Elton R Smilie

The Manatitlans

A Record of Recent Scientific Explorations in the Andean La Plata, S.A.

Elton R Smilie

The Manatitlans
A Record of Recent Scientific Explorations in the Andean La Plata, S.A.

ISBN/EAN: 9783337414689

Printed in Europe, USA, Canada, Australia, Japan

Cover: Foto ©berggeist007 / pixelio.de

More available books at **www.hansebooks.com**

THE MANATITLANS;

OR A

RECORD

OF

SCIENTIFIC EXPLORATIONS

IN THE

ANDEAN LA PLATA, S. A.

R. ELTON SMILE,
PRO-SCRIPTOR.

𝔅𝔲𝔢𝔫𝔬𝔰 𝔄𝔶𝔯𝔢𝔰:
Calla Derécho, Imprenta De Razon,
1877.

DEDICATED

AS A

MEMORIAL TRIBUTE OF AFFECTION

TO THE EVER PRESENT ANIMUS OF MY

PARENTS, SISTERS, MRS. HIRAM HOLLY, AND
MRS. SOPHIA VISCHER.

PREFATORY INTRODUCTION.

BY THE HISTORIOGRAPHER.

IN the following record of the explorations of the Teutonic corps of the R. H. B. Society of Berlin, dispatched for the classification of parasitical animalculæ peculiar to the vegetable productions of the tropics, I shall confine myself exclusively to the revelations of the day until the culmination of the corps discoveries, and then to Manatitlan dictation, either direct or through the medium of thought dictation.

The discoveries, as verified, will undoubtedly tax "public credulity" to its utmost stretch; but as the absorptive power of human instinct for the marvelous is unlimited in its superstitious gullibility, it will peradventure receive — with perhaps an awkward spasm from the novelty of goodness — the practical experience adduced as worthy of disputatious consideration. Still we feel assured that there is a reasonable minority who will adopt the practical suggestions with joyful avidity. The facts related — although at present stranger than the instinctive fictions projected from the unreason of the stomach's rule — will prove, to the affectionately disposed, of easy reconciliation with healthy digestion, and in every respect worthy of universal adoption by our race. Assuming the privilege of narrative relation in recording the progressive events, I shall only advert to the leading adventures of the scientific corps while en route toward their ultimate field of exploitation. But while in progress shall endeavor to render the characteristic peculiarities of the members sufficiently conspicuous for the clear exposition of their national traits, that the reader may realize the obstacles opposed, in degree, to their assimilation with the practical teachings of the Manatitlans demonstrated by Heraclean example.

Lucenhouck, in prophetic forecast, says, "Man, in the arrogance of his pride, believes that he is of a race separate and distinct from the lower orders of the animal creation. Assuming attributes of deity he has constituted himself arbitrator of his own destiny. Yet, with all his affectation of superiority, there is an approximation in his form and physical conformation that distinctly declares his relationship to the simia species; among which there is as great a variety in form and racial intelligence as with those of the genus to which he stands confessed. With the full development of microscopical power, future generations will learn that the wonders of Creation are beyond present conception, and that well defined organic humanity may yet be revealed on the utmost verge of atomic divisibility."

THE MANATITLANS.

BOOK FIRST.

CHAPTER I.

IN the month of January, 187–, M. Hollydorf was selected to conduct an exploring corps of the R. H. B. Society to the head waters of the Paraguay and its tributaries, for the purpose of observing the habits and classifying the different species of animalcular life native to the trees and plants appertaining to those regions. The Royal Society had supplied him with able assistants, and the most complete set of instruments ever constructed for botanical or other research in the fields of natural science. Among the instruments of recent invention, was one of Lutsenwitz's solar reflecting microscopes, especially designed for field explorations. This was of the highest concentrated power yet attempted by that artist, — the intensity of its magnifying capacity being capable of showing the facial contortions of the most minute animalculæ. Attached to the focal platform was one of Phlegmonhau's highest grade of responsive tympanums, with reflecting auricle for magnifying the articulation of sound. The corps arrived at Montevideo on the first day of April, and was fortunate in finding a small trading steamer, under neutral colors, ready with quick despatch for a barter voyage up the Paraguay and its tributaries, without a specified port of final destination.

The captain was sole owner, and proved to be a man of rare intelligence, which had been cultivated by travel and study. To his love of adventure was added a strong amateur predisposition for the pursuits of natural history. These qualifications led to a speedy agreement, with conditional arrangements for a charter of the steamer open to variations suited to the requirements of the corps.

On the 15th of April the members of the corps, instruments, camp utensils, and travelling gear, were safely stowed on board the little steamer *Tortuga*,— a name that implied slow progress, which to our satisfaction her speed decried. At eleven A. M., having bid farewell to our newly acquired friends, we left the anchorage with their " Good speed," and after threading her way among the vessels in the roadstead the little steamer puffed her way up the broad expanse of the La Plata estuary. The balance of the day was occupied in arranging instruments for river observations, the while listening to praises lavished by the captain upon the " worthy " qualities of his little propeller, of which he was the architect and builder. During the evening he regaled us with incidents of his life in California and the East Indies. His adventures in California received occasional illustrations from a genial individual introduced as Padre Simon, the prefix having been conferred — as we afterwards learned — from his zealous support of the Catholic dogmas, theoretically. As the padre was eventually enlisted in our corps, we will foreshadow some of his peculiar characteristics. In form he was of medium height, with a rotund outline visibly inclining to jovial obesity; his face was indyed with a complexion blending with the Roman auburn of his hair, which gave a warm glow to his expression when lighted with a smile. In the first generation of descent from Irish parentage, he retained the full impression of inconsistency in the practical adaptation of his habits to the faithful index of goodness ingrafted from the maternal stock. Guileless in thought, when free from temptation, he possessed a ready facility of excusing his habits of excess with the plea of saving grace administered under the seal of confession. With this hint, in forecast of development, we will proceed in the relation of events transpiring during the river voyage.

On the morning of the 21st of May, after having been subjected to our full share of vexatious delays, incident to the provincial *poco pocoism* of the guarda and custom-house officials, the steamer gained the river post of Santa Anna on the Pilcomayo, two miles above its mouth. At Santa Anna they found the well-known American naturalist, Diego Dow, waiting for an opportunity to obtain sufficient aid to attempt the exploration of the Pilcomayo as far as the reputed settlement of Tenedos, which rumor located on a confluent stream rising and flowing eastward through the valleys of the Andean spur that reached into central La Plata.

The ultra-savage disposition of the wandering tribes on the banks of these rivers, having defeated every previous attempt made to establish trading-posts, but few had been found willing to incur the hazard proposed by Mr. Dow. Even the indomitable Jesuits had been foiled in all their endeavors to conciliate the Indians in degree sufficient for the establishment of missions preliminary to their subjugation.

The magnet of Mr. Dow's desire had been drawn thitherward by the reputed existence of a walled city inhabited by a white race of great beauty. He considered the report sufficiently well authenticated to warrant the adventure of his life for its discovery and relief from the constant siege to which it had been subjected by the savage tribes from time beyond date. His chief authority, which had incited him to engage in the emprise, was his Auraucanian servant, who had, in his wanderings and progress northward, served in an Indian marauding expedition, which invaded the valley of the city for the purpose of lifting the cattle of the inhabitants, who were in seasons of drought obliged to protect them while feeding beyond the walls. As Indian forays were expected, the herds were well guarded by shepherd escorts, whose persons were safely protected with defensive armor, so that with the exception of the face the other parts of the body were proof to the poisoned arrows. In addition they were armed with a bow which in their practiced hands sent the arrow sure to its mark far beyond the range of their savage foes' weapons, so that in the open valley they were safe. Besides, their tactics embraced so many precautionary variations that the Indians were almost invariably decoyed and blinded from real intention. These feints caused the savages to become over wary, never venturing an attack unless with the advantage of overwhelming numbers. The party with which Aabrawa, Mr. Dow's servant, was engaged, met with a severe repulse that indisposed them to renew the attempt, notwithstanding an opportunity was offered on the succeeding day. So well managed were the citizens' plans of protection that they rarely lost either men or cattle, and without being aggressive frequently administered well merited punishment upon their foes, who were inspired with wholesome fear from a superiority so manifest in deadly effect. Unable to cope with their white antagonists in the open field, they, with constant wariness peculiar to the savage, neglected no opportunity to harass, hoping at some time with constant worrying to catch them off their guard. The cause of this implacable hatred was hereditary, reaching, as Aabrawa learned, far back to a time when the forefathers of the citizens abused their supremacy by enslaving their Indian benefactors. The Indians having surprised and overcome their oppressors, a remnant of the whites obtained refuge in the present city, which had since been kept under constant espial. As the city was overlooked from an adjacent height, but little passed in the streets unknown to the besiegers, who were quick to discover any relaxation of vigilance; and whenever from pestilence or other cause it did occur, couriers were dispatched to summon aid from distant tribes.

Curiosity and love of exciting adventure had enlisted the members of the corps in favor of aiding Mr. Dow's projected enterprise, and through their continued solicitation. M. Hollydorf consented to waive the strict interpretation of his commission, designating a particular field of operation, by using his discretionary power in favor of the proposed scheme for raising the siege of the beleaguered city. Captain Greenwood without hesitation tendered the aid of his steamer, and being one of those peculiar persons who are accustomed to take the head of time by the forelock, he immediately commenced the precautionary labors to protect his vessel from the wily tricks of surprise practiced by the savages. The commandante of Santa Anna, being well acquainted with the methods of attack that led to the defeat of the various expeditions directed against the Chacas, proved of great use in suggesting precautions. The chief dread arose from the poisoned arrows of the savages, which inflicted incurable wounds, adding to death the horrors of lingering putrefaction. The fears anticipated from this source were relieved by the confidence inspired through the energetic character of the captain, whose experience with the superior cunning of the North American Indians prepared him to cope with the lower instincts of their southern congeners.

On the morning of the 23d of May the *Tortuga's* bow was turned against the swift middle current of the Pilcomayo's bayou expanse, then at its height from the copious contributions of the rainy season in the high lands and mountain sources of its tributaries. Night still found us in the broad sea of waters, baffled in search of the interior mouth which was made more difficult from the confluent branches uniting with it near its Paraguayan embouchure. The commandante, anticipating the difficulty likely to be encountered, had been particular in giving directions; but although strictly followed, from a calculation of the steamer's speed, twice the distance had been run without discovering the described landmarks. Uncertainty was rendered still more uncomfortable by the shallowing of the water, showing plainly that we were inland from the river's channel. At midnight, while anchored, a hurricane, heralded by a thunder-storm, made the waters seethe with its force, causing our little craft to careen and bob with a politeness to the gusts that impaired our confidence in its self reliance. Padre Simon declared that the lightning set his teeth on edge, prompting him from its dazzling flashes to pray, but that the thunder so startled and confused him that he was unable to think, and as a dernier ressort was obliged to drink. This remedy finally rendered him proof to the best efforts of Jupiter Tonans; but on waking in the morning he complained that he could still hear the roll of the thunder in his head.

On the morning of the 24th the sun rose bright and clear in a cloudless sky, compensating with its splendor the discomforts of the night; its reflected light glancing upon the waters discovered far to the south a broad ripple, indicating the sought-for channel. The river's stream was soon gained, and followed in a southwesterly course until the river's limits were defined by partially submerged trees growing upon its banks. Having at Santa Anna filled every available portion of the vessel with fuel, sufficient for a run of four days, the boat was enabled to keep on her course under a full head of steam, without anxiety from the dull prospect offered for replenishing.

May 25th, at sunrise, after a good night's run, we discovered a headland above the surface of the water covered with fire-scathed trees, from which the captain, for a surety, concluded to add to his diminished supply of fuel. The labor of taking in wood from this source was by no means pleasant, but the sailors with good-will made the "virtue of necessity" cheerful with songs and jokes, the "passengers," suitably clothed, contributing with the zest of energy their labor for its stowage, so that by eight o'clock we were again under way. With the exception of this wooded bluff nothing but sky, water, and foliage had met our eyes since leaving Santa Anna, the monotonous compound making us well content with cabin associations.

On the 28th at sunrise, our ears were gladdened with the cry of "Land ho!" Rushing on deck, with the expectation of a greeting from well defined banks, we were disappointed, as the contrasted elements of the previous day still prevailed. Seeing that we were a little inclined to be vexed, at what we considered to be an ill-timed joke, the man at the wheel, an old river navigator, pointed to a mud bank that closed our view with the bend of the river, at the same time directing our attention to the eddy cast from it far out toward a line of trees on the opposite shore. From these indications he assured us that in a half hour's time we should hear the songs of birds to make us lively. Doubling the muddy cape we were greeted with the screams of parrots, while other birds of gay plumage were crossing and recrossing the river singly and in flocks, causing, in apparent salutation, a lively line of demarcation between the land enclosed current and the smooth waters of the flood below. The welcome sight raised our spirits into a sympathetic mood of song, which was unfortunately too nearly allied to the screaming discord of the parrots to evoke other than a mirthful disposition for repartee which expended itself in humorous comparisons, favoring the advent of genial omens.

Mr. Welson, a prominent official of the Panama Railroad Company, had accepted the freedom proffered by the steamship lines plying between the maritime cities of the eastern coast of South America, for his recuperative vacation of three months, and on his arrival in Montevideo had been induced by Captain Greenwood to extend his voyage up the river.

A Scotsman by birth, he possessed in an eminent degree the predilection of his people for dry, caustic humor; and in his position of commercial agent had cultivated the art of extracting fun from the vagaries of migrating humanity in their transit across the isthmus. Scientific whimsies were especially adapted to his quizzical vein, and a happier combination of material could scarcely have been conjured for his entertainment, than he found on board of the *Tortuga*. Padre Simon was his especial favorite as a stimulating provocative. Won by his naïve simplicity, he had soon interested himself to learn the object of his river voyage, with the intention of rendering him assistance. Greatly to his surprise the padre informed him that he had no other expectations in visiting Entre Rios than the chance one " of hitting an opportunity to make a strike." Amused with his vernacular, and the easy carelessness of his manner, which seemed to defy disappointment, he was delighted to discover his growing fondness for polemical disputations, which was gratified by a kindred disposition cultivated by Dr. Baāhar, the naturalist of the corps. On the steamer's arrival at Entre Rios, the port of his destination, the padre's thoughts were absorbed in the dogmatic discussion of the soul's material identity with the body after the resurrection, so that he gave no heed to the frequent repetition of the name of the town. Aware of his total abstraction from all thoughts and anxieties connected with the business responsibilities of life, necessary for material sustenance, Mr. Welson connived with the doctor to hold him in argument until after the steamer's departure, well assured that no material harm could arise from the derangement of plans so lightly impressed as to give place to chimerical argument. For a characteristic illustration of the disputants' peculiarities we will give the burden of their colloquial subjects of exposition.

Padre. " My conscience' sake alive, man! Why, you might as well set us down as beasts at once, as to argue that in resurrection we shall assume the form of animals whose habits we most affect in life! Surely your naturalistic learning has run mad with your orthodox catholic ideas, for, upon my soul, they are rank with transmigration, and if confessed, you would be denied absolution by every ecclesiastic in the Christian world. Look you! the very fact, if admitted, would controvert all that we hold sacred. Why, man, it would render absurd our reliquary faith in the efficacy of sainted bones and vestments for healing the sick and lame, for the marrow-bones of swine and the hair of dogs would hardly serve to enlist belief in the Christian doctrine of divine transubstantiation?"

Dr. B. " As we claim that reason has been bestowed as an endowment to distinguish us in reality from the brute creation, its possession presupposes preordination of intention in decree for its use. Now, if you will devote your share of this human endowment to the demonstration I am about to give of cause and effect, you will not fail to perceive the distinctions upon which our faith is founded. Humanity possesses omnivorously, in its varieties of genera and species, all the habits of the lower orders of the animal creation in their separate representation! But superadded to this resemblance in the community of instinct, man has a discretionary power inherent with his endowment of reason, which enables him to profit by experience in shaping his course for the avoidance of consequent evils which follow from the transgression of natural laws. This power presupposes accountability that directs itself to Creative Cause. Upon this innate feeling of responsibility, impressed by repentance from transgressions, and joys imparted from adherence to the monitor indications of our superiority, man has founded his religious distinctions of vice and virtue. In furtherance of this natural division man has volunteered to represent vice, and woman, unprejudiced by his influence, would have naturally assumed the rôle of virtue in truthful vindication of her vocation as the mother of our race. Now, as you well know, it is impossible to harmonize vice and virtue, even with the instinctive coalescence of the sexes? Hence, as you must acknowledge, there will be a constant struggle for ascendency. Man as the stronger of the two, in representative selfish determination, and the moral force of muscular strength, is as full of devices for the beguilement of woman from her sacred trust as the variations of his ability admit."

Padre. "Yes, all that may be true; but you don't talk at all like yourself, and I can't see what you have said has to do with revealed religion."

Dr. B. "Why, its connection is self suggestive; virtue and vice in sexual array, for the supremacy of example, naturally oppose to each other their attractions and temptations. Fortunately, the harmonizing beauty of woman, with loving affection, impressed on the rude selfishness of man the preferred happiness of a home subject to graceful refinements, and with her sex in the majority held his passions and appetites of instinct in abeyance. To overcome this tacit rule man devised a series of temptations to hold her in subjugation to his control. These were addressed to her vanity and envy, incited by the jealous instigations of man's preferment on the score of beauty. This led to artificial adornment, which placed the means of temptation in the hands of man. Then, as a plea for the encouragement of virtue, religious revelations were instituted under the conjurations of mystery to control, with fear, superstitious simplicity."

Padre. "Perhaps I don't quite understand you, for I can scarcely account for my own thoughts as they seem to be so mixed with new impressions; but if I understand what you express in words, I will answer for myself that the revealed way of salvation is to use all the blessings of life with moderation."

Mr. Welson. (Amused.) "With the doctor's permission, you will perhaps appreciate an illustration that occurs to me? Woman's naturally unselfish affection, unbiased by the temptations of vanity and envious curiosity, exerts with gentle forbearance a restraint upon the more brutal appetites of man, softening asperities provoked by over indulgence. Theodosius, the emperor champion of Christianity, opened a way for the incursions of northern barbarians by patronizing the intolerant sway it usurped over the more primitive and lenient rites of paganism, as it weakened, by the introduction of effeminate luxuries which allied the sexes for degeneration."

Padre. "I have never been much of a book-worm, but it appears to me if man, as Dr. Baāhar says, represents vice and woman virtue, your college learning directly tends to the cultivation of a vicious course by keeping before the people the barbarous acts of the ancients derived from their own language, which gives the scholar a directing power, from a studied understanding of the corruptions practiced in past ages. So you see, it's far better for woman, and the world at large, that she's denied the means of classical study; for from your own admissions, her curiosity and envious vanity rages so greatly at the present day she'd be more likely to play the part of a Cleopatra than a Zenobia. As the world runs, I think the less we know of the past the better it will be for our salvation."

Mr. W. "But you forget church history, padre, from the record of which you derive your knowledge of the fathers?"

Padre. "Well, but that is different from profane, for it teaches us the way of salvation by saving grace."

Mr. W. "Yes, through the tender mercies of the Inquisition."

Mr. Dow. "As a listener I must acknowledge that you have each with good arguments strangely confounded your former selves."

The above colloquial rejoinders will serve as an illustration of the attraction that beguiled the padre's attention until the second day after he had passed his port of destination. Then inquiring of the captain the distance that still "intervened," the supposed number of miles being given. he relapsed into his usual routine without suspecting that it was calculated from the stern instead of the bow. When informed at the port of Rosas that the town of "Three Rivers" had been passed some days previous, he exclaimed, "My goodness gracious, there was where I wished to stop; my conscience' sake alive, what shall I do?" The captain, to whom he appealed, answered by asking, "What did you intend to do at Entre Rios, padre?"

Padre. "A brokerage business of some sort, real estate or sugar, whichever offered the best opening."

Captain. "But, padre, you cannot speak the language, which would render your expectations abortive, for a bargain is never closed in these countries without a great deal of word chaffering. A clear understanding of the language is absolutely necessary, for the inhabitants of the river towns are very apt to "fly" from a bad bargain when they find themselves caught and lightly held, so that the only safe way to secure them is to clip their wings and hood-wink them in black and white. But I can send you back without cost when we meet the next downward bound steamer; then you will have the advice and assistance of Mr. Welson, who perfectly understands the habits and customs of the people."

Padre. " Well, I declare to gracious, I hardly know what to do ? "

Captain. " Would you like employment on board ? I think that there is a berth that would suit you ! Besides it will afford you an opportunity to convince Dr. Baāhar of his errors ; at the same time you can perfect yourself in speaking Spanish."

Notwithstanding the captain's quizzical looks and speech the padre thankfully accepted the proffered position of second officer, with the expressed hope that he might perform its duties in an acceptable manner. Captain Greenwood, although somewhat crispy in speech and austere in address, had a strong undertow of humorous appreciation when the shafts of irony were not directed against himself. His disinterested disposition, prompted by the padre's kindly *vis inertiæ*, had suggested the offer ; nevertheless he really desired a person capable of superintending small matters that would relieve him from a responsibility not greatly to his relish. The duties imposed by the captain were as follows : " You must be the first up in the morning and the last in bed at night. While on duty, see that everything in the way of labor is well done, and never interfere with advice when a helping hand is required. Lastly, never report to me necessary changes until after they have been made."

Padre. " But, captain, if I am never to speak how am I to improve or correct to suit you ? "

Captain. " With the moral influence of your head and hands, when you see anything necessary to be done ! "

This settled the question of the padre's new vocation, and he was forthwith introduced to the crew, who greeted his installation with marked approbation. At night, when he became genial in confessional overflow and dogmatic in argument, he was the source of humorous repartee and good-will among the passengers on the quarter-deck. His American birth having toned down the quarrelsome disposition legitimate as an inheritance to the native-born Irishman, when under the influence of whiskey, he indulged in quaint disputations, peculiar to his Yankee ingraft, in freedom from ill humor.

With this insight descriptive of mood foreign to the members of the corps, we will now resume our narration of events transpiring in the daily progress of the steamer's river voyage.

May 28.—The banks of the river are now clearly defined, but the water still submerges the undergrowth that margins its lower stages in the season of drought; the more matured growths are already peopled with the smaller species of birds delighting in the bushy retreats overhanging the waters. Our naturalists' eyes are now greedily engaged in busy search for new specimens of the feathered species.

May 29.—This morning we reached a sand-spit formed by a confluent stream, upon which the receding waters had left a wood-drift well suited for the steamer's use, having been forced by the jam of flood-tide high out of the current. The eddies and back-water of the Pilcomayo's stronger flow had carried the raft and lodged it high up above the mouth of the lesser stream, leaving an extension interstayed by the roots that reached into deep water; alongside of the raft, in the smaller stream, the steamer moored. The axes of the firemen and sailors were soon busy, wakening for the first time the forest echoes to the chucking sound of their strokes. The more active members of the corps volunteered their services in aid for speedy replenishment, deriving in recompense the invigorating novelty of exercise. While actively engaged with ready hands and merry voices they were suddenly startled with the scream of the steamer's whistle, simultaneously accompanied with a flight of arrows from the ambush of the forest screen above the raft. Fortunately distance and trepidation from the unearthly screech of the whistle rendered their aim harmless; the check it afforded enabled the woodcutters to scramble up the sides of the steamer before the savages recovered from their surprise. When they realized that the shriek was harmless in effect, the Indians rushed forth from their concealment to secure the axes which had been abandoned by the men in their sudden fright, but were again momentarily intimidated by the rumbling sound of the gong, which Antonio, the steward, had seized to increase with concerted din the scream of the whistle. The savages' hesitation was but momentary, seeing that like the former the steward's overture was harmless in effect, then with a counter whoop of defiance they sprang forward to secure the coveted prizes. But the second diversion brought with it presence of mind and time for the use of more effective weapons than empty sound. One of the two howitzers, which had been taken as freight to Santa Anna, the commandante loaned to Captain Greenwood for the voyage; this had been loaded as a precautionary measure the day previous, and intrusted to the charge of Jack and Bill, two sailors who had " shipped " on the river voyage for a " lark." With thoughts trained to the duty of their charge they were the first that reached the steamer's deck, and before the savages recovered from their second hesitation sighted the gun and answered their whoop with a discharge of grape, with an effect that left five of their number stretched on the logs, killed outright, the others in quick retreat leaving a trail of blood showing from its copious flow the infliction of dangerous wounds. The retreating savages in their turn dropped clubs, spears, blow-pipes, and arrows, so that there was but little danger of their return. But the premonition caused the captain to place a guard in a position to command the isthmus, accompanied by two hounds belonging to Mr. Dow. The dogs following the bloody trail soon gave intimation that they had discovered the wounded savages. Proceeding cautiously into the thicket beyond the abattis they found near together, an elderly savage and a boy of seventeen or eighteen years, both severely wounded. The padre, with heedless but kindly intention, attempted to raise the head of the old Indian upon his arm to relieve his uncomfortable position, while the others stanched his wounds. In a second from the time the padre's arm came within reach of the savage, his teeth were fastened upon the arm above the elbow, while with working tenacity he used his utmost energy to penetrate the sleeve of his coat. His intention was evident from the greenish slaver that oozed from the corners of his mouth, betraying in appearance the characteristics of the dreaded poison. Bill, who was near at hand, relieved the padre from the danger of poisonous inoculation, before the teeth of the savage had penetrated the cloth, by the introduction of a marlin-spike with a decisive force that showed but little care for their preservation. The boy was more tractable, permitting his captors to handle him as they pleased. Two other savages were overtaken dragging themselves from bush to bush. When surrounded they were still defiant, threatening all who approached with spear-heads attached to short staffs; these were finally struck out of their hands, but they still repelled peaceful overtures, making a formidable show of resistance with teeth and nails. We had been specially warned against coming into close quarters with them by an old trader, who had frequently encountered their ferocious tendencies in his travels. Finding all our conciliatory attempts futile the wounded savages were left to their fate. Adopting the padre's suggestion, the young Indian and his savage companion were taken on board, with the intention of trying the effect of kind treatment, but a lasso in the practiced hands of a guacho was required to persuade the latter to accept the proffered hospitality of the boat. Aside from the comparative docility of the boy, his lack of resemblance in feature and general conformation plainly declared that his subserviency to the will of his companion did not arise from parental affection. Shackling them to the windlass they were placed under the guardianship of the dogs, whose favorite lounge was on either side of the bowsprit heel beneath the shadow of the chocks. After they were secured, all hands, with the exception of the engineer, steward, and cooks, resumed their labors on the raft. As the padre insisted that it was a barbarous shame to throw the bodies of the dead savages into the water to become the food of alligators, when a few minutes' labor would make them a decent grave in the sand, he was allowed the privilege of extending to the defuncts the rites of burial. As the spade in his hands had not been a favorite specialty during the more elastic periods of his existence under the benign influence of temperate heat, the torrid glow of the morning acting in concert with a stimulant he had taken to steady his nerves, caused a sweltering perspiration that in no way accelerated the progress of his pious undertaking. The sands having become quick from recent saturation were constantly caving, so that in addition to aggravation he was in danger of becoming a victim to his sextonic benevolence. While trying to extricate himself from the caving sand, the while vainly pleading for assistance from the laughing spectators of his disaster, his attention became fixed upon an array of yellow nuggets which he had overlooked when thrown from their bed with the sand. His silence and curious investigation with hands and eyes extorted the inquiry, "What is it, padre?" The laconic answer, "Gold!" brought the whole party to his rescue, including the sentinels from the logs above, while the engineer, steward, and cook deserted their posts in greedy haste. When the truth of his announcement was verified they with some difficulty dragged him from his grave, then oblivious to thoughts of savage surprise and poisoned arrows, they consigned the dead to the river, without remonstrance from the padre, and with flushed avidity commenced with spade and pan to unearth the precious metal. Mid-day, with its heat, found them still engaged, heedless of danger from the sun's rays and the miasmic current converging upon the spit from the confluent streams. Silence alternating with wild bursts of hilarity, caused the captive savages, chained to the steamer's windlass, to gaze with wondering looks of amazement.

Through the day, until darkness precluded the possibility of detecting the golden grains, the wild search continued, then when collected on the steamer's deck they bethought themselves of the dangers to which they had been exposed. Although resolved to be more cautious in future while gathering their golden trove, its tangible presence banished fear; still as a thoughtful precaution the steamer was dropped into the stream as a guard against surprise.

CHAPTER II.

At early dawn on the 29th all were on the alert, anxious to recommence their gold-gathering labor, but obedient to the captain's request the steamer was first supplied with its full allotment of wood. This was accomplished with a despatch that betokened an earnest desire to resume their yesterday's toil in the sands. The captain and padre explained the most approved methods for the economical saving of the smaller particles, which brought into requisition the steward's and cook's wares. The tableau of the second day, although lacking in the wilder excitement of the previous, incident to the impressions of first discovery, would have afforded a novelty unparalleled in scenic variety for the study of an artist, but unfortunately our own was too much engrossed with interest to heed the rare advantages of the absurd comicalities of selfishness. In truth all were so moved by an acquisitive spirit, but little thought was given to the ludicrous groupings of the parties engaged, or the solitary wildness of the surrounding scenery, contrasting so vividly with the pretentious civilization of the laborers.

On the morning of June 3, the spit was left in the wake of the steamer, exhausted of its free surface gold, and much to the surprise of all there was a general expression of relief when it was lost to view, and the discomfort it had caused began to disappear with the revival of order. But a still greater surprise was in store, which removed all the barriers of distinction bred by the pride of birth and station from the standard of laboring vocation, inasmuch as they debarred in exchange kindly equality in reciprocation. Unusual alacrity and kindliness of feeling had been observed in " putting" the vessel to rights by the hands, which was explained, when accomplished, by Jack and Bill, who came aft with hats in hand. After bowing all round, Bill the prompter nudged Jack the spokesman to give way, which he essayed to do, but from confusion was unable to get a running bight of phrase, until aided by the captain's inquiry, "Well, what is it, my man?"

Jack. "You see, Bill and I started up the river to freshen our jints, which had grown stiff and creaky with salt junk and hard tack. Well, after we had loosened our barnacles with the treacle of a Spanish skipper we took French leave and laid low until you hove in sight. Now you see after we entered with you it took us some time to get the run of the fair weather you made for all hands. Expecting to be taken aback with a sharp squall we kept our eyes well to the wind'ard, for you see on this river with cannibals on the lookout and no vessels there was no chance of skulking on shore for a down-river craft. To be sure, we soon found that we were out and wide in our calculations, so when brought to our bearings we began to take kindly to the lay of our watches in scrubbing and wooding, as there was no hand-spike snubbing or squeak of hard words. Then comes this gold lay, and when you says, ' Boys, here's your chance, pitch in, every man for himself without envy,' we were taken aback with a fair wind. When we came on board to empty our hats we began to take our bearings, and ses Bill to me, after an observation, 'We've shipped and signed the papers, and this gold is way freight, so you see it's not right to tap the cargo on full rations.'

"There was the p'int clear, and we said 'Never a bit!' So you see after the flurry was over we put the question to the others and they took the bearings at once ; so you see that we've concluded that we're only 'titled to prize money at most, just as you valer the danger we run with the savages."

This construction, regulated by the sea usage of man-of-war's men, who had grown gray and poverty stricken in "service," was so generous in the sincerity of honest proposition for revoking the captain's liberality that he asked time for consideration. In submission the procession, headed by the two honest tars, retreated to the " for 'ard " hatch, on which they placed their well-filled hats to await the captain's decision. A consultation with the members of the corps was immediately held to decide upon a method to insure an equitable division of the gold suited to the emergency. After a variety of propositions had been made and rejected, the padre advanced one that proved the most acceptable. His suggestion was that the passengers and officers should abide contented with their own gatherings, as they were proportionately less than those of the crew ; but that an equal division of theirs should be made to avoid envy. When this equitable measure was made known to the men, Jack, with the advice of Bill, objected that the most important persons had been left out, which in their opinion were the vessel and captain. As this amended consideration met with general approval, it was adopted. Then Antonio, the steward, said, that the men for'ard, from being accustomed to work, had gathered so much more in proportion than those aft, he would propose to " lump " the whole for an equal division, after one fifth had been deducted for the vessel's and captain's share. This was acted upon, notwithstanding the captain's protest that all should share alike. The division accomplished, there was a hearty shaking of hands that opened a sympathetic current of reciprocation void of selfish envy, which as an omen heralded a happy result for their adventurous voyage. After the parties to this happy arbitration had resumed their usual avocations, Jack and Bill — to whom had been assigned the duty of "freshening up" the trimmings of capstan, binnacle, and other extras aft, usually attended to in their watch below, to save time — entered upon their duty during the siesta hour of the day. While engaged they ruminated in silence until the deck was cleared of chance listeners, then the rapid change of tobacco quids from side to side of their mouths, and an unusual flow of the green ooze from the corners gave indication of thought's supremacy. At length when they " supposed " the coast was clear, Jack gave an expressive tug at his waistband, then after blowing his nose with a clarion note, he sputtered, " Blast my buttons, Bill, if this fresh-water turtle of a captain hain't sounded and found a salt-water leak in the water run of my eyes!" Bill without answering, except with a suppressed sniffle, found it necessary to expectorate and blow his nose over the bulwark nettings. A prolonged effort having relieved his emotions he shuffled back, and shyly exclaimed, with a whispering sob, " Don't, Jack."

Woman's distress, from the period of youth and beauty, through all the gradation of cause, to its decline with the influence of age and ugliness, when haggish distemper engendered from selfish disappointment makes it repulsively loathsome, I have felt with impulsive variations, but never experienced the like choking sensations of affectionate sympathy, from the evidences of gratitude, that held me bound during the enactment of this short scene, so truthful in expression. Probably during their long term of service they had never felt a like cause, foreign to themselves, for the revival of emotions so nearly allied to affectionate reciprocation; for it was evident that the gold of itself occupied a minor impression in the ruling of their thoughts. Indeed, in the after detached rehearsal of their seafaring experience, they declared that a glass of grog was the only compensation they had ever known a sea captain to bestow upon his sailors for extra labor. The representatives of tropical countries, of which a majority of the crew was composed, were more open and volatile in their expressions of gratitude; but like the English sailors attested that the self-denial of Captain Greenwood was the only exception in their experience in which the master of a vessel had failed to exact to the uttermost the fruits of their labor.

From the Tortugian era of the third of June Captain Greenwood became a deity of adoration to his crew, who offered daily sacrifice of labor for kindly propitiation, which from promptness in anticipation rendered the padre's official vocation a sinecure.

The sun of June 4 found the *Tortuga's* decks neatly scrubbed and washed in readiness for its rays; the two savages having participated in the cleanly overture, the elder receiving his somewhat copious douche with a grateful show of teeth; but the younger's eyes were used with such an evident desire for pitying sympathy that Antonio volunteered his tonsorial service as an initiatory introduction to civilized habits. This act won the young savage's first love; while it added another count to the special hatreds of the old, who bestowed upon Antonio a toothful longing to recompense his civilized barbarity. The improvement of the young savage was so marked from the use of soap, sand, and scissors, with the grateful expression produced, that Antonio was fain to crown his morning's missionary labor, and his neophyte's satisfaction, with a hat.

CHAPTER III.

WHILE the events related in the preceding chapter were in progress, which gave advent to the new era, the manacled savages would have fared poorly but for the ever mindful benevolence of Padre Simon, who ministered to their relief after depositing with his traps his godsend, which he averred came from the source of their misfortunes. His arm warned him to be cautious in his approach to the old savage, but he could not refrain from the pitying exclamation "It's a shame," when he saw him bound to the links of the cable with its coils for his bed. Placing the food he had brought cautiously within reach, he left with intention of pleading for some aid in mitigation of their painful position, but the question of an equitable division of the gold trove diverted his thoughts. But after the ablutions of the succeeding morning, and Antonio's improvement of the younger savage, his derilection occurred to his thoughts under the stimulating inspiration of a somewhat copious oblation to memory, which served to render the sincerity of his repentant remorse heedless. Under the sacrificial impression he hastened forward to make amends for his forgetful inhumanity. Without observing the change already made for the ease of the savage, he attempted to place an oakum fender between his back and the cable. Exposing his arm the brute again seized it with a vicious energy that bespoke his determination of obtaining recompense for his morning's aggravations. With the pain, caused by the working teeth of the savage, the padre's terror of the deadly poison was revived, which caused him to cry for help in frenzied accents, alarming all on board. Again English Jack was the first to reach the struggling victim of misplaced pity. With a sailor's promptness he forced his sheath knife between the back teeth of the cannibal with a delicacy peculiar to the tar when called upon to repel boarders; working the blade, with a prying motion, hither and thither with the edge directed toward the ear the backward capacity of the mouth was insensibly enlarged, which produced a diminution of muscular tenacity and consequent release of the padre's arm. His release was not effected until the teeth of the savage had penetrated through his linen coat and sleeve of his shirt, inflicting bruised punctures beneath the skin sufficient for the absorption of virus. The general consternation was greatly increased by the exultant gleams darted from the eyes of the bleeding savage. Dr. Baāhar had just prescribed whiskey to be taken in copious draughts for *ad diliquium* effect, which the padre, with a sense of relief, said he had premised, when the young savage attracted attention by pantomimic gesticulation, at the same time producing from his mouth a small sac of an acorn's size and shape. From the pleased honesty of his expression and the scowls of the old savage, it was apparent that it contained an antidote for the poison. Aabrawa having caught some familiar words, he was soon able to add his assurance in verification of the boy's ability to counteract the effects of the poison with a sure antidote. The padre with fear hesitatingly submitted his arm to the boy's mouth, the old savage regarding the operation with looks that boded ill to the savior and saved if by accident they should come within his reach for injury. The padre, when impressed with the kindly intention of the boy, apostrophized the old wretch in this wise: "You ungrateful venomous old serpent, upon my conscience you ought to be made to crawl on your belly all the days of your life with a rattle tied to your — well if you haven't a tail, you are a vile reptile all the same, and I dont believe all the purgatories in creation can change you! Upon my soul, it's a shame and an imposition for you to pretend to be a man with a soul to be saved!" Here the padre observing the smiles provoked by the earnestness of his address to an object as incapable of appreciating as he was of understanding the language in which the anathematizing sentence was couched, apologetically appealed to his auditors, "You know that what I have said is as true as there is a day of salvation for man to sin away."

"Are you not assuming," asked Mr. Welson, " the privilege and understanding of a judge without knowledge sufficient for the condemnatory sentence you have pronounced as a penalty against this savage?"

" By their works ye shall know them," replied the padre, looking wofully at his arm.

This retort placed the padre's star in the ascendant, and it was immediately proposed that the mouth of the old savage should be rid of its poison, a task which Jack and Bill volunteered to accomplish. Preparing a running noose they slipped it over his arms, pinioning them to his side, and then proceeded with sheath knife and marline-spike to open his mouth for investigation, but not without strenuous efforts on the part of the subject for revengeful retaliation. Beneath his tongue they found two sacs, or bladders of the river whiting, attached to the cuspid teeth, which by the tongue's pressure could be made to eject their contents into wounds inflicted with the sharpened teeth, which were pointed like fangs, verifying the padre's estimate of his reptile instincts. Above, attached to teeth upon either side, were the sacs containing the antidote in position to be pressed by the cheeks. Rid of these venomous appliances the nozzle of the steamer's hose played the part of a purifier by injecting a bountiful supply of water into his mouth, regardless of the published restrictions of the humane Society for the Prevention of Cruelty to Animals.

During the passage of the two days succeeding that of the padre's mishap, parties of savages were discovered tracking the progress of the steamer, the while with opportunity holding communication by signs with the captive chief. As he did not appear to be in the slightest degree amenable to kind treatment, and his presence on board was neither safe, agreeable, or ornamental, a consultation was held for the best means to be used for his disposal. As no feasible method appeared for his immediate transfer to the shore with beneficial effect upon his kindred, Mr. Welson asked the privilege of retaining him on board as a subject for instinctive experiment. The savage chief having, in the thoughtless zeal of the two sailors, — bred from automatic education on board of a " man of war," — received gratuitous injury, they lost caste in the captain's favor, which caused them to " overhaul " their thoughts for a restorative. Bill sagely remarked that " What's done's done, but now we see the drift to smooth water we must kedge for the current and a fair wind; so we must try to make the old shark as comfortable as we can." This opinion meeting with the hearty approval of his mate, they at once "set about" rectifying the effects of their brutality, without fully realizing in thought the extent of their own culpability. Still there was a vague remonstrance that " loomed up " from youthful impression which admonished them of the source of the captain's silent reproof. While engaged in their propitiatory labors the Indian boy, or " cub " as they styled him, watched, and apparently detected the source of the kindly influence wrought in the mood of the sailors. His looks of grateful appreciation attracted the sailors' attention, which caused Jack to exclaim, " I say Bill, the young un's throwing out signals of distress; odds, we were too hard on the old brute. P'raps we can take the young un in tow; suppose we give him an outfit, he seems to take kindly to his head-gear."

Bill bestowed an " observation " on the boy, and became convinced that no treachery was meditated, but that all was fair and above board, so they resolved to rig him out ship-shape in their watch below. Their intention being discovered while in progress, there was a general overhauling of kits, so that the originators were obliged to accept contributions in excess of their requirements. Aabrawa, while the metamorphosis was in progress, discovered that he was an adopted prisoner of the old savage, and that his name with his own tribe was Waantha. To all the trial changes in the process of clothing him, Waantha submitted with unmistakable evidences of gratification; and when fully dressed to the satisfaction of his impromtu guardians he was escorted by Antonio and the sailors aft for the captain's inspection and approval. The pleasing expression of his joyfully bewildered face won the kindly confidence of all, and he was voted his liberty. When asked by the captain if he would like to be employed, he expressed his desire to help Antonio, who with permission cordially adopted him as an apprentice in the culinary department. When duly installed, as a dish-washer, the concentrated ire of the old chief was fully aroused, causing his eyes to fairly scintillate with fury as he readily understood that his plans would be exposed. The sailors' thoughtful endeavors to win back the captain's favor gradually proved successful, and when fully reinstated showed a careful regard for its retention.

Mr. Dow in his naturalistic wanderings had acquired a keenness of perception for the detection of danger from premonitory indications that exceeded, from his natural endowments, the sagacity of the veteran trappers of the North American wilds, so that with Aabrawa and his two well trained dogs he had felt himself proof from surprisal. In proof of his cultivated superiority he instructed the members of the corps in the various causes inciting the flight of birds along the banks of the river and over the distant forests, which invariably proved to be correct in inception. The flight of water-fowl disturbed by alligators or other causes, birds by serpents or monkeys, or like inimical foes, he could detect the intruding species with unerring certainty while distant to the utmost reach of the eye. Early in the afternoon a flight of parrots rose over a distant headland, settling again in the same place; this was repeated frequently with upward impetuosity, which with irregularity in rise and descent indicated some vengeful cause. In explanation, Mr. Dow said, " You will find on rounding the headland a settlement of Brazilian apes, of a different species from any you have yet seen, also in the neighborhood a plantation of sugar bananas. These the natives believe the apes plant, as the spot selected is always adapted in a special way to their growth, and in close proximity to a grove of trees suited in spread of limbs for their arboreal habitations. The parrots have likewise a great fondness for the luscious fruit, which is known as the ape banana, and gather in flocks for poaching depredations, in which large numbers lose their lives, for they are no match in quickness of flight for the nimble quadrumanal defenders of the rights of freehold proprietorship, who have acquired considerable skill in the use of projectile weapons. When we reach the plantation you will find them engaged in defending ' the fruits of their labor,' unless the unusual appearance of the steamer alarms both parties."

Doubling the headland a well protected cove opened to view with a crescent shaped hill sloping to the southwest, enclosing in its semi-amphitheatre a tamarisk grove with a banana patch upon the rise of the hillside. As the parrots had taken flight on the approach of the boat, and there were no signs of Indians or apes, the members of the corps proposed an exploring party for the verification of Mr. Dow's descriptive sagacity. Mr. Dow excused himself from joining the exploring party, on the plea that he had once visited a settlement on one of the tributaries of the Amazon, of which he still retained a vivid impression, that was too recent to require revival. His ambiguity in describing the peculiarities of their domestic economy and defensive resources we had occasion to recollect. After precautionary measures had been taken to avoid surprise from the tracking savages, we landed, directing our steps in the first instance to the banana plantation. Its appearance well sustained the popular traditions of the Indians, as the plants were separated by well defined paths, and around their stalks not a weed or spear of grass was to be detected. This at least denoted care in grubbing, which of itself is an initiatory indication of cultivation. The plat was continued within the slope of the hillock; at one time the bluff bank of an inlet from the river which had been filled up by the drift debris and alluvial deposit caught in its curve, intermixed with the wash from the highlands. After completing our survey of the banana garden, and in our progress selecting and cutting unbidden the ripest bunches of the golden fruit, which were sent on board, we descended into the basin of the tamarisk grove to inspect the community habitations of the apes. Supposing, from the universal silence, that the inhabitants had fled in alarm on the steamer's approach, we were admiring the high order of architecture displayed in the arrangement of their habitations, at the same time questioning with wonder their unnatural desertion despite the prevailing curiosity of the species in the presence of mankind, when a guttural challenge was reëchoed from hundreds of mouths in answer to our query. In a moment the branches above were alive with the hosts we had excluded from our reckoning, who in chattering response tendered us the hospitalities of their aerial city in a shower of cocoanuts, stones, clubs, and other missiles rank with the "reverence" of ordure, prostrating three of our number outright, while they bewrayed all with an unendurable odor, that would have rendered the stink-pots of ancient Greece worthy of being esteemed pouncet-boxes for relief. These tokens of high admiration, designed for the distinguished reception of allied humanity, were accompanied with a jabbering outburst which could only be likened to an explosion of Chinese tongues. To save ourselves was impossible, for in a moment after they had discharged their weapons, pendant from every branch above was an ape ready to fall upon us. At this threatened juncture, when our lives depended upon the drop, the screech of the steam-whistle saved us. Some of our late assailants, paralyzed with the fearful shriek, dropped nerveless to the ground; others upon us, and clinging to our persons grinned beseechingly for protection. But the majority swung themselves from limb to limb in wild panic, disappearing over the brow of the hill. Without waiting to test the permanency of their fears, or courage for a rally, we shook off our personal attachments, and assisted the wounded on board, under cover of the still sounding whistle. In candor I must confess that our reception by those who remained on board ill accorded, from a lack of pitying sympathy, with our narrow escape from imminent peril. Yet I will as frankly acknowledge that there was ample cause for the levity of their manifest disgust at our approach; but when the old savage added his grin to the measure of our disgrace it was more than human nature could bear, and we thankfully accepted a warm bath, in our clothes, proffered by the engineer, while standing on the outjutting portion of the gangway plank, which he administered through the nozzle of the deck hose. Even Jack, who had received an ugly gash which had sounded the depth of his scalp, was obliged to submit to purification before Doctor Baāhar would bestow upon him the rites of absolution conferred by adhesive plaster, notwithstanding his plight was equally abnormal. But the sailor, in the spirit of his invincible good humor, provoked by the novelty of the encounter, declared that he knew the fellow who had barked his head-piece, and would have his revenge. Although we failed to appreciate the mirth of our scathless "friends," we were exceedingly thankful for our escape, for we realized in the cool moments of reflection the peril we had encountered too vividly for the capital of a laugh at our own expense. Neither did we wish for a second trial of Mr. Dow's skill in apeing practical jokes. Bill, in expressing his gratitude for his friend's escape, said, "There you lay, Jack, knocked on the head, and them fellows just ready to drop on us tooth and nail; well, I can tell you our lives wern't worth the flutter of a gaff to'sel in a gale of wind, when the whistle brought them up with a sharp turn. But what's food for one's fun for another; the squall just took the wild ones aback like the wink of a gib in a luff, so they turned tail and scuttled away, and we hauled off for repairs, mighty glad they didn't grapple."

While the explorers' ablutions were in progress ape sentinels were seen in the tree tops above their habitations, in which position they continued until a curve of the river concealed them from view.

June 8. — Large parties of Indians have been seen inland on both banks of the river during the day. The swiftness of the river's current has greatly increased, giving indication of an upward incline to a more elevated plateau. Open glades reaching to the river are now of frequent occurrence. The left or eastern bank is less defined than the western, and bears stronger evidences of alluvial deposits in its arboreal growths.

June 10. — Our redeemed captive boy begins to show many pleasing traits, among which grateful fidelity is not the least. His attachment to Antonio, who first bestowed upon him pitying kindness, is prominently manifest and touching in the simplicity of its promptings. He desired Aabrawa to ask the captain to allow him to remain on board, promising that he would try and speak and make himself useful when recovered from his wound. The captain received his professions of attachment with a warmth that made his eyes glisten with joy. Mr. Welson suggested that it would be necessary to christen him, proposing that Padre Simon should officiate in administering baptismal rites. But the padre objected that he was not in orders, and for a layman to assume the solemn responsibility of baptizing was in his opinion but a grade less than presumptuous blasphemy. M. Hollydorf referred him to the example of John the Baptist when in a similar position, exhorting him to do his duty fearlessly, as the act of consummating the conversion of a heathen would be esteemed a meritorious service by the most bigoted of the sects. The padre still urged, " He does not understand our language, and consequently the effect of redeeming grace necessary for the consecrational rites of Christian adoption fulfilled by baptism." Mr. Welson said, he need have no scruples on that score, for Xavier, Ricci, and other missionary apostles of the Church boast, each, of the baptism of five thousand and more heathen Chinese in less than a month after their arrival in the country, and without being able to communicate with their catechumens by the aid of interpretation. Having a strong reverence for the opinion of Mr. Welson, he reluctantly consented to officiate. Antonio standing as godfather, he was christened "Tortuga Waantha." Scenes of this description were a source of renewed vitality to Mr. Welson, as it afforded him special delight to expose the vagaries of the three professions founded upon theoretical science. In fact, the very chairs of his Panamanian office were made available for startling effects in support of his specialities; indeed, his reputation had obtained such distant recognition, that strangers en route preferred to stand isolated in his presence. From these experimental essays none of his friends escaped; sensitiveness, dignity, and reserve, were in fact special invitations for the exercise of his curative skill, if in the slightest degree morbid in tendency. After meridian, when his books had been laid aside for the day, it had been his custom to indulge his quizzical humor in trolling for fun, and it was a rare occasion that did not offer a European or American gudgeon, isthmus bound, ready to take his bait.

As before mentioned, it had been his intention to return from his river voyage by a Brazilian steamer, but the varied characteristics of the members of the scientific corps, with the chance additions, made him resolve to forego the obligations of his business relations for the indulgence offered to his humorous inclinations. Meeting unexpectedly with his old friend Dow at Santa Anna, he eagerly seconded the exploring adventure of the Pilcomayo, from the prospective novelty it offered for the cultivation of his humorous studies. In addition to the incompatible whimsies of scientific association, the questionable reports of an undiscovered inland city provoked a second incentive. With this more explicit introduction of Mr. Welson, who from accident and inclination became one of the most important aids in directing and harmonizing the attainable objects of the expedition, we will resume the thread of our narrative.

CHAPTER V.

NOTWITHSTANDING the confirmed assurance of the sufficient efficacy of the antidote applied by Waantha for counteracting the poisonous inoculation of the padre's arm, he still continued the use of whiskey with the thoughtless lack of consideration that fosters habits of indulgence and self-imposed penalties. In verification of the advanced statement, that artificial stimulation gave birth to war and the three curative professions, the padre, in common with his paternal ancestors, became polemically disposed when subject to the influence of his imposed habits. Waantha's happy manifestations of "regeneration" caused him to urge dogmatically, " You must acknowledge, Mr. Welson, that the Jesuit fathers have done much good, for of all nations and sects they alone have suceeeded in bringing tribes of Indians under the influence of civilized control."

" Yes," replied Mr. Welson, but with the reprobating clause, that " they have manifested in all their missionary labors a paramount zeal for the selfish aggrandizement of their partisan order in the extension of its power for enforcing the control of a hypocritical despotism ; the real welfare of the heathen converts being held as a blind of nominal consideration. Indeed, the Jesuitical method enacts the part of whiskey in its habitual rule over the faculties of civilized society ; in conjuring for the subjugation of reason superstition for the supremacy of fanatical instinct."

The padre startled, exclaimed, " Upon my conscience, Mr. Welson, I am afraid you are little better than an infidel!"

Mr. Welson left the padre with an ill-concealed show of disdain. Finding M. Hollydorf engaged, with the assistance of Mr. Dow, in removing a powerful electro-magnetic battery — one of Shockwit's best — from its case, it occurred to him that amusement, if not more permanent advantage, might be derived in trying its effect upon the savage chief. This proposition was readily adopted, with the resolve that only those necessary for the working accomplishment of their purpose should understand the nature of their occupations. The experiment, under the experienced management of Mr. Welson, promised some rare developments of motor effects, in the production of instinctive superstition, without committing an act of cruelty beyond the wholesome excitement of animal fear. As it was necessary to keep the instrument out of sight to secure the full impression of supernatural effect, the captain offered his stateroom as the best adapted for the preservation of secrecy and the effectual working of the instrument. With the aid of the two sailors, the wires were passed out of the port and run unobserved outside of the bulwarks, and so arranged that the old savage could not escape the full force of the electrical shock. When completed, the connection of the circuit was tried in the absence of persons from the neighborhood of the intended victim. The result was a prolonged yell, that not only surprised the uninitiated on board, but brought inquiring heads forth from ambush on shore. To the wonder and alarm of all on board excluded from a participation in the secret, the old savage was found writhing in an agony of fear entirely bereft of stoicism. Various explanations were suggested to account for the startling phenomena. The padre admonished Mr. Welson that it was, without doubt, the working of the spirit of repentant regeneration, as the Fathers had recorded numerous instances where the self-convicted had cried out in anguish, "What shall I do to be saved?"—the fact being made known after they had acquired a knowledge of missionary language. He averred that there could be but little doubt that it was the workings of the spirit of conviction, from the agony of his expression. Thereupon he desired Aabrawa to inquire into the cause, as it had all of the appearance of a miraculous conversion. But the old chief stared at Aabrawa, helplessly unable to speak through an excess of fear. Mr. Welson then counter-admonished the padre, that as a professed follower of the Church it was his evident duty to point out to the convert the appointed way of salvation. As all supported this suggestion, the padre remonstrated, while looking wofully at his arm, " I once offered him my sympathy and aid for his relief, but he repulsed me so brutally, upon my conscience, I am afraid to try him again."

His attention being called to the helpless condition of his late antagonist, he was finally persuaded to adventure one of his hands upon the head of the savage in the way of benediction. Answering to a given signal the battery claimed the padre as a victim through the chief, whose yell was accompanied with the exclamation, " My conscience' sake alive !"—then his fears became as vivid in expression as those of his intended convert. Mr. Welson, addressing himself somewhat scornfully to the padre, said, "You accused me of infidelity when I endeavored to use my privileged endowment of reason bestowed by the Creator for human direction ; now you will see how much better it serves as an exorcist than your faith in a religion that ignores man's duty for the fulfillment of intention in its bestowal." He then made a few passes over the Indian, and when he had gained the full attraction of fearful awe with mumbling incantations, the padre was reluctantly induced to replace his hands on the chief's shoulders and remove them without alarming impression. Then assuming an awful aspect and tone, as if addressing the powers of air with the spirit of invocation, he implored their aid to convict the reptile savage, and civilized devotee of a blind infatuation, of their willful errors alike dangerous to the well-being of humanity. When made sufficiently impressive he commanded the padre to take the chief's hands. Overawed by the majestic impersonation of sublime authority enacted by Mr. Welson, the two joined hands, both keeping their eyes fastened in blank wonder upon his face and movements. The conjuration having fixed their attention, he pronounced in a loud voice the magic word "Letonnow!" Immediately the two commenced a series of contortionate grimaces, directed toward each other, accompanied with spasmodic handjerking. The actors were so engrossed with their fears that the spectators were fain to have recourse to a variety of succedaneum vents to suppress the outburst of laughter, the sailors adopting the novel expedient of revolving their quids around the tips of their tongues, which ejected a jet of saturated decoction from the corners of their mouths with every revolution. But for Mr. Welson's practiced command of his emotions, subject to the control of judgment, the ludicrous scene might have been continued to the extent of injury, for his associates were, from spasmodic action, to all intents speechless. When at length the larger fraction of a minute had been exhausted in husky attempts to command his voice, he managed to stay proceedings with a sign evoked from head and hand, faintly sustained with a vocal negative. When the current was checked the last vestige of ferocity had departed from the face of the savage, leaving the vacuum unsupplied, as it was his sole dependence for the facial expression of his emotions. The padre's face was confounded with a blending of superstitious dread and suspicion, for with all his phantasmic nervousness provoked by the excessive remedial use of whiskey and tobacco, he could not fail to detect the covert effort of restraint that prevailed. Indeed, with his natural powers of perception free from their imposed embargo, he would have detected the means employed for the production of effects known to the most illiterate members of scientific academies. To dissipate his suspicions the padre had recourse to Doctor Baāhar, of whom he anxiously inquired whether Mr. Welson derived his power from a legitimate source compatible with the apostolic faith inculcated by the tenets of the Church. The doctor, as instinctively absurd when out of the scholastic thills of antiquity, found especial gratification in teasing those subject to the common frailties of his kind. So, taking his cue from the padre's necromantic suggestion, he explained that Eusebius, and other Fathers of the primitive Church acquainted with the practice of Egyptian astrology, had confirmed the prevalent belief that in certain families, under peculiar conditions, there was a power developed similar to that exhibited by Mr. Welson.

Here Mr. Dow interrupted their conversation by calling the attention of the padre to the savage, who was following Mr. Welson with the docility of a spaniel. Observing his emotions of superstition he asked, " Are you in reality so blind, padre, that you are unable to detect the agency of Mr. Welson's power over the savage ? You seem to be impressed with the belief that Mr. Welson has been enacting the part of a magician in producing these effects upon the savage, whose ignorance sympathizes with, or rather reciprocates your superstitious delusions ? How is it possible for you to overlook, with thought, an impression so familiar to your understanding, and in fact, place yourself on a level with this savage from a lack of intelligent perception ? Really, padre, you confound me with astonishment. Time, place, and circumstances, with certain abetting aids, have thrown you off your guard." A shake of Mr. Welson's head prevented Mr. Dow from revealing the means employed, as he wished to confound the padre with further evidences of his simplicity and heedlessness. Beckoning the sailor satellites of the savage, he was led back to his place of confinement, and secured in contact with the wires of the battery; then, when the padre's attention was otherwise engaged, a glass of whiskey from his bottle was administered by Mr. Welson to his experimental victim. But a short time had elapsed when attention was called to an unusual disturbance forward, in which the fierce snarling growl of the dogs was commingled with the guttural "ughs" of the savage, whose face was contorted with an expression of demoniac rage, causing his mouth to froth, exposing through its slaver his pointed teeth, while his eyes gleamed with a ferocity that prompted the padre to flight. But when assured that he was securely confined, the padre asked Mr. Dow what he thought of the source of Mr. Welson's agency now! Mr. Dow led him to the captain's room; with a glance at the instrument the nature of his ludicrous position began to dawn. But when his whiskey bottle with diminished contents was produced and proclaimed as the magician of ferocity, his face mantled with the scarlet dismay of shame, which with his ejaculation of " My goodness gracious, what a fool I have been!" filled the cup of mirth to overflowing.

Since the morning of the 9th the strength of the current had increased so rapidly that the captain feared we were approaching impassable rapids; but at nightfall we entered into a broad expanse of water resembling a lake. Keeping beyond the range of arrows, Mr. Dow and Welson in the punt succeeded in killing sufficient wild fowl for a week's supply. Shortly after nightfall the dogs with their muzzles primed over the chocks kept up a warning cry. Waantha with a crutch, the gift of the carpenter, hopped about the deck with eyes on the alert, and ears primed for sounds from the water and shore. Through the night his vigilance was sustained, until in the darkness of the morning hours he aroused Jack's attention to floating objects on the water just visible to his sight, but while peering the whiz of an arrow interpreted the source of danger. The angle of flight enabled him to judge with tolerable correctness the position of the foe who discharged it; the yells which answered the report of his escopeta loaded with buckshot bespoke his success with others if not the one whose intention provoked retaliation.

June 11.— Jack's morning salute awoke all on board, causing a general muster to learn the source of provocation. While Mr. Dow was taking his coffee in the dawning twilight, Waantha hobbled to the place where he was sitting and after directing his attention to an approaching swan, took one of the dead ducks hanging under the awning and placed it on his head, at the same time imitating the movements of a man decoy. Understanding his meaning, Mr. Dow took his rifle from the rack and sped a bullet with sure aim; the unfortunate bird extended above the surface a black pair of arms, then with a gurgling cry sunk out of sight. Flocks of ducks which had been gradually nearing the steamer on all sides made for the shore without taking wing, showing by the wake the nature of the fowl before the submerged Indians clambered up the banks. The undaunted perseverance of the savages in tracking the steamer, despite of our superior weapons, showed an indomitable determination, proof to danger and disappointment, which detracted greatly from our prospective feelings of safety when exposed to the disadvantages of land travel.

The steam-whistle and gong had startled them at first, but they had tested their harmless natures, and evidently thought the howitzers relatives, whose destructiveness could be avoided as easily as the poison of their arrows when they had obtained a knowledge of the antidote. The forbearance of the captain had favored this impression, and it was determined in consultation to use our weapons to the full extent of their destructiveness. An opportunity was soon offered, for in passing a raft lodged on the eastern shore Waantha pointed out a rampart of logs ready poised for an overthrow, with insterstices between in which were seen the protruding muzzles of their blow-pipes. One of the mountain howitzers loaded with solid shot was discharged point-blank against the upper tier causing it to fall inward, catching the lurking savages in their own trap, while it exposed those in the rear to the full effect of grape and our small arms, which caused the river echoes to resound with the yells of the wounded. Without stopping to learn the extent of the slaughter, the steamer kept on her course. In passing a glade reaching to the water the plain was seen covered with panic-stricken savages on foot and horseback, directing their course to the foot-hills. Although surprised at the large number collected, we felt safe with the impression that the wood rafts of the left bank would be left free for our acceptance thereafter.

June 14.— While collecting wood from the scattered lodgments of the western bank, parties of mounted Indians watched our movements from the opposite plain. These Waantha informed us were of his own tribe. When asked if he would like to be set on shore to rejoin them, he expressed, with signs, a reproachful negative, blended with fear and sorrow. After a moment's hesitation he seemed to understand that the proposal was made to test his feelings, then with a pleased look of Indian cunning he pointed to the old chief, who had been regarding him with a revengeful look of ferocity. Understanding his meaning as a proposal of substitution, Mr. Welson asked, through Aabrawa, if they would kill the old chief if set on shore? This was answered with a decided negative, and the pantomimic addenda of labor as a substitute for death. As the captive was sufficiently recovered from his wounds to control his own movements, Mr. Welson took him in charge for initiatory preparation in presage for association with his foes on shore. That it might not, in form, be considered an arbitrary expedient for riddance, after Mr. Parry had fitted to his neck a brass collar, proof to Indian appliances for removal, he was freed from his bonds under the supervision of Mr. Welson, who offered him his choice between the continued hospitalities of the steamer, or liberty, such as he might be able to secure from his congeners on either bank of the river? The speedy announcement of his choice was urged by three strong shocks of the battery. When his agitating consternation had sufficiently subsided from the last talismanic touch to his neck decoration, his head disappeared over the bulwarks with his heels in reversion, giving farewell nods to his civilized entertainers. When last seen beneath the water's surface he was making for the eastern shore with a frog's exampled despatch.

The kind-hearted readers will be unnecessarily excited, if from the foregoing relation they are inclined to think our enactments were dictated solely for the gratification of instinctive mirth. Mr. Welson's object was to obtain a clear demonstration of instinct in the rudimentary foundation of habit as the source of progressive inclination in its bearings upon the present standard of civilization. The participation of the padre in the vague terrors of the savage from a reciprocation in kind, from the two extremes of cultivated progression, offered absolute evidence of a common origin and source of provocation, the variations in expression being dependent upon practiced habits and customs. The padre attempted to offer his own experience to subvert the ferocious testimony of the old savage while under the effects of whiskey, pleading that it had ever exerted an opposite influence with him, exciting in its action a genial flow of sympathy. This partial testimony was overruled by the acknowledgment that in social whiskey bouts, indulged in as night passatiempos, he had invariably been obliged to act as a peaceful arbitrator. With the impression made from the effects of whiskey on the savage, all our habits of indulgence were curtailed, greatly to the advantage of kindly reciprocation which had often been chilled by theoretical disputations that ended as they began, in the void of instinctive mutation.

CHAPTER VI.

THE constantly increasing perils of the voyage from the pertinacity of our savage foes, recalled the warning words of an old priest of Santa Anna who had engaged in one of the Jesuitical expeditions. He advised us to keep at a safe distance from the shore, and never attempt to hold friendly intercourse with the savages, or endeavor to conciliate them with presents, as it would expose us to their deadly treachery. "You must be constant in your guard or they will board you in the night, for they are as familiar with darkness and water as the land. If they come within reach of your guns kill and spare not, for fear, if you can inspire it, will be your only source of safety." Our daily experience had thus far confirmed the prudence of his advice, and it was yet a question of extreme hazard if we should attempt to land. Each day afforded additional evidence that the tribes were banded together in a defensive alliance against the whites, with a politic foresight that made intertribal jealousies secondary to their exclusion. When partisan ferocity, so deadly in manifestation with the aboriginal races of America, could be made to coalesce for protection against the agressive tendencies of a race in customs and habits inimical to their own, it seemed an act of desperation to attempt the farther prosecution of our Quixotic enterprise. This feeling had perceptibly gained strength while the ferocious characteristics of the old savage remained unsubdued, under the impression that our vitality was held with a lease as precarious as his own. The padre's exhibition of fear had established him in the belief that in stoical courage we were inferior to his own race. This impression he had evidently found means to convey to his tribe. But Mr. Welson had, by a seemingly chance train of humorous experiment, dissipated his reliance upon the savage hypothesis of instinctive sagacity. The fancied superiority of his exaltation realized to the old chief the attributes of deity, while the padre became reduced, in his estimation, to a kindred caste with his tribe. In train the gyved circlet of his neck, as a talismanic badge of investment, would be likely to afford material evidence in proof of Mr. Welson's deitistical power. These impressions, which it would be natural for him to impart, would prove omentious as a prestige of awe, similar in effect to that afforded by Moses to the Israelites. The contrast between the old chief and Waantha discovered a marked distinction in tribal caste, dependent upon the miasmatic influence exerted by local impressions derived from degrees of purity in the sources of exhalation that gave birth to kindred habits and customs. The former devoted his attention to engendered animosity, while the latter eagerly searched for some token of kindly sympathy, and when it was bestowed his whole being became instinct with grateful pleasure. Even the dogs evinced an inherent perception of Waantha's higher grade by fawning acknowledgment, while with the old chief the defiant acrimony increased rather than diminished. In habits the same characteristic features prevailed. In eating the old savage used as little ceremony as the dogs, and far less in the modest observance of the other requirements of nature; while the younger seemed to derive intense pleasure from cleanly imitations. With these instinctive demonstrations we will resume our descriptive course.

June 17. The three previous days passed without any active indications on the part of our tracking foes,

THE MANATITLANS. 47

but during the twilight dawn of this morning Waantha discovered parties crossing the river in advance from the right to the left bank. With every safe opportunity fuel was renewed to guard against unforeseen emergencies. At noon large bodies of Indians were seen watching our progress from eminences inland, and the trees of either shore. Their appearance caused M. Hollydorf to question his duty in opposition to the prospect the adventure offered for the fulfillment of his commission. All, with the exception of Mr. Dow, expressed themselves in terms of discouragement. Dr. Baāhar depicted the horrors of a death from putrefactive poison, which entailed in life the lingering corruption of bodily decomposition, which even the vultures would disdain to hasten. Mr. Dow was obliged to acknowledge that the preoccupation of their thoughts, while engaged in field avocations, would expose them to certain surprise, and inevitable extermination. But he had set his heart upon the venture and pleaded the advantage that would accrue from the river's exploration, hoping for some chance interposition for the furtherance of his enterprise. Captain Greenwood, for the relief of Mr. Dow, proposed that the exploration of the river should be continued as far as admissible for the safety of the steamer. M. Hollydorf accepted this proviso, notwithstanding the loss of time it would cause.

June 19. While the captain and M. Hollydorf were engaged with the calculation of their meridian observations, just as the steamer was closing a long reach, Waantha hobbled aft in great excitement, pointing with energetic gesticulation to a headland we were approaching, and then to our guns on the forecastle deck. Interpreting some new emprise on the part of our savage foes, the boat was kept in the centre of the current, until the view opened beyond the headland, when in melée encounter were seen parties on horseback. On nearer approach women and children were discovered huddled together within a barrier of mules and horses. Parry, the engineer, always prompt with his weapon, sounded a parley, which caused a momentary cessation of hostilities, allowing the boat to gain a position commanding a full view of the parties engaged. A glance, aided by the imploring gestures of the women, whose garments and other indications bespoke an approach to civilized origin, at once enlisted the inclination of our sympathy. The novelty of a scene so unexpected, rendered us for a moment undecided how to act, but the sound of Antonio's Chinese weapon restored our presence of mind. The Indians quickly recovering from the momentary panic, caused by the shriek of the whistle and clangor of the gong, engaged in a renewed charge upon the unfortunates, who were defending their families with the desperation of despair, and in numbers seemed scarcely one to ten of their foes. The charge of the Indians was accompanied with a derisive whoop, this was almost simultaneously echoed back by the bray of the mules opposed in forlorn hope, which revived Mr. Dow's with a realizing perception of the ways and means for the achievement of his ambitious project. His rifle had reported the death of four Indians before a general volley put the survivors to flight. The rescued, when they saw the Indians fall and themselves spared, hastened down the bank that they might not interpose their bodies as shields to the savages. The panic of the Indians who were in flight over the pampa was increased by a shell, the report of the gun startling from the western shore a party lying in wait for the issue of the battle on the eastern, with the probable hopes of a chance advantage to themselves. Acting upon the hint that there were among them those who had witnessed the effect of shot and shell on a former occasion, the opportunity was embraced for reviving the impression. When satisfied that all able to molest had carried their bodies out of range, preparations were made for landing to succor the rescued with food and raiment, for they appeared to be in a deplorable condition.

Before landing for the personal expression of sympathy, the punt was loaded with provisions and dispatched to allay the immediate cravings of hunger. The steamer in the meantime was moored to a woodrift, from which the captain and members of the corps gained the shore. They were received by a man past the middle age, whose face was exceedingly attractive, although wan with fatigue and anxiety. Momentarily embarrassed, as if with doubt of his capacity to make his emotions of gratitude intelligible, he bowed himself down with the intention of prostrating himself at the feet of the captain, but this act of humiliation was arrested by the grasp and hearty shake of his hands. As distress evokes compassionate emotions with the kind-hearted, the captain's eyes were not alone mindful in the reciprocation of the stranger's outburst of grateful tears. Quick in demonstration, when his generous impulses were aroused, the captain exceeded the cautious discretion that usually guarded his movements, from fear of imposition, by bestowing a hearty embrace of sympathy upon the careworn guardian of the rescued flock. This act caused, with one exception, a general prostration accompanied with a grateful outburst of tears. The exception to this indicative act of eastern humiliation, bestowed alike in reverence to the tyrant and benefactor, was a maiden who had probably numbered eighteen seasons. Tall and erect in stature, she stood unmindful of the prostrate throng, but not unmoved by the scene enacted between the representative leaders of the rescuers and rescued. The clear transparency of her skin, with the healthy purity of its texture, combined with a graceful form, exceeding in height those with whom she was associated, declared her at once alien to them by birth. Seemingly aware that grateful expressions confined to pantomimic enactment would at the close of the introductory scene prove embarrassing, she advanced, after securing with touch the companionship of two young maidens who had prostrated themselves beside her. Approaching Captain Greenwood, she addressed him in an unknown tongue, which M. Hollydorf with surprise recognized as an idiom of the Latin language. His wonder was augmented by her confident assumption that there were among us some who would be able to converse with her, and through her interpretation would be enabled to hold communication with her protectors, her companions speaking a dialect in remote correspondence with her own. The captain, although gratefully recompensed for his lack of language by the eyes of the fair vision, felt himself unaccountably moved in his isolation, notwithstanding she continued to bestow upon him from those members sympathetic admiration exceeding the compass of speech. The maiden announced herself as a native of Heraclea of the Falls, a walled city but a few days' travel remote.

"My name," she continued, "is Correliana Adinope, daughter of the Prætor Adinope, in body deceased, and step-daughter to Adestus the present Prætor. The city has sustained a constant siege for centuries by the savages in revenge for the wrongs committed against them by our ancestors. Its inhabitants are at the present time in the extremity of distress from pestilence engendered by famine. While endeavoring to obtain remedial plants without the city walls I was made prisoner by a band of our besiegers, and was rescued immediately by these fugitives, whom in turn you have saved from destruction."

Having satisfied in outline the curiosity of M. Hollydorf, she begged that safe means of rest might be afforded her protectors, for they had been constantly harassed for weeks without an hour's undisturbed sleep. But long before the preparations were completed for a comfortable resting place, Correliana and the wounded were the only ones that remained awake.

Waantha, assisted by the guacâcioes of the crew, collected from the hair and mouths of the dead Indians antidotes, and from the growths of the river bank counteractive remedies, which relieved the excruciating pain of the wounded, and stayed the progress of gangrenous putrefaction. At sunset all the rescued were in a deep lethargic sleep, and as the night was pleasant, and the glade where they lay was open to the river, with a day draught that freed it from miasm, but little fear was apprehended from their exposure, notwithstanding the tattered condition of their clothing. Fortunately, before the evening was far advanced, the captain bethought himself of his trading stock, from which he soon obtained fabrics well adapted for their protection. Mr. Dow, restored to the full vigor of ambitious vitality, busied himself in organizing a guard for the protection of the mules and horses, listening the while to Aabrawa's relation of their owner's source, for he had recognized them as belonging to a colony located far to the eastward of his place of nativity, who were known to his people by the name of Bamboyles. Mr. Dow viewing his night charge as the keys destined to unlock the gates of his New Jerusalem, he picketed them in the most verdant portion of the glade. When morning dawned his fears were startled to find them still prone with scarcely a sign of vitality; and as his attempts to arouse them failed to elicit more than a drowsy snort he feared that with all his vigilance they had been poisoned, but was reassured by Dr. Baūhar, who pronounced their immobile condition as lethargic, induced from hunger and fatigue.

While the night dew was still on the foliage Waantha pointed to a long line of animals approaching the river from the plain, which proved to be llamas. Upon this hint the three marksmen took the steamer's boat to find their " toch," or path to the river, and were successful in securing a supply of game sufficient for several days' consumption. Before his guests awoke the captain had prepared a tent for the reception of the women and children, and an abundance of food for all. In addition he was able to furnish from his trading stock dresses, which, with a little alteration, would supply the requirements of the women.

The mayorong, or chief of the Bamboyles, and Correliana were the first to awake in the morning; the latter, with her two companions, were conducted to the tent and there presented with the means of renewing their garments. In communicating the kindly expressions bestowed, with the gifts, her companions, in returning thanks, used the Spanish idiom, which startled Mr. Welson with pleasurable surprise, as it opened to him a direct avenue of speaking intercourse, for its varied provincialisms were as familiar to him as his patrial mother tongue. After the agreeable confusion, occasioned by Mr. Welson addressing them in Spanish, had subsided, the eldest introduced herself as Cleorita and her sister as Oviata Arcos, daughters of Don Santiago Arcos, a native of Madrid, the chief city of Spain. On hearing this announcement he became joyfully elated, bestowing upon both a fond recognition, as they were the daughters of a personal friend of former years. After a long conversation, in which they gave him an outline history of their people, and the cause that forced them to become wandering exiles from their loved country, with the distressful mishaps which had attended their search for a new home, they separated reluctantly for the day. In answer to Mr. Welson's sympathetic desire to render burial assistance in the regretful disposal of their dead relatives, the mayorong replied, that unless their preservers especially wished to be present they would prefer to indulge in their sorrows alone. Readily understanding the motive, Mr. Welson and associates returned to the steamer while the ceremonies were in progress.

As Waantha had discovered Indian scouts lurking above and below upon either bank of the river, Mr. Dow exercised his engineering skill in forming on the pampa a defensive redoubt for the night protection of the horses and mules. Dr. Baāhar theoretically explained the Latin nomenclature of the different departments of the Roman castrum, which possessed from his natural and cultivated innocence from mechanical attaint the supreme "virtue" of novelty. Mr. Dow submitted to his classical dictations, but stoutly refused to adopt his method of fortification, which the doctor styled *fossa cingere internus*, or moating inside of the redoubt, notwithstanding the strongly urged advantage of its stratagetic intention of concealment, that would lead the savages, on gaining the summit of the embankment, to take a blind leap into it. Fortunately the padre was present to divert the argument, which enabled him to render practical assistance to his Bamboyle aids for the completion of the inclosure in time for the night's occupation. The absence of the doctor and padre from the supper table caused the captain to inquire where they were? Mr. Dow said that he had left them but a short time previous seated on the sods of the embankment engaged in a dogmatic discussion of the feasibility of the various methods adopted by the ancients and moderns for citadel defense, the doctor quoting from "Plutarch's Lives," and the padre from Bunyan's "Holy War" as the best English authority. Aggravated by the heedless lack of sympathy shown in the use of their tongues, the while withholding the useful aid of their hands, the captain, on their appearance, reprimanded the doctor over the padre's shoulders with tart severity, which caused both to give heed to the practical suggestions of Mr. Welson in train for the outfit of the overland expedition. From the direction of Correliana, who seemed to have an innate perception of her entertainers' dispositions, the captain concluded to continue the voyage up the river to a point she described as more favorable for debarkation, as it was nearer the southern passes of the mountains that opened a way to the city of Heraclea.

June 21. After the morning meal a majority of the women and children were brought on board of the steamer, and of the males all that would be likely to impede the progress of the land party having in charge the horses and mules. When ready for the start, the doctor joined the shore party equipped in naturalistic costume, which, in defiance of the recent sad experiences of the Bamboyle women, excited a mirthful inclination; even the more sedate demeanor of Correliana was moved in despite of her efforts to suppress her risible emotions. With his nether bifurcations disappearing, in extremity, within the capacious leg receptacles of boots, a blouse surcoat, or smock frock, elaborately supplied with Sanskrit labeled pockets, depended loosely from his shoulders, reaching to his knees, his head being surmounted with a bell-crowned hat, bestudded with impaling pins, technically called the kaleidoscope. Protruding from the larger pockets were seen the mouths of a pistol barrel, powder and drinking horns, with various articles for insect preservation.

Aware of his uncouth presentment, he pleaded that its adoption combined usefulness with policy, for he had noted in his travels that all tribes and nations bowed down in reverential worship and awe to ugliness; and he felt certain that he had often been indebted to the contributions of his costume for the preservation of his life, while sojourning among the natives of the Polynesian and Ladrone Islands. When fairly mounted upon a mule, who seemed to be affected with emotions peculiar to his species, but seemingly averse to awe and worshipful respect, Mr. Welson could not refrain from commending the happy conjunction as talismanic for the rider's preservation from savage attacks.

It required much coaxing on the part of the mayorong to reconcile the mule to the novel eccentricities of its rider, but in the course of the forenoon he seemed to enter into the humor of his direction with unusual zest. When fully reconciled to the swaying of the doctor's net, with the sharp turns and checks to which he was subjected in the chase of insects, the Bamboyles left them to the full sway of their own moods. Fortunately the saddle was well adapted to secure the safety of its occupant. As they were crossing the opening of a glade, when the day was well advanced, a splendid specimen of the pampa *Nyctaloide* hovered over the cavalada long enough to attract the doctor's attention, then floated away, leisurely, over the plain. In a moment the insect-hunter's net was in hand and, before he could be checked with warning caution, was under full headway in pursuit, and, when fully engaged in following the doublings of his quarry, he became deaf to the mayorong's calls. Feeling secure in being able to keep within hail of the boat, the erratic movements of the doctor had been a source of amusement to the Bamboyles, but as the distance was narrowing between the foot-hills and the river, and withal hummocky, his danger increased. Still he was armed, and little fear was entertained for his safety, for while within call his mule could be brought back with a whistle. As he still kept heedlessly on, the mayorong sent a party of young men to bring him back. They had scarcely started, when a shrill shout from the mayorong urged them on, he and Mr. Dow following at full speed. The cause of these movements was a pursuing Indian close in the wake of the doctor. Unheedful of the danger, the doctor and his mule — who seemed to enjoy the novelty of the chase with his rider's gusto — neared the foot-hills, where a band of Indians were seen watching the strange scene. His frantic gesticulations had undoubtedly impressed them with the belief that he was bestraught with madness; a condition held in especial reverence by aboriginals, — as they continued to regard the movements of the Indian in pursuit with negligent indifference; indeed, from his frequent hesitations, when within the cast of a spear, he seemed to be subject to the restraining influence of the same fear. The mayorong, who had allowed Mr. Dow to overtake him, had twice discharged his rifle in hopes that the report would apprise the doctor of his danger, so that he might use his pistol. But these offensive demonstrations only aggravated his danger, for the band of Indians moved rapidly forward for the rescue of their scout; he at the same time, warned by the rifle reports, cast a calculating glance backward to determine the extent of his own danger. At this juncture the butterfly rose and doubled just without the range of the distracted enthusiast's net, then coquetted backward and forward with all the instinctive blandishments of its human type, showing as little concern for threatened danger as its pursuer. This tack brought the doctor face to face with his foe, who had sprung upright upon the croup of his horse, holding his spear poised ready for the cast. The cool indifference of the doctor to this offensive act, although within reach of the spear's thrust, caused the savage to pause, backing his horse out of the way, as if still doubting the sanity of his meditated victim's self-possession. In this act a bullet with the mayorong's novice aim startled the savage from the close proximity of its whizz, as he started suddenly aside. A quick glance turned toward us determined the doctor's fate just as he succeeded in capturing the tantalizing object of his chase. While in the act of lowering the staff of his net to remove his prize, he received the blow from the cast of the spear aimed at the unprotected portion of his head; the point glancing upward upon the skull divided the scalp on the forehead, reflecting it backward over the crown. The blow forced him backward from the saddle to the ground; at this stage Mr. Dow brought his rifle to bear, which caused the savage to bite the dust, just as he was about to finish his victim with a spear thrust. The blow and report brought back the doctor's scattered senses in time to anticipate with his pistol an attempt upon his throat from the teeth of the Indian's no less savage horse, for the completion of his dead master's unfinished work. This instinctive impulse of self-preservation announced the presence of the doctor's mind, and that he still survived, but the horse, deprived of life, fell forward over his prostrate body, as if to accomplish in death his defunct master's intention. When dragged from beneath the horse Dr. Baāhar looked as if he had been resurrected from a slaughter-house, but he was a naturalist still, for his first thoughts were directed to his captured butterfly. A more striking contrast could scarcely be imagined than that presented by the captor and captured, the former being clothed in blood and the latter in beauty, for it had escaped injury in the conflict. After the doctor had examined the condition of his hat with its contents and garnish of insects, he submitted his head to the mayorong's treatment, with the proviso that his restored scalp should be swathed without washing. When mounted on his mule his appearance was as fruitful of humorous mirth as those attending the most ludicrous mishaps of the valorous knight of La Mancha. The Indians, after the mayorong's party left, held a consultation over the dead body of their scout, which seemed to result in a determination to avenge his death, for the main body, which outnumbered ours in the ratio of three to one, followed, standing on croup in a menacing attitude, occasionally making a dash forward, and as suddenly retreating. These maneuvers were continued for an hour or more, serving to retard the progress of the cavalada, until Mr. Dow, our rear guard, getting out of patience with their annoyance, proposed a long shot with his Spencer rifle which in effect astonished the Indians by dismounting one of the most defiant. This caused evident dismay, for they immediately retreated with all speed to the foot-hills, leaving us unmolested for the rest of the day's stage. Notwithstanding the delays of the land party, they were obliged to wait at the first open glade until night-fall for the arrival of the steamer. After the doctor had submitted to a thorough ablution of body and head, administered by the Bamboyle women, the cause of the steamer's delay was explained.

The steamer, after an hour's progress from her night's moorage, entered a broad expanse of water of lake-like dimensions formed by a confluent tributary from the west. The strong eddy caused by the making out of a spit from the eastern bank forced the boat to the opposite shore covered with the rank growths common to extensive alluvial deposits in semi-tropical latitudes. While the engine was exerting its utmost power to stem the current and cross the walled strength of the combined streams, Waantha, who was at his post with his canine friends, called Mr. Welson's attention by signs to a broad spreading mangrove banian peculiar to the tributary deltas of the large South American rivers, which bear a strong resemblance to kindred growths in India. Among the pendant hybrid limbs, which had taken root in the muddy deposit, there appeared one that seemed to vibrate to and fro, coiling upon itself. With the glass the captain discovered that it was a huge amphibious anaconda hanging pendant from one of the horizontal branches by the prehensile attachment of its tail. The waving excitement of its corrugations and swaying reflection of its head from side to side, within circumscribed limits, aroused the spectators' curiosity to learn the nature of its attraction. A nearer approach discovered, prone upon the interwoven platform of mangrove branches, a huge alligator with his head inward from the river. The reptile relation of the parties foreboded an instinctive encounter of sagacity and strength, which excited in Mr. Welson a strong repulsive desire to witness, as a comparative study, the result of a duel between individual representatives of species so nearly allied on the cold blooded verge of vitality. The captain, in order to afford him the privilege of recording the result for future reference, directed the bow of the boat cautiously toward the scene of encounter.

When sufficiently near to witness the movements of the monsters, who were engaged in preliminary tactics, one to prevent insinuating surprise; for the alligator, from his shrinking contractions, was evidently aware of the impending danger, if his foe was allowed to gain his object, and the other to excite the advantage he wished to gain, the headway of the boat was checked. As the distance intervening was shortened, the scaly tail, back, and immense snout of the alligator, were exposed to view in sidelong reflection within the umbrageous shadow, proclaiming him the patriarchal champion of his species, and well matched in strength to contend with his ophidian foe, should he, from tantalizing banter, proceed to actual hostilities. Gradually the serpent's curves and retractions grew more energetic in gliding movement as its head darted hither and thither, now disappearing on one side of the saurian, then retracting over his back for an investigation of the opposite side, with the evident object of seeking a passage beneath. The alligator, although passive in his defensive movements, was observed to crouch closer to the underlying branches whenever the head of his foe touched a part beneath the scales of his armor, his apprehension being made manifest by a nervous twitching of his tail, as if aware of the fatal vantage sought.

The captain had requested the engineer to keep the steamer in position until the victor in the duelistic contest was determined; but the wariness of the alligator, who was not in a position to accept the wager of battle, made the result of the siege doubtful, as it might be prolonged until they had tested their respective powers of total abstinence to the extent of endurance. With the thought of his own culpability should the gratification of Mr. Welson's curiosity prove fatal to the hopes of Correliana, who had placed her reliance in his direction for the relief of her kindred, he was about to request the engineer, who acted as pilot, to proceed, when the pagan exclamation proh Jupiter! from the object of his thoughts called attention to the cause. The alligator had attempted to gain the advantage of his preferred element by a backward movement, this act had opened to the head of his foe the sought for advantage, which had already passed underneath his body between his dwarfed legs before his hind quarters reached the water. In a twinkling two coils had involved the saurian's body just behind his fore legs, the part most susceptible to wounds and compression. Then came a fearful struggle that swayed the tree attachments through the wide expanse of its reach, causing in the minds of the beholders a loathsome interest devoid of sympathy, offering the test of instinctive strength and endurance as a meagre source of gratification. Still, to Mr. Welson, the contest was not altogether devoid of useful application for parallel deduction when compared with the animal traits of human instinct. The tightening of the prehensile coil of the anaconda's tail on the limb of its attachment, and upward retractile corrugations of his body with corresponding attenuation, disclosed the difficulty he encountered from the elasticity of his leverage, which prevented the concentration of muscular strength necessary for the strangulation of his victim. To the elasticity of the limb the alligator owed his prolonged existence and chance of advantageous retrieval. At this stage of doubtful emergency the instinctive "wisdom" of the serpent became meditatively apparent in the darting movements of his head and gleam of his watchful eyes, which were engaged in alert study to advantage his position, while guarding his straining body from the frantic strokes of the tail and distended jaws of his antagonist. The anaconda's intention was soon made manifest, for we could plainly see his corkscrew tail traveling with insidious progress toward one of the main trunks of the tree; this once gained the moments of the saurian's existence could be numbered, for it would afford the required resistance for crushing his body in its armor of proof. The "spectators" had watched the conflict with a superlative degree of indifference, inasmuch as favor for either of the contestants was concerned, hoping that both would be fatally disabled. But the moment the alligator began to manifest symptoms of exhaustion in the weakened strokes of its tail, and gasping throes, the human instinct of a guacho fireman sided with the weaker party in the struggle. Yet the object of his championship was scarcely a shade less repulsive than the symbolic cause of man's squirming meanness and disposition to involve in his folds of treachery all that adventure within the reach of his cupidity. The alligator's champion, born and nursed in the saddle with the lariat and bola for his rattle, asked the captain, in an undertone, for the skiff, with permission to terminate the combat. This granted he soon gained a footing upon the mangrove thicket and in a few seconds the quick gleam of a machéte was seen, then with the accompaniment of a prolonged hiss the serpent's writhing body fell separated from its tail. Relaxing the portion inclosing his nearly lifeless victim, he strove with instinctive energy to release his folds, but his efforts were vain, for the retractile power of his muscles had departed with his tail. Helplessly retained by the dead weight of the alligator's body, the serpent seemed at a loss to account for the futile result of his efforts, for he continued to retract his bereaved stump, while investigating with darting head the progress effected by vermicular contraction beneath. The reviving spasms of the alligator increased the anxious rapidity of the anaconda's movements, but as with the fabled flight of Samson's strength shorn of his locks, he was held for sacrifice bound in the toils of his own instinctive intention. His helpless condition was aggravated by the guacho, who, after cutting away the intervening branches, was seen struggling with the writhing tail until he had drawn it to an overreaching limb, from which he dropped it within reach of the head of its late owner. Its detached appearance seemed to impress upon the majority, of the relict anaconda, the diminished extent of his misfortune, for it was seized with its late mouth and bitten with impotent rage. While engaged in inflicting punishment upon its supposed traitorous tail, instinctive caution was made blind with rage, and its coils, released by the recovered consciousness of the alligator, convolved athwart between his open jaws which seized and severed the serpent's body while its head was endeavoring to execute ultimate vengeance by swallowing its recreant tail. M. Hollydorf and Mr. Welson closed the scene and the alligator's repast with their rifles, the bullets taking effect in the soft parts which were exposed in his endeavors to regain the water. With this humane addenda to the reptile duel, the serpent's head was left to shuffle off from its mortal coil. Correliana Adinope and the Bamboyle women had screened themselves from the revolting sight under the awning aft, from which they could not be induced to look backward until the scene of the duel was left far behind. The steamer, to make good the time lost, was urged to her best speed. With the relation of these retarding incidents of the day, Antonio announced his readiness to serve the evening meal.

CHAPTER VII.

CLEORITA ARCOS, at the request of her grandfather, the mayorong, gave the following relation of the causes that led to their exile:—

"As Aabrawa has informed you, our people have received the name of Bamboyles from the Aurancanoes. This was derived from the noise of our workmen's hammers in mending their utensils. But our transmitted, and more pleasing name of designation, which we hold in reverence as an evidence of remote ancestry, is Kyronese. Our late place of residence is called Pompolio, which is also of remote hereditary origin. Mendoza was said to have been founded by our ancestors, from which their more recent descendants were driven by the Spanish half-breeds who coveted their vineyards, which produced excellent grapes for the manufacture of wine, of which they were fond to excess. Their envious hatred followed the victims of displacement to Pompolio, their new home, and still continues. Our ancestors were also beset by wandering tribes of savages in their new home, as determined for our destruction as those from which we were rescued by your timely arrival. But as they were constantly at war among themselves it gave our people an opportunity to build walls and gates to defend the passes.

"The Aurancanians were always friendly, for our people never exacted more for their labors than their employers were pleased to give in exchange; and until the event occurred that caused us to become outcasts from our dearly loved homes, they were ever more ready to bestow than we were to accept. But the same cause, from the same source, has reduced them to a condition worse than our own, for they can no longer command themselves in their own country, being constantly at variance in their own households. We are so unlike our neighbors, and their visitors from other nations, in personal appearance, habits, and customs, our curiosity has labored long and patiently with the transmitted emblems, but they refuse to unravel the secrets of the past.

"My father gave our people much information, which they supposed to be reliable. First, he said that Kyron, from which our name was derived, was an ancient Assyrian department, which gave birth to the city of Sidon, famed in its day for the boldness and enterprise of its navigators; and that the vessels portrayed by our ancestors were similar to theirs. But he said that our short bows, and spears, as well as our defensive armor, afforded the strongest confirmation of Assyrian origin. In addition, he found utensils designed for household use which corresponded exactly with pictures in the books he obtained from Europe; and furthermore, he made a journey to Peru and brought back vessels of pottery exactly similar. From these evidences he naturally concluded that our ancestry, and those that inter-married with the aboriginal inhabitants of Peru, were derived from the same source. However, you will understand all these things better than ourselves; for he said your learned men devoted their lives to the study of the past, and were skilled in tracing vestiges, and conjectural probabilities.

"From what I have related, you can judge of the past, and from what I shall now relate, whether we have acted prudently, and are worthy of the interest you are disposed to take in our welfare. We lived happily according to our knowledge, neither eating or drinking what we considered to be impure, or indulging to excess beyond the body's requirements for the gratification of taste. Our amusements were harmless and serving as a vivacious warmth for affectionate love. Those who visited us, like my father, were kindly entertained, and not one of the few has disdained to accept our friendship. The cause of my father's departure was not that he loved us less, but the wish to induce his father and brother to come and see that he had succeeded in finding a people who were content to live without money, in freedom from want and envy, with the security of a common affection to make them realize a more perfect existence after the separation of vitality from the body. It was our misfortune to lose him when his advice was most needed, for we feel assured, if he had remained, he would have averted our calamities, for he claimed that our goodness and simplicity invited imposition, which we had not the diplomatic skill to avoid."

Mr. Welson, with a humorous twinkle of the eye, interrupted Cleorita, questioning whether her father explained the meaning of the word diplomatic. To which she replied with blushing trepidation, " My father gave his own version, but he was so chioptic (jocose) in his way, and inclined to speak disparagingly of his people's sincerity, we did not press him to asservate the truth of his interpretation, for we could not wish to believe that civilization consisted in the art of successful deception. As you knew him well in former years, I will not withhold his exact definition. He said the word was a comprehensive cover for all the variations of lying evasion practiced in the adjustment of national encroachments, as a pretext for more extended impositions. The immediate cause of our exile was the reappearance of a tribe of Indians who had been expelled by the Aurancanians for their atrocious acts. The return of the Abacknas (marauders) was announced by their sack of the settlement of Guaspe. When pursued by an avenging party they fled to the mountains. Their leader, named O'Grady, a sailor who had escaped from a vessel in the straits of Magellan, betrayed them to the vengeance of their pursuers, so that few escaped. By this act of treachery he gained admittance into Aurancania for the introduction of a destructive cause more insidious in its perfidy.

"In all the valleys of Aurancania the apple and pear grow to perfection, and, as with us, those bordering the countries on the north and east are well adapted for the culture of the grape and fruits kindred to the peach. The extracted juice of these had been used as a pleasant and harmless drink. O'Grady, although mistrusted, proposed to make the juice more pleasing in its effects if suitable vessels could be procured. As these were to be made of copper, of which we had an abundance, and were skilled in reducing it for the manufacture of utensils, he was referred to us. Unfortunately, on his way to visit us, he met one of our most ingenious workers of the metals at Muloa, who comprehended the kind of vessels and attachments he wanted. Insisting upon accompanying our brother to oversee his labors, he gave him abundant reason on the way to regret the chance that made him responsible for the stranger's introduction to our people. On their arrival within the gates of the pass, he would not accept the hospitality provided for strangers, on trial, — outside of our Douang, or walled town of defense, but insisted that he should be received as a guest within. This act of aggressive presumption was firmly but politely opposed by his sponsor, which from his slight stature led to a trial of strength, with a result seriously unfavorable to O'Grady, who was glad to accept assistance from his antagonist and a bed in the strangers' quarters, which he kept for a month, until a fractured leg and an arm were again serviceable. Nevertheless, he was kindly attended;

and after his recovery never attempted to overawe any of our people with threatening overtures provoking personal encounter, having seemingly lost confidence in the accounted advantages of superior size ; but the revengeful leer of his eyes boded us ill if the opportunity of exacting it should ever occur. The vessel, with our troublesome visitor, were transported back to Muloa as soon as he was able to travel ; he neither offering, or his conductor requiring aught for the labor or material bestowed, other than the desire, on the part of our people, never to see him again. But the hopes entertained that our parting would be final, were void ; and in view of the calamity which the heedless fulfillment of our brother's stipulation wrought upon the friendly Aurancanians, we have questioned whether our own misfortunes were not justly merited."

"Were you aware," inquired Mr. Welson, "that the vessels your artizans were fabricating were intended for the transformation of a beverage juice into a fiery distillation, that in product would reduce your friends to the condition of enemies to you, by the introduction of 'civil' discord into their own households ? "

"The only information our people had upon the subject was derived from my father," replied Cleorita, "who had often described the misery it had caused among your people. But his habits were abstemious, and his example prevented the full impression of the danger, for we did not forethink that others lacked his discretion, and would pervert actual blessings for their own destruction. Alas, we soon found that the track of our heedless labor was marked with the blight of provident affection. To controvert our own agency in the misery inflicted upon the families of our ever kind neighbors, the mayorong sent those abroad who mingled substances with the ashes beneath the vessels that in burning destroyed the metal. But the O'Grady had gained the means before this was effected, of obtaining others from Mendoza of larger size, after we had refused to supply his loss. These we also felt warranted in destroying, which aroused his suspicions and his third enterprise was carefully guarded. When its product exceeded the demand, he sent a still over to Pompolio and seized our fruit for its use, which caused our people to destroy it openly, expelling his aids. This provoked his bitter emnity, and he swore that he would exterminate our people root and branch.

"Two years passed without cause for alarm, when, with a morning's dawn, we were aroused by the boom of a great gun and a loud crash in the midst of our houses. When rushing forth to learn the cause the gatekeeper gave the mayorong a letter written in Mendozean Spanish which I translated. The missive was a demand for the immediate surrender of the Douang, unconditionally. In the event of refusal, the lives of all the males were to be sacrificed. This was signed, 'Patrick O'Grady, Commander-in-chief.'

"Of course, without hesitation, our people put on a bold face and sent him back a defiant answer. In less than an hour our gate became a mark for the cannon. This we had anticipated, and a second gate prepared for an emergency of the kind, was closed inside of the outer, the interspace being filled with faggots of osiers and tough mountain moss. So that our second gate was well protected, for they kept prudently out of reach of our spring-engines which were almost as effectual as their guns, but could not be directed as easily. But our people were sorely disheartened, for he had brought with him a large band of the guachos and Indians of the plains, who had often attempted similar enterprises. Finding, after many days, that their guns were breaking through our strong walls, our people determined to conceal in the mountain caves all that was held valuable, leaving in charge of a band of our young men the old and infirm, with our cattle; while the mayorong, with the majority of the able-bodied of both sexes, should set forth to seek a new home farther north. When all the arrangements were completed a passage was opened in the southern wall opposite and in concealment from the besiegers' encampment, for the outgoing of our cattle, through the heap of litter that had accumulated from our stables overthrown from the wall. After our departure for the mountain strongholds, the way of escape was again closed and concealed as before. When everything was made ready for the departure of the mayorong's party northward, they resolved upon another night attack upon their foes for intimidation, that they might not seek to molest the mountain party in reserve; but with such precautions as could be used to prevent the loss of life on our part. The success of our people, if it had been followed up, promised a complete rout, so great was the panic they caused, but it sufficed to render their guns useless, with the destruction of their munitions, and such other damage as we could accomplish without hazard to ourselves.

"With a sad farewell we set forth in search for a new place of habitation. Encountering many hardships, we finally succeeded in reaching the fruitful valleys to the north of Mendoza without the loss of life, where a new race of foes have driven us hither and thither as relentlessly determined upon our destruction as the O'Grady. When we started, our men numbered an hundred and eighty, and our women and children two hundred; these have been reduced by death and capture in our long wanderings among savage foes, to ninety men, and an hundred and twenty women and children. Twenty days ago we rescued our loved companion, Correliana, in sight of her city, while her guards were fighting bravely for her defense against overwhelming odds. For many days we hovered in sight of the city, hoping to regain for her an entrance into the gates; her friends understanding our intention endeavored to render us all possible assistance, but it availed naught for her advantage, but caused us great distress. Yet that she has been the means of our preservation we doubt not; for without the support of her undaunted courage and device, we should scarcely have been able to elude the many schemes planned for our destruction and her capture. When she found it was impossible to gain an entrance into the city, and we were fainting for the want of food, she led us by devious ways to Indian villages, left in charge of old men and women, where we obtained an abundance of food without causing other injury. From that time we have had no rest, except what we gained in the sillia while our horses were moving. Her desire to keep the river in view has been so urgent that we saw clearly she expected succor from it in some way. Although her language corresponds with Spanish so closely as to furnish me with a ready understanding in other matters, she was not disposed to impart the nature of her hopes from this source. We are not greatly given to superstition, nevertheless, we cannot rid ourselves of the grateful belief that you were in some way overruled for our rescue."

When Cleorita closed her relation the Kyronese women bowed themselves down in grateful acknowledgment for their preservation. This act of humility caused the padre to utter a remonstrance coupled with the declaration that prostrate humbleness for human aid seemed to him an affectation that smacked strongly of hypocrisy. But when reminded of the obeisance paid to the pope's toe, and similar absurd acts inculcated by Christian doctrine in the education of youth subject to the bias of sectarian supremacy, he was silenced. But all joined in expressing their strong sympathy and proffers of aid in solace for the unmerited sufferings of the Kyronese.

CHAPTER VIII.

WHILE Mr. Welson was engaged in listening to the rehearsal of the proposed plans of Correliana for the speedy rescue of her people, a falcon in the act of stooping from its poise attracted the quick eyes of Mr. Dow, who raised his rifle, but before he could secure his aim the Heraclean maid uttered an exclamation of alarm which arrested his destructive purpose. In explanation and apology for her impetuous words and act, the falcon settled from his waft upon her shoulder with a flutter of glad recognition, coaxingly pecking at her ear with side glances for accustomed caresses. In a few moments the fair perch became so abstracted with varying emotions hovering between sorrow and gladness, that her pet was fain to stoop to her wrist for the mechanical recognition of the right hand; yet, as if unmindful of neglect, it plumed itself in the pride of feathery vanity, seemingly confident notwithstanding the reserved affection of its mistress.

At length, as if suddenly made aware of her preoccupation from the silence that prevailed, she asked the privilege of retiring to the cabin for a few minutes for the recovery of her composure. During her absence Cleorita said that she had been similarly affected on several previous occasions from falcon visits. Nearly an hour passed before Correliana reappeared, then, with the pleading animation of anxiety, she requested M. Hollydorf to urge all warrantable haste in preparation for the overland journey from that

THE MANATITLANS. 73

point, if they proposed to rescue her people, as they were in extremity from the increased virulence of the pestilence aggravated by famine, of which the besieging savages were preparing to take speedy advantage. Naturally supposing that the bird was the carrier medium of communication, all their energies were exerted for the accomplishment of her affectionate solicitation.

Mr. Dow, with Jack's and Bill's assistance, drilled the Kyronese in the art of loading and discharging the howitzer, with effective aim, also in the use of rifles and pistols. During the day hampers were filled with prepared munitions and rations, and the party selected for the expedition. Having assisted, with wonderful tact, during the process of packing, just before night-fall Correliana dispatched the falcon in homeward flight, with encouraging promises of speedy relief. When with the approach of darkness, and fatigue, the labors of the day were suspended, she pronounced herself anxious that we should become acquainted with the history of her people, that we might judge of their worth before venturing the hazard of our promised aid. With an assurance of unwavering determination to adventure their lives for the rescue of her kindred by all, she commenced her narration.

"The transmitted written history of our people, derived from our ancestors of old Heraclea, has not been esteemed reliable by the later renewed generations of our present City of the Falls, inasmuch as the historians, of the middle period, were invariably inclined to ascribe the partial prejudices of degeneration as evidences of progression in their assumptive decisions of right and wrong. With self exaltation they did not hesitate to extol the most arbitrary and licentious acts of persons in power, which in accommodation for the selfish retention of favor were constantly subject to reversion. These sources of selfish contradiction, serve to impeach the veracity of the whole, so that from the adventitious impressions of truth we have been obliged to make conjectural deductions to subserve our desire for the preservation of a probable outline record of the causeful events that led to ancestral translation from the Pontine to the Iberian Heraclea. However, in my prompted relation I shall endeavor to give a simple rendering agreeable to the expressed judgment of our advisors, without attempting to force your concurrence with reasoning similitudes. Your knowledge pertaining to coincident history will certainly attest to the correctness of the alleged source from which our remote ancestors were derived.

"Our original stock, in translation, might well be represented in the variations of caste by the contingent elements with which I am at present surrounded; for the place from which our ancestors embarked was a central point for the fermenting commixture of the peoples and septs of Asia, Europe, and Africa. Our patrician historian states that the original stock were all derived from noble Roman families who were emigrating, with collateral provincial branches, from the Euxine Heraclea, in a Macedonian ship, to an Iberian city of the same name, situated a short distance inland from the ocean opening of the straits of Gades. After touching at the African port of Rusander Gaditarius for supplies necessary for support during the interval of planting and harvest, they set sail for their port of destination.

"When in sight of the landmarks of Heraclea, while offering sacrifice to the gods of their worship, for the prosperous termination of their voyage, a sudden tempest arose which forced their vessel out into the broad Atlantic. For days the storm raged, while before it their bark was driven heedless of mortal control, every moment threatening destruction. At length, after hopeless despair had held them bound in shadowy darkness through a lapse of time unmarked by the full distinctions of day and night, the sun rose clear over a limitless expanse of waters. Still they feared to offer thankful oblations, for they were drifting they knew not whither. In the listless inactivity of despair they had allowed the waters of heaven to accumulate in their vessel mixed with the briny wash of the ocean. As the sun rose in the firmament to its meridian, the heat parched their mouths with thirst, then they recognized the providence of heaven for the supply of water tempered with salt to make it unpalatable for excess.

"'Again hope began to dawn, which was strengthened on the following night by a flight of fish seemingly attracted by the altar fire, which had continued to burn through the fearful tossings of the vessel when impelled by the merciless tempest urged by the god of the ocean. Revived, with the second sun, the sailors spread the vessel's sails to a favoring waft of the ocean wind, showing their recognition and resignation to the decrees its providence had ordained. There was no lack of food, for the supply obtained at Rusador for anticipated wants between seedtime and harvest, more than sufficed for prospective requirements, unless the ocean proved boundless. Of luxuries there was also a bountiful supply; dates, dried figs, grapes, and Chian wine. Strange as it may appear, with the revival of our hopes, a large portion of the wine was sacrificed to furnish vessels for treasuring the water preserved by the ship. But with the rising of the eleventh visible sun all the supply of water having been exhausted, — for there were many mouths and great thirst, — despair, which dried up the moisture, began its reign of terror, from the moans of mothers who freely offered their tears to still the wailing cries of their children.

"'In this condition, when all coveted death to relieve the tortures of thirst, there came on the sixteenth of its rise upon our forlorn hopes at sundown, a waft that made all murmur thanks in their weakness. This was followed with genial showers which brought a reviving consciousness of an overruling presence inspiring a love of life and the blessings of kindred affection. When the clouds, to whose timely benefactions we were beholden for our preservation, were dispersed by the rising sun, our eyes and hearts were gladdened with the sight of land, which called forth tears with whispered rejoicings, and wan smiles of congratulation bestowed with embraces, and hand pressures in thankful praise that we had been once more permitted to see the element from which we had been so long divorced by cruel fate.

"'Borne onward by a gentle wind from the ocean, we entered a broad estuary whose banks, or shores, were bordered with a forest verdure of trees exceeding in magnitude our previous conceptions. Far off in the interior, as the sun declined, were seen mountains whose summits were clothed in fleecy mists while beneath the varied descent appeared dressed in rainbow tints of moving light and shadow. The banks of the mighty river, or arm of the ocean, became more distinct in the approaching twilight, until darkness with its pall withheld them from view. Again another day dawned; refreshed with the dews of the night we bethought ourselves how we might bring the vessel to land where we could obtain water to quench our thirst, when lo, with the first feeble dip of the oars the trickle of the water inward discovered to a child its freshness. The faint struggles of the oarsmen strengthened with the fear of again being carried out into the ocean, for the current was forcing the vessel backward, were at length rewarded with the stranding of its keel beneath the steep bank of an inlet.

"'In vain our eyes, from the mast, searched the shore for evidences of man's habitation; neither smoke from hamlets or signs of cultivation could be traced. Weary and weak, but composed in spirit, from our now secure attachment to land, which, although foreign, seemed afar off fruitful, all sank into a deep and refreshing slumber, lulled by the familiar sound of the cicada's shrill vibrations, which continued unbroken until the dawn of another day, when we were awakened by the sound of strange voices speaking an unknown tongue. Surprised, but not alarmed, when we discovered that the utterances were from a collection of human beings who were viewing us and our vessel from the bank that overlooked the transtra, our own curiosity was in like manner attracted by the novelty of their appearance. In stature they exceeded in height our own, but were gracefully formed, with expressive features inclined in color to a brownish red. With eyes of vivid blackness they seemed capable of giving intensity to the two extremes of passion — expressed by revengeful anger and dalliant softness. The covering of their bodies was so slight that it failed to afford the shadow of concealment or restraint to their persons.

" 'While we were sleeping they had drawn our vessel into the inlet so far that with slight assistance we could raise ourselves to a footing upon the bank, this with signs they proffered and we accepted. When seated upon the grassy plain, the women with native grace prepared in shell a thick paste compounded of milk and fruits, exceedingly palatable and refreshing. For a drink they pierced the eyes of large nuts from which flowed a milky fluid that found special favor with our women and children. These tokens of kindly regard were presented with timid gentleness and solicitude that won our confidence.

" 'When our appetites were appeased in their craving for the *novus res* in freedom from the ocean's savor of salt, signs of mutual curiosity began to flow in pantomimic gesture. First, they questioned from whence we came? We answered by pointing over the ocean. But when they pointed to the sky in its descent to the horizon, we saw that they would ask whether we were descended from the gods. Humoring their implied belief, we answered truly by uprooting a stalk of grass, then holding its seed filled follicles dependent we in addressing the roots to heaven shook the semina from their receptacles to the earth, therewith, to their apprehension, acknowledging our heavenly origin.

" ' Communing among themselves, with a defferential review of our persons they seemingly acknowledged the superiority of our pretensions, while questioning the cause of our forlorn condition when found. At length in their doubt they appealed to an aged man whose appearance augured wisdom, who answered sagely by addressing, for our comprehension and approval, his symbolic exposition of the cause. Selecting two tall spears of grass, overtopping the heads of their kind, he pointed to the eldest parents of our group, then reversing the stalks with the roots upward, he forced the symbols apart by introducing a younger female blade between, adherent to the tendrils of the paternal branch, causing the mother and her seed to fall to the earth. This disruptive demonstration so clearly defined his knowledge of the human passions, in accordance with the experienced injustice of our own race, that a blush of shame suffused, with its evidence of conviction, the faces of some of our elders whose withers of frailty had been touched. Taking these symptoms of assent as evidences of conviction, the oracle, with a self-complacent air, relapsed into silence, his kindred mingling their admiration for his ability in prescientia with reverence for our supposed paternity. Having arrived at the Ultima Thule of their curiosity we endeavored to satisfy ours without lessening the kindly reverence we inspired from our presumed descent from the gods. But learned nothing beyond the impression that the land extended, in the three opposed directions to the ocean, to the horizon, and that their country was the full of a moon nearer the setting sun, to which they invited us warmly to accompany them.

"'Although still fancying that we were in a remote division of our own land, yet hopeless of regaining our homes, or intelligence of our people, we concluded to avail ourselves of their invitation, for an attempt to return by the ocean augured sure destruction. Nourished with fruits and wild game, which nature furnished and sustained without the aid of human labor, and nursed with the tenderest care we soon regained our strength. Signifying our readiness to accompany them, and desire to take with us our household lares, utensils, harvest, and fruit seeds, they brought, after the lapse of days, diminutive beasts of burden, which seemed united in equal relationship to the camel and goat. When the day of departure came, we bid tearful farewell to our vessel, then with the ready aid of our benefactors buried it from vision that it might escape desecration from wandering tribes.

"'Many days were occupied in our inland journey before we reached the valley of our destination. When at length, after surmounting many difficulties, it opened to our view we were overjoyed with its beauty and the bounteous prospect it afforded for the fruitful recompense of our mischance in original intention. In the sincerity of our joy we could not withhold our thanksgiving for the divine direction that had conducted us through so many perils to a land, where, as demi-gods, we could live in freedom from the dread of invasion and corrupt oppression of imposed tyrants. Our advent brought peace to our benefactors, who had been forced into wandering exile by the neighboring tribes; who instead of opposing their return solicited the privilege of bestowing their labor as a willing sacrifice in atonement for their injustice in expelling the Betongo tribes from their lands while under the favor of the ruling spirit.

"'Season after season followed the advent and propagation of our Latin generations in the Betongo valleys, each more bountiful than its predecessor, until years were multiplied into centuries. The reproduction of the exotic grains, fruits, and vegetables yielded tenfold returns in excess of their rates from native soil; and while our people preserved their original prestige as a race of superior beings, dealing with arbitrary justice free from forced oppression, they prospered and were reverenced by the aboriginals for the happiness they conferred by kindly example. During the first century, the castaways and their descendants did not disdain to give instruction to the natives with the exampled labor of their own hands; and through the adoption of their children in allied direction with those of the Latin race, easy communication in language was held.'"

Correliana here remarked, that in the first part she had adhered closely to the rendering of her Latin ancestor, Marcus Adinope, the Prætor of the castaways in their first settlement of the Betongo valleys. "I will now," she said, "append his apology for practicing duplicity in accepting the homage of the aboriginals as their due in the assumed character of demigods.

"'In the first instance, we felt constrained to accept their proffered reverence paid in fealty to our supposed descent from the gods, not from the feeling that the assumption would offer us the means of practicing arbitrary oppression in safety; but as a necessary composition for an exampled restraint of gentleness in association among ourselves, as a secure hostage for imparting its godlike virtues to our trusting neophytic benefactors. Aided with the harmless reverential impression, we were able the better to control the plebistic democracy incorporated with our element of self command over the thoughtless impulses of the subservient oarsmen and hinds of our vessel. Our memories were kept on the alert with the monitorial revival of insurrections and massacres, which had their origin from impositions exacted in the conquered Roman provinces by plebeian officials who had paid a price for their promotion. Indeed, the cause of our transmigration had had its birth from that illegitimate source of instability.'

"After the passage of many centuries, another of our family has recorded the result of the democratic majority's usurpation of the power of equalizing self-command, evidently in readmonition of his predecessor's apology: —

"'How void of self enduring forethought are the uncontrolled instincts of youth, when reckless of experienced premonitions! It is with painful emotions that I am obliged to record that the descendants of the aboriginals who succored our forefathers in their castaway distress, and preferred them to their own hereditaments, with the reverent homage accorded to the gods, are now subject to the cruel exactions of the taskmaster. The hardships to which they are now subjected by the multiplied progeny of the sailor, — who in thoughtless frenzy attributed their thirst upon the ocean to exact equalization in water distribution, — will prove the sure precursor of our common destruction. The frailty of our godhead assumption has been long since exposed, engendering hatred from the enslaved in abhorrence of their own submissive weakness, so that with the opportunity they will destroy every vestige of their humiliation.'

"This prophecy indicates the period when the defense of a walled city was required for sustaining the exactions of the taskmaster. The traditionary scenes enacted by the old Heracleans, as the inhabitants of the first city were styled, would be as painfully oppressive to your kind-hearted generosity as they would be to me as relator. It will be sufficient for me to state, that the 'City of the Falls' was built by Indian labor, enforced by the cruelty of the taskmaster, as a place of recreative resort during the heated solstice, for the old Heracleans. When remonstrance failed to abate the oppressive exactions enforced from the accumulating slaves, and stay the wild orgies enacted by the democratic rule of the city's majority, the kind-hearted stipulated for the cession of the new city for their seceding occupation, subject to their own governmental rule. In less than a decade of years, after the separation, the inhabitants of the old city were surprised, during the celebration of nocturnal orgies, dedicated to mythical patronage, by the uprising of their slaves; and with the exception of a few, who had been forewarned, an hour previous, in time to make good their escape to the City of the Falls, all were massacred, and the old city has continued a tenantless ruin to the present day.

"Unsatisfied with the partial success of their vengeful retribution, the Indians entailed upon their successors the unlimited enforcement of a constant siege, until the inhabitants of the new city were exterminated, a result that without your effective interposition in our behalf would be well nigh accomplished."

CHAPTER IX.

LONG before daylight on the morning succeeding the narration of Correliana Adinope, the busy sound of preparation was heard on board of the *Tortuga*, and on shore. Food and hclotes for raiment were bestowed in hampers and bales, by the Kyronese, in quantity sufficient for the easy carriage of the mules; while Captain Dow and his subalterns, Jack and Bill, marshaled the Kyronese guard in preparation for rifle, pistol, and howitzer, defensive and offensive practice. At sunrise, when nearly ready for the start, Correliana clapped her hands with a joyful exclamation, and, in a moment after a messenger falcon stooped in perch upon her wrist. This was of the species *Falco peregrinus* of the pampas, but much improved in size and plumage from culture. Its greeting, as with the first, was replete with pleasurable animation, extending its wings in impulsive sway to the voluntary and involuntary action of its talons, peculiar to birds and beasts of prey, when subject to intensified sensual gratification. As with the cat kind, who make their "friendly" satisfaction manifest by extending and contracting the sheath muscles of the claws, the falcon unconsciously closed its talons upon the wrist of its mistress, causing her to utter, with the painful punctures, "Soh, soh, Merlin, mon brachiale!" Captain Greenwood, observing the flow of blood from her wrist, quickly supplied her with a pair of gauntlets. Merlin, when again restored to her wrist, seemed to understand the intention of the buckskin proviso, for he used his talons in the expressive ruffling and extending of his wings; succeeding with his coquetry in attracting her attention from the train of meditation in which she appeared to be engaged, he raised his wings upright, exposing beneath parchment scripts; these removed he leisurely commenced a survey of his surroundings. After their perusal she wrote a few words in reply upon some French tissue paper furnished by M. Hollydorf; this secured in Merlin's sacks, he desired Captain Dow to take note of the bird's course, before it rose to its poise, as it would guide him to the opening of the pass in the foot-hills. After the bird in floating flight had reached the point of designation, it soared to its poise and in descent quickly disappeared from view.

When the train was fully in motion, Correliana beckoned Captain Greenwood apart, and then to his surprise addressed him in English, with slow, measured enunciation the involumed supplication " Will-you-come-to-us-if-we-are-successful ? We-are-happy-among-ourselves,-and-if-you-love-happiness-as-we-en-joy-it-in-our-simplicity,-and-your-educated-habits-will-permit-you-to-love-me,-without-regret-from-other-cause-than-my-own-demerits,-there-will-be-great-joy-in-store-for-us."

The captain's faculties, notwithstanding his bewildered amazement caused by her sudden acquisition of power to express her thoughts in English, and with such clearness his most coveted desire, in terms so agreeable to his perception of her worth, answered with prompt energy, in quick imitation of her method, " If-my-life-is-spared-I-will-visit-you-soon ! "

After a moment's hesitation, as if to realize the full comprehension of his reply, she, with a sudden flush of joyful animation, exclaimed, " I-am-certain-you-feel-that-my-happiness-depends-upon-the-consummation-of-our-love-in-Heraclea!" Then with the proffer of salutation, she answered to the hastening call of Captain Dow.

This parting scene between Captain Greenwood and Correliana caused M. Hollydorf's countenance to become overcast with a rueful shadow of dismay. At nine o'clock the train reached the foot-hills where they exchanged their last farewell signals with those left under Tortugan protection. On the fifth day after their departure from the anchorage of the *Tortuga*, the train had gained the eastern slope of the highest mountain pass that opened to their view the Betongo valleys, with but one interruption to their progress from Indian opposition, which was quickly turned aside.

On the first of July, while in midway descent to the valley, the falcons returned after a short flight over a wooded district to the left of their course, which was interpreted by Correliana as an indication of danger from an approaching party of Indians. This startling news caused the greatest activity. While Captain Dow reconnoitred with his glass the descent for a point of advantage for their reception, his two cannoniers prepared the howitzer charges for immediate action. Fortunately they were able to reach a comparatively level plat that offered for their train's protection the vantage of a natural rampart, which was improved for the reception of the gun with a wall of stones serving as a mask. When the defensive preparations were completed, the pack train, under its guard of women, was sheltered behind it as far in the rear as possible.

While yet engaged in strengthening our position for their reception, a large body of Indians on horseback debouched from a wooden pass upon the plateau below. It was evident from their movements, when collected for consultation, that they were aware of our near approach, and when discovered would be set upon immediately. That the crisis might be hastened, and the obstruction to our progress removed as speedily as possible, the weakness of our party in numbers was exposed outside of the temporary walls of the fortification as a temptation for speedy onset. Their eyes were soon directed toward us, at first with silent curiosity, then after a short consultation they sprang upright upon the croups of their horses, and commenced brandishing their spears and clubs, with the evident intention of intimidation. Accessions to their number were constantly appearing from different quarters showing that our progress had been watched. Nearly an hour elapsed before a forward movement was attempted. Their waiting delay enabled us to strengthen our position. They commenced their approach with feats of equitation that would have delighted a circus audience, seemingly determined to entertain us to the death. Indeed, their evolutions, which were timed to a war song and dance with a display of acrobatic agility as they advanced at a gallop, attracted our admiration. When within six or seven hundred yards they came to a sudden halt, then after a short "palaver" they reformed in sections, which commenced an involved circle dance, the horses performing their parts without prompting from bridle or lash. The object of the entertainment was soon apparent in the narrowing space between the outer circle and our rubble stone wall. Jack, although amused with the nearing foes' tactics, nursed the fuse fire of his linstock with watchful care, Bill keeping the howitzer in range with their rising advance to the point intended for the discharge of their spears. While yet without the bounds of their spears' range, quick as thought the whole band were in full career toward our cover, the foremost launching their spears at everything human exposed. The ducking and dodging on our side was naturally and skillfully executed, but not in every instance gracefully. Jack reached the ground in the style of turtles sunning themselves on a water log, when surprised by urchins with a flight of stones, but in his descent did not lose his presence of mind, for the report of the howitzer was simultaneous with the report of the rifles. The massing of the horses in the onset caused a fearful havoc. The effect produced upon the survivors, from the turmoil of bewilderment, subjected them to a second and third discharge of the cannon and rifles; then in view of the slaughter the mayorong's pity was excited, and with imploring signs he petitioned Captain Dow to withhold the fire of his men. The cessation allowed the Indians to recover from their daze, but panic succeeding, they dispersed wildly in flight, giving expression to the tumultous effect of fear in attitudinal variations, which, in equestrian display, exceeded in diversity those improvised as a prelude to the battle.

When the last of the fugitives had disappeared, it was discovered that Correliana had sustained the only injury inflicted from the cast of spears. Fearing that her protectors, in amused scorn for the unwarlike antics of the foe, would allow them to attain their intention of securing, with the impetus of onset, an effective range for their weapons, she had risen to caution Captain Dow, when in the act a spear grazed her shoulder inflicting a flesh wound. This had been immediately cared for by the Kyronese women, and her anxiety and pain were so slight that she rallied the two sailors, who were sincerely affected with sympathy for her safety, on the speedy methods they adopted in avoidance of the spears.

Jack with a humorous smile, rendered comical by the perceptible movement of his tongue, as if in the act of revolving a quid from side to side of his mouth, replied: " To be sure it was sum'ut lubberish to your ledyship's eyes, but it's a way we learned at sea to draw the enemy's fire."

The effect of our arms had been terrible, the dead

and wounded Indians greatly outnumbering the shots fired; the predominance of the latter bespoke in plain terms either the unpracticed skill of the Kyronese in the use of firearms, or their more probable instinctive humanity. Captain Dow, anxious to retrieve lost time, had the wounded and dead bodies of the Indians removed for the passage of the train. The mayorong caused the former to be tenderly carried into the inclosure, and when the train had passed beyond the human obstructions, he requested permission to remain with the elder matrons that they might bestow some relief upon the suffering until their companions recovered from their panic, promising to overtake them before they encamped for the night. Although the objects of his delay received but little sympathy from the members of the corps, and its male adjuncts, they could not refuse the request, but insisted that he should retain a sufficient number of his men as a guard for their safety. When the moon rose we had gained the valley of the Betongo, and the rare beauty of the scenery, under its resplendent light, invited us onward; but the mayorong's party had not overtaken us, which caused some anxiety, but this was soon dissipated by their appearance. Urged on by the delightful prospect, heralding the speedy attainment of our journey's object, we were enabled to encamp in a shaded nook upon the banks of the Betongo river. Notwithstanding the lateness of the hour, the Kyronese added game and fresh fish to our delayed repast.

With the morning's dawn we moved onward over a paved causeway, with its massive stones still intact after untold centuries of wear from Time's detrite usage. Inland from this shaded causeway, we passed Indian villages at intervals of a few miles, pleasantly located upon knolls surrounded with banana, corn, and vegetable plantations. One of the largest we entered, but found it deserted; there were, however, abundant evidences of its recent occupation. Finding an abundant supply of roasted corn, dried fish, and other edibles, an equal quantity was taken from each house, the hampers of the mules furnishing cloths in exchange. The site of each village was connected by a branch causeway with that of the river's bank, confirming the relation of Correliana.

To kill, rout, and destroy, is the orthodox inculcation of civilized progression; so in view of relieving the inhabitants of the besieged city from the besiegers' stores of provision, we resolved to visit all the villages in our route, and mulct from their abundance as much food as we could transport with our limited means of carriage, leaving with each an equivalent. Dr. Baāhar advocated the total destruction not only of the provision left, but of the plantations and villages, in opposition to the mayorong's pleading expostulations for their preservation. But the doctor urged the curative plan of extirpation of the sources of vitality, as the only authorized means sustained by classical experience for rendering the enemy's efforts nugatory. "For," said he, "it will be neither consistent or prudent to leave your enemies the means of prosecuting their unrelenting siege of Heraclea." The mayorong, with sad deprecation, pleaded that acts of revengeful destruction would only enrage, and in naught avail the beleaguered; as they would increase the inveteracy of hatred, with justice, against the white race, that so not only the lives and peaceful happiness of the Heracleans would be sacrificed, but others with like kindly intentions. For in making others suffer needlessly, we cannot hope through futile intimidation to be spared ourselves, if an opportunity for revengeful reprisal should occur? This half soliloquized questioning appeal of the mayorong, seemed to be addressed to all, and from the impression conveyed by his intonations in speaking, its truthfulness, when interpreted, was sanctioned with general approval. Still, although manifestly grateful for the appreciation of the majority, his countenance lacked the fullness of satisfaction that the hearty concurrence of Dr. Baähar would have afforded. But the doctor, with the proverbial fatuity of the precedentalist, substituted for the required solace the revised saw, " they thought themselves wiser in their generation than their forefathers," evidently with the intention of reproving his associates for their defection from the transmitted creed of warful usage. That there might be no lack in the practical support of the mayorong's behest, Correliana left as equivalents, in exchange for food, a large proportion of the cherished gifts bestowed by Captain Greenwood.

Determined to reach the besieged city before midday on the morrow, we did not halt until the dividing range of hills, that separated the upper and lower valleys of old Heraclea, had been surmounted. Upon the shaded summit overlooking the vegas we encamped for the night. The cool refreshing breeze that swept over the hill, and an abundant supply of sweet grass for recruiting the strength of the horses and mules, lured us to delay our start on the following morning, until the sun had dispersed the mists from the valleys. When the fleecy veil was at length dissipated, an enchanting view was presented upon either hand extending as far as the eye could reach. Paved roads or causeways followed the windings of the river and canals through all the alluvial districts. These were of easy detection from the checkered overgrowth of brambled weeds, which ever delight to erect their prickly domes and spires above the ruins of palaces, churches, monumental tombs, and the most splendid mechanical achievements of man, as if in derision of his instinctive claims to immortality, after a life spent in arrogant oppressions, and thorny assumptions, opposed to the kindred sympathy of reciprocal goodness. While the Kyronese were bestowing their kindly attention upon the animals,

M. Hollydorf, with barometrical aid, calculated the altitude of the valley above the plain of the Tortuga, and found that its elevation exceeded four thousand and nine hundred feet. But with heat lessened by only a few degrees from the tropical zenith, the valley, from its still continued facilities for irrigation, appeared to be the scene of perpetual verdure. Its altitude gave a climate, from mountain inclosure, especially adapted for the cultivation of exotic fruits and cereals, of which, in wild growth, there were abundant specimens.

While Correliana was in thoughtful meditation, overlooking the beautiful scene, her attention was attracted to the labors of Mr. Welson, who was engaged in writing out his diaretic observations upon the developed phases of instinct. With Dr. Baāhar's aid, he, at her request, imparted in outline the result of his labors, which he styled, "A Relative Exposition of Instinctive Traits Common to Animal Life." Under this head he had classified those common to savage and civilized humanity, in the following order. Poison, Material and Immaterial. The lowest grades of savage life use material poison almost exclusively, as a destructive agent in their intercourse with each other. Representatives of civilized nations compound with speech vituperative venom, which is as deadly in its effect upon happiness, as material poison upon the body. Its insinuating use, in language, is a speciality of women who have suffered in reputation from its taint, and in turn, to conceal their own frailties, use it as an imperative means of counter irritation to blind the censure of their kind. Illustrative examples of savage and civilized superstition, compared by an experiment upon savage and civilized representatives of the human family. Both submissive to instinctive fear. The savage is dubbed a knight with the collar and conferred order of Bath. His departure, after the ceremony of consecration, in pursuit of adventures.

Reptile duel between a Boisdean serpent and an Alligator. Instinctive tactics of displayed strategy. Guacho " sympathy " enlisted for the weaker party. Reverse. Result of civilized arbitration. Correliana readily interpreted the satirical import of Mr. Welson's comprehensive method of illustration; but questioned if the women of civilized society had ever in fact given truthful cause for the expressed venom of his satire. In answer, he referred her to M. Hollydorf, as a more ready exponent, who would truthfully inform her whether he had by insinuation libeled the market value of female "virtue" as a negotiable article of appraisement in the gossiping marts of fashionable society? Still puzzled, in the absence of the referee, she applied to Dr. Baāhar for a direct elucidation of the word "virtue," which she had so often heard him make use of in conversation. The doctor in explanation said, that in the highest caste relations of female association, termed fashionable society, the word virtue was used as a compendious cloak for the concealment of instinctive gratification, which remained unblemished in its sanctity of expression, while it remained impenetrable to the searching eye of scandal.

At this stage of her sophistic bewilderment, the mayorong directed their attention to the nearest village. The Indian women having discovered their encampment, were waving their trophies, obtained from involuntary exchange, with jubilant manifestations of happy elation. At this exhibition, after a suitable recognition had been made by Jack and Bill, who waved aloft, from their gun carriage, bunches of bananas, all turned with thankful expression to the mayorong, who had so earnestly advocated the conciliatory means adopted, so that he was fain to have recourse to his animal charges to conceal his emotions. Dr. Baāhar, however, could not withhold a disdainful expression of chagrin, that the chief of a wandering tribe, without a pedigree, or a home, should presume to plume himself upon his approved controversion of national usage that had been revered from time immemorial as the sanctified source of wisdom.

Correliana turning to the two sailors, whose countenances were moved with joyful emotions from the Indian women's grateful demonstrations of pleasure, asked how it happened that they were able to retain their destructive presence of mind when forced to evade the Indians' spears by a disordered movement? Her slow enunciation of English gave Jack time to work up his "reckoning" for an answer, which he gave with the blush of shamefacedness peculiar to the British sailor when accosted by a "lady," deepened by the reminder, that to his sensitiveness implied the "white feather." "You see, your ledyship, those Indian chaps had been cutting up their anticks so long, we sort o' lost our lay, but they brought us too with their spears, so we returned the compliments of the season in our fashion. Th'of as Bill says, we 'd much rather had the dig of the spear than it should have touched you by our ducking."

This new source of sensitiveness they had conjured through self-reproof, from the impression that their bodies might have averted the course of the spear. But when assured that she was out of their range when she received the wound, they were greatly comforted. Jack expressing his relief in the phrase, "things being as they were, it could n't be helped!"

As we proceeded on our way, along the eastern margin of the broad southeastern valley, our progress was overlooked by the women and children of the villages, who waved as we passed, our "forage" exchanges of yesterday, with an evident civilized interpretation of gratitude expressed in favor of their neighbors. But our supply of provisions being accommodated to our means of transportation, we could not gratify the desire that prompted the acceptance of their overtures. Evidently interpreting the cause, we found upon rounding a hill in advance a herd of cows panniered with bunches of bananas, plantains, and other edibles waiting for our acceptance, the donors watching us from the leafy screens of the hill plantations. The contraband gift — for their male protectors were evidently absent — was too acceptable, for the prospective relief of want, to be refused, and the recompense was suited to the full gratification of the womanly promptings suggesting bestowal.

In descending from a hill in advance, the valley proper of old Heraclea opened to our view. The plain, under the golden light of the morning sun, exceeded in beauty of variegation as in extent the famed vega of Granada, when clothed in the productive vestments of Moorish culture. At nine o'clock we passed the field fortalice commanding a view of the valley, and through the river gate those below. It had evidently been designed for a signal station and barracks for those employed to guard the ripening crops; the necessity for its erection bespeaking the inaugurated reign of oppression. The rock used in its construction, as well as of the bridges, dykes, and bank supports of the canals, was basaltic. Unlike granite, marble, and other stones used for building, it had withstood the disintegration of friction and chemical action through the lapse of ages with scarcely perceptible change. The style of architecture bore a strong resemblance to that inaugurated by Cestius, and introduced some sixty years before the Christian era. Our way from the tower to the hill city of old Heraclea, was a paved roadway overshadowed with relict growths of trees, whose ancestry had probably "ennobled" it with shade as an avenue of recreation for the citizens. Reaching the headland of the city esplanade, its level was gained by a zigzag ascent of the same breadth with its connecting avenue, its gradations being easy and of curious construction. Gaining the esplanade we were surprised to find its dimensions so extensive, as from below we scarcely conceived its plain would exceed an acre in area, whereas in reality it afforded a promenade that appeared to approach in length and breadth a half of a mile. As in the avenue below, the remains of parapet seats, and protected spaces for trees, were everywhere apparent. Entering from the esplanade, which extended in narrowed proportion to the gateway, through the single broad street of the first walled inclosure built for its protection, we passed to the fora, around which were the houses of those preferred to its distinctive advantages from the ruling qualifications reverenced, as godlike, from the fluent flow of speech. Built in an ampitheatre its walled defense could be made certain against the united tribes of the aboriginal race without, while the system of construction combined economy in space and in labor, giving evidence of emergency from doubtful crisis. The first inclosure had probably furnished ample space for the accommodation of its founders. Passing from the nucleus by the nether street of the fora, we entered the second surrounding, which corresponded in breadth with the original. The third and last, bespoke the disruptive reign of sensual gratification, heralding dissolution. Its expanded breadth from wall of circumvallation to nucleus, must have exceeded the distance of a mile, the palaces being detached from it by gardens and outhouses, the latter subserving the purpose devised from original intention. The structures retained, almost unimpaired, their original perfection; while within many of the heavier household utensils were found in place, touched lightly, from the comparative dryness of the climate, by an age of centuries' duration. These indications proclaiming the sudden calamity of successful insurrection, and extermination, were to be seen in every direction.

Leaving this city solitude, once peopled by the instinctively thoughtless and "gay," we gained the summit of the dividing ridge separating the Betongo from the Vermejo valley. A glance sufficed in answer for the question of causes that led to the selection of the "New City's" site as a safe place for recreative resort. Limited in extent, and remote from the larger cultivated district, it could not be made available as a permanent place of residence for the guard of growing crops; but was naturally adapted for the indulgence of luxurious ease in a revoltful country, as its walls inclosed sufficient arable land for the support of a limited number of inhabitants, while its natural and artificial aids for defense rendered it impregnable against aboriginal weapons, without taxing the energies of the citizens. Our introductory glances of admiration were arrested by tokens of recognition which greeted us from the citizens, who had assembled along the guard walk of the southern parapet in waiting expectation of our appearance. Their signals soon informed us of the enemy's position, which was in a grove surrounding a temple, and reaching from it to the road of descent at its escarped junction with the level avenue leading to the city gate.

In consultation for the devisement of means for dislodging them, Dr. Baāhar, and the curators of sound, still urged the precedent of classical experience, which advocated the greatest possible destruction of life when engaged in war with barbarous nations and tribes. Notwithstanding the pleading appeal inspired by the sight of her distressed relatives, Correliana manifested strong emotions of repugnance against the wanton destruction of life, even when the advocates strengthened their advice by quoting the padre's experience on board of the *Tortuga*. Turning to Mr. Welson and the mayorong for their support, she was relieved by the former's humorous expression, as he asked Dr. Baāhar to enumerate the number of generations that had passed, since his ancestors could urge equally well merited judgment for their own destruction? Then turning to Mr. Dow he asked whether he would prefer to seal the fruition of his hopes with slaughter, or the more lasting effect that would be insured by arousing their superstitious fears. Although urgently impatient of any delay to the full realization of his historical source of fame, his respect for the pungent elements of his questioner's resources caused him to offer his willing acquiescence if an effectual plan could be suggested for insuring their dispersion. Correliana asked the sailors through Mr. Welson if they could not think of some way to frighten the Indians without injury, as she could not bear the thought of exposing to death and mutilation the husbands, fathers and brothers of the women who had bestowed so gratefully of their means for the relief of those who were descended from their oppressors. After the two sailors had "put their heads together to overhaul their lockers," Jack said, if he knew exackly where the enemy lay, he could in a giffin fix a shell so that it would scream like a broadside of devils before it burst; and th' of they were civilized, and not up to the thing, they would scud like swallows caught in a gale at sea. The sailors' invention was adopted, and when everything was in readiness for all the emergencies that could be anticipated, the descent was commenced; but notwithstanding the eminency of danger, admiration gained the sway, attracted by the natural beauties developed at every turn in our downward course. The skill displayed for the artificial improvement of the natural advantages, would also have received like commendation if the means employed had not excited emotions of abhorrence. For the Indians who accomplished these labors of Heraclean devisement were in fact the benefactors of their oppressors.

Having arrived at the desired position for the essay

of Jack's "devilish experiment" the shell was belched forth from the howitzer upon its frightful mission. Its screaming powers had not been overrated by the projectors, but it exploded before it had accomplished half of its intended distance, seemingly in the very midst of the concealed foe, for the grove became swayingly alive from the panic imparted to its wooded growths. Moving rapidly forward, a second shell, true to its intention, accelerated the flight into a rout as wild with dismay as was ever enacted by congeneric warriors with civilized instincts.

Advancing to the bridge spanning the river moats to either bank of their conjoined stream, the city gates were open and the parents of Correliana stood upon the threshold waiting to bestow with tearful gratitude their acknowledgments for opportune deliverance from the manifold perils to which they had been subject. After they had bestowed upon their daughter tokens of affectionate welcome, in which all present joined with kindred sympathy, we were ushered in and made the centre of grateful attraction. It soon became painfully apparent from their wan features and tottering steps, that their vital energies were reduced to the lowest ebb from over anxiety and the want of suitable nourishment; so we at once mustered our prepared resources, and became their directing entertainers. Even the saturnine dignity of Mr. Dow, and the patronizing sagery of Dr. Baāhar, relaxed under the beneficent influence imparted from their ministering attentions. When the prætor and tribunes requested an introduction to the patriarch of the Kyronese, his absence was first noticed by the members of the corps, Correliana, and his granddaughters; when in the act of apologizing for his absence and the elder matrons, they were seen issuing from the temple grove; with their welcome the gates were closed and the sailors placed in charge. Then the Heracleans were placed upon the sillias of the horses and mules, — notwithstanding their earnest protests of ability to walk, — while each, as they proceeded up the avenue of the latifundium, was attended with the sympathetic support of the Kyronese and members of the corps. At the oppidum vera gates, nearly a mile distant from the cinctus, or outer wall gates, the Heracleans insisted upon dismounting, thankfully accepting the Kyronese proffers of assistance in rendering service to the sick. Correliana then directed us to the quarters prepared for our use, expressing the hope that the condition of her people would afford ample explanation for whatever was found lacking or amiss for the assurance of comfort in their accommodations? Having unpacked and disposed of our instruments and personalities in the house prepared for us, an evening consultation was held to devise means for the purveyance of supplies for the nearly famished inhabitants. Feeling certain that the besiegers were effectually dispersed, the hunting of wild game was proposed as a dernier for present support.

CHAPTER X.

At daybreak, of the morning following our entry into Heraclea, the prætor and Correliana paid us a visit. After salutations of renewed welcome the prætor addressed us, in substance, as follows: —

"You are already partially aware of the means of communication which have been employed to advise us of your presence, and the deliverance of our daughters' rescuers from their extreme peril! Through the same source we have been advised of your daily progress for our relief, now happily consummated. When the health of our families shall have ceased to tax your anxious care, we will then endeavor to make you sensible of our gratitude through the warmth of affectionate reciprocation. For the present we will ask you to assume the responsibility of your own entertainment, for we are utterly powerless for the fulfillment of that duty so inseparably imposed by our obligations. But with our energies restored we shall claim the gratification of reassuming the privileges of our natural charge. Until this sum total of our past indebtedness shall have been fulfilled, please accept the keys of our city in token of our submission to your direction?"

In reply to this tender, M. Hollydorf said, "We will accept the keys, but only in the light of a necessary facility to render our sympathetic aid more readily effectual, and will certainly feel more sincere gratification when your own, and the health of your associate citizens, will admit of their restoration.

Until then we shall rely upon your advice for direction, for we have already learned to prize its transmitted agency beyond measure, as it exceeds the power of material recompense." Then taking the prætor's hands, he continued with glowing warmth and tearful emotions: " Indeed we feel assured, beyond the possibility of selfish reflection, that in preserving your people, we have acted as agents for the opening of a way through which the children of our race may exalt themselves above the gregarious instincts of animality. We have already realized premonitory emotions, which bespeak an assured glimpse of immortality, albeit our habits and customs intrude their practiced grossness to mar the beatific visions inspired from the influence of your exampled reflection."

Here the tremulous cadences of a pæan˙ hymn caused the prætor to beckon us beyond the threshold, and from thence we saw gathered in groups, before the portals of each door, the residents uniting in the choral anthem of thanksgiving to the Creator for blessings vouchsafed with deliverance. At its close, we were apprised that it was their morning and evening custom to offer grateful salutations of praise with the rising and declining sun. Then, in the fullness of their grateful joy, they left to engage in the nursing avocations of the day.

After their departure we engaged in preparation for our first hunting expedition; when nearly ready the mayorong appeared accompanied by three Indians whose bearing proclaimed them upland chiefs. With their introduction, he stated that they had visited him while he was attending the wounded of their tribe in the temple of the grove; and as they evinced kindly emotions while endeavoring to make him understand the chief object of their visit, he followed them to the margin of the wooded growth, and he there beheld a train of horses loaded with panniers containing a plentiful supply of grain, so much needed by the famishing Heracleans. "Unable to withhold the elation of joyful surprise I embraced them, and could not resist the pleasure of bringing them to receive your personal acknowledgments for their timely supply of food." The prætor and tribunes, having been informed of the Betonges chiefs' introduction into the city by the mayorong, with the supply of food they had brought as a voluntary peace offering, hastened first to the hospidium, and then to the quarters of the corps to give them a fitting reception. To the surprise of all, Correliana, who accompanied her father, addressed the chiefs in their own language, with expressions of such grateful warmth that the eyes of the savages became tremulous with tokens foreshadowing the impressions of a moisture as nourishing to unselfish sympathy as dew to plants. When these exotic emotions had subsided, the Indians in turn tersely expressed their regrets for the unmerited sufferings their tribes had caused from the remote acts committed by the old Heracleans, who paid the penalty of death for their oppressions.

Correliana, in explication of what appeared mysterious in her ready use of the Betongese idiom of the Quichua language, said that she had learned it from children taken from the Indian villages, and adopted as hostages to be educated with those of Heraclea. "You have been puzzled," she continued, "with many mysterious passages since our first introduction, which have appeared more unaccountable to reasonable suggestion than this, still in due time they will be as readily solved." After a lengthened conversation with the Indian deputation, Correliana proposed that the gates of the cinctus wall should thereafter be left open for the free ingress and egress of their Indian allies, in trustful confidence as leal as though mutual faith had been kept from the beginning.

Mr. Welson suggestingly asked, if the river Indians, or in more truthful expression, the reptile savages, would not avail themselves of this open invitation to wreak their poisonous vengeance? To which Correliana smilingly replied : " Our benefactors have informed me that the river Indians, when in dismayed flight from their repulse, met the old chief who had been retained as a prisoner on board of the *Tortuga*. Holding them in check while he described the power you had exercised over him, and one of your own kind, he urged that any further attempt against the city would result as in the battle they had just fought. His collar investment was, in their panic, a sufficient verification of authority, and although a victim to your sorceries they proclaimed him an embogator, or prophet itinerator of the tribes. His description of the effects you produced upon him, conjoined with the padre's fears, has established your reputation as a magician capable of filling their bodies with tormenting scrouls, or demons ; this increased their panic, causing the tribe to disperse in all haste to their swamp feudalities. We are fully assured from the Betongese recital, — and they are not altogether free from the fear you have inspired, — that your presence will prove ample security for their absence from the highland valleys, as well as a protection to the *Tortuga* on her downward passage. In pledge of their fidelity, the Betongese have volunteered an escort for the Kyronese remaining at the anchorage of the *Tortuga*."

After the chiefs had partaken of food prepared by the Kyronese matrons, they were escorted without the cinctus gates by all within the city able to walk. When the gate keepers, Jack and Bill, were notified that from thenceforward the gates were to be left unclosed, they fired a salvo of a single discharge, then limbering up the howitzer stowed it, with munitions, in the keep of the gate tower ; but asked permission to retain their quarters, with the more than probable inducement of having their rations brought and dispensed by two Kyronese maidens, with whom they had been on "signal" terms from the day of their rescue.

Cleorita, after the Indians' departure, expressed to Correliana the hope that her grandfather had not been by her judged over-presuming in caring for the wounded Indians, or bold in assuming the responsibility of introducing the Indian chiefs into the city? " For with truth, he says," — she urged, " he would not have hazarded the venture, if he had not felt certain that they were trustworthy. Indeed we have seen many worse who have been grateful for kindness!"

"Say to your grandfather," returned Correliana, " that we justly merit the punishment he has inflicted, and I feel more sincerely indebted to him for the last service than the first. I will own frankly in self-reprobation, with the belief that the self reproof includes all except your own kindred, that my thoughts were altogether diverted from the possible sufferings of our wounded foes; and I will not pretend to assume even the merit of feeling sufficient solicitude to inquire whether any were injured."

The mayorong, who, with Mr. Welson, had overheard this plea of his granddaughter in his behalf, and understood its import, said to Cleorita, "you have spoken according to my desire, but you must not forget that the members of the corps were fewer in numbers than ourselves, and were expected as the sponsors of the expedition to present themselves for the relief of the famished citizens, so we each acted the parts of our allotment."

But Mr. Welson expostulated: " You need not attempt to say anything in our extenuation, for we turned a deaf ear to the groans of the wounded, and passed them with as much indifference as we left the severed serpent. Now that we have seen the effects of your unselfish sympathy, we cannot withhold from ourselves the fact that you are the real deliverer of Heraclea. You have merited and will receive the untutored homage of the Indians."

"You forget," replied Cleorita, prompted by her grandfather, "the eminent services of the most favored of the magicians, who has controlled the savage 'instincts' of the river Indians?" With Correliana's asservation, that the Heracleans were so universally indebted to the united members of the corps, and its adjuncts, the personal distinction of preference was resolved into the grades of adaptation for the parts enacted, they separated, with mutual congratulations, to engage in the alloted avocations of the day.

In view of their peaceful prospects, enhanced with food bestowed by their late foes, the Heracleans recovered rapidly from the pestilential flux, so that in a few weeks they were able to enjoy the liberty of the open country, and enter upon the reënjoyment of the boon of self dependence. The households enlivened by their reappearance, assumed the renovated impression of a happy vitality breathing outward for the kindred invocation of reciprocal goodwill. Correliana, with renewed vivacity and mysterious facility, had conjured the ability for conducting her own correspondence with Captain Greenwood in English, also for ready communication with the members of the corps. Her Kyronese companions, Cleorita and Oviata, had with her revived a speaking impression of the language derived from their father.

On the morning of the 7th of October, after the journey had been prolonged far beyond the time set for its accomplishment, from the grateful desire of the valley Indians to honor the people of the mayorong, the Kyronese remainder arrived under the conduct of Abdul, his grandson, and padre Simon. Their reunion and reception was joyful in the extreme. The compendic ejaculation of the padre, in sanction of the corps' expressions of happy satisfaction, will prove ample for the exposition of the prevailing impressions of renewed goodwill. "By my soul's salvation," he exclaimed, "I have by the same tokens come to a belike conclusion! For surely I would have as soon thought to see the lion and lamb lay down peaceably together, as to have been entertained as I have been by these same Indians. It was so unnatural, for you know the delegations from the tribes brought on to Washington are exhibited as specimens of wild beasts indigenous to the soil? But I can tell you, I never was treated more kindly in my life, bating I could not speak their language, nor they mine."

Mr. Welson inquired, whether in the item schedule of good treatment they asked him to take something, or smoke a weed? The padre happily averred, with a blush, that he had neither tasted of spirit or tobacco since his departure from the *Tortuga*. In testimony of the improvement from his abstinence all bore witness.

"But," asked Mr. Welson, "had you no fear of being bitten again?"

The padre smilingly expostulated, "I see that you have not left off all your bad habits, yet, notwithstanding the good example of the Heracleans! Why not let bygones be bygones? My own thoughts are a sufficient torment, without having my friends poke fun at my lameness."

"It is from no ill intention that we keep the crutch in view, but rather to prevent the necessity of its future use," suggested Mr. Welson. The padre closed the sally port of banter, by quoting the saw, "Sufficient for the day is the evil thereof."

CHAPTER XI.

WHILE the chiefs of the valley tribes of Indians were entertained in the city, one from the Vermejo petitioned the prætor for permission to settle with his tribe on the vega of the lake expanse of the Boetis below the temple grove. This petition awakened a pleasing smile in the expression of the prætor's face, who, without consulting his associates, requested his daughter to proffer his fealty to the united chiefs and their tribes of the valleys, in behalf of the citizens of Heraclea, with the hope that they would trustfully extend their permission for the continued occupation of lands alienated by the cruel oppressions of their ancestors of old Heraclea. When, with some difficulty, Correliana was able to make them understand the nature of this request, they pondered, and looked upon each other in bewildered silence. At length, one of the oldest Betongese chiefs, "saw the approaching ends of the long severed thread of unity that had caused the siege of hatred, and the concession offered by the prætor for uniting it with the durable bonds of privileged equality." His explanation was received by his compadric chiefs with smiling assent, assured that it was for mutual behoofment. With united sanction, evidences of mutual understanding were passed in tokens of goodwill, until the rays of the sun, in decline, were cast aslant from the brink of the precipice of the falls, covering with its bright canopy the shadows that enveloped the city beneath, then in strengthened concord the pæan hymn of thanksgiving rose in unison from every Heraclean threshold, and after it a responsive refrain repeated in swelling harmony from group to group. Of its import M. Hollydorf gave the following rendition, —

> "Neighbors good-night, good-night;
> A day of right,
> Without a wrong,
> Hallows our evening song."

At an early hour of the day, succeeding the arrival of the Kyronese detachment, Indian women brought fresh fish and fruits as presents, then volunteered their service for clearing the houses, colonnades, and patios of the accumulations consequent upon the sickness of the Heracleans, and were made happy by the acceptance of their proffered aid. Gradually the cheerless gloom which had held sway in the depopulated portions of the city for ages, from the harassed anxiety of its defenders, passed away under the active hands of the Kyronese and their Indian aids. Fountains, whose conduits had become choked, were opened and cleaned, causing the house gardens and latifundium to rejoice in primal gladness from water distribution above and below the surface of the ground. The loving sympathy of the Heracleans made manifest in the tender care bestowed upon the reviving sick, brought forth the latent gentleness of the corps, which had been suppressed from childhood by the civilized decrees of fashionable folly and vanity, begot from the precedental inoculation of habits and customs derived from the heroic ages of classical brutality. Indeed the members of the corps were so often moved to express genial emotions with glistening tears commingled with smiles, they seemed to have developed a new inherent combination as necessary for the joyful expression of happiness, as sun and showers for the behests of fruitful vegetation. The padre, in his quaint emphatic style, expressed the prevailing influence in an evening salutation addressed to his

compadre Dr. Baāhar after even song, in this wise:
"Well I declare, doctor, upon my soul, I have passed
such a happy day in useful labor that it seems as if I
had just emerged from a life's nightmare of torpid inactivity. Really, upon my hopes of salvation, I believe that I could live and thrive upon the joys of
others, without material food."

But the doctor, who was impaling the insect game
obtained from his day's hunting excursion, replied
sneeringly. "So, so, h-m — I see, you are taken in,
with the others, by this humdrum life of these Heracleans, with their puling, wishee-washy affectations
of caring more for others than they do for themselves.
The long and short of the matter is, that you are all
subject to an unnatural influence, and if it is not
thrown off immediately, from whatever source derived, you will shortly forswear manliness, and your
hopes of heaven."

This baited injunction caused the padre to exclaim,
"My goodness gracious, doctor, you frighten me! I
hope you don't truly think there's anything like
magic or sorcery used upon us here? To be sure,
now that I remember, I have had strange thoughts,
to which I have never been accustomed to before!
But they have been in motive pure, urging the necessity of controlling the appetites and passions, if we
would attain the abiding confidence of a trustful
affection, that outreaches self. But then, as you
know, the devil can preach, and practice too, if it so
minds him, self-condemnation?"

"Certes, the fact is," replied the doctor, "you are
subject to vagaries when your stomach is empty, and
require to feel the force of sound German philosophy
that urges substantial fullness as the source of generous impressions, eloquence, and heroic deeds, and for
exorcism thorough fumigation with tobacco smoke."

M. Hollydorf, from the intervention of multiplied
causes, had procrastinated the inauguration of his

scientific explorations, until compelled to enter upon the duties of his commission through fear that inquiries would be instituted to learn the cause of his long silence. Fully aware that the manifold attractions of Correliana had served to abate his professional enthusiasm, and urgency of his desire to fulfill the trust reposed in his discretion, he resolved to make a test of his naturalistic occupation for the diversion of his thoughts from an object of hopeless attainment. Notwithstanding his knowledge that her affections were irrevocably fixed, he could not withhold the manifestation of a hopeful desire in her presence, within the limits of reverential respect. Correliana, on her part, seemed to fully understand the import of his attentions, but was in no way embarrassed by their indulgence, which with her frankness appeared inexplicable. When he expressed his intention of commencing his microscopical field investigations, she asked the privilege of assisting him when free from the indispensable duties of the household; promising, if her request was granted, to be diligent for advancement in scientific knowledge. She was promptly accepted as a catechumenic aid, notwithstanding the promptings of his judgment which suggested that with the ever present cause of his disquietude, his remedy would prove of little avail for relief. But he determined, with a lover's infatuation, to converse with her as an abstract divested of material embodiment.

On the first day of November, while engaged in preparing his instruments after evening song, M. Hollydorf was surprised with a visit from the prætor and family. Observing that the unusual hour caused fear that some mishap had occurred, Correliana hastened to relieve the anticipation of evil tidings by stating the object of the visit. "My father," she said, "has been delegated to proffer you the perpetual hospitality of Heraclea. Not, however, with the design that you should hold it as an acknowledgment of service rendered, but rather as the promptings of affectionate esteem for your companionship. As you are aware, we have no practical knowledge of the world beyond our city walls, and feel that in winning from you a reciprocation of our affection, we shall be advised of a course that will avail us as a protection against the grasping cupidity you have described as the inherent motive power of civilization. To be forced to adopt habits of corruption, in defiance of local option, because your enlightened civilization holds that the power to enforce their arbitrary despotisms with brute strength, aided by destructive mechanical adjuncts, is right; would, with the introduction of 'luxurious' poisons which frenzy and degrade the human instincts, make us regret with anguish our liberation from the deadly intent of our savage foes. For their speedy poison, with its putrefactive torments, does not degrade the animus of goodness, but relieves it from material bondage in purity for immortal association with those who have gone before. We feel self-conscious that we are in intention pure and free from cupidity, which assures us that we merit the affectionate interest that you have bestowed for our liberation and welfare. This much we will advance for initiation without infringing upon the more matured wisdom in store for your direction. With the full development of our loving resources, we feel confident in securing your permanent residence among us, as advisors, in warding off those who would, for the gratification of craving instinctive cupidity, sacrifice our well-assured happiness, from which we realize in life a foretaste of immortality (*smiling*). Fortunately, the sage suggestion which led Mr. Welson to confer the honor of knighthood upon the savage for the indomitable bravery of his instinctive propensity to inflict deadly wounds with his teeth, has relieved us from anxiety from his kindred; and if we can persuade the grand master 'Lobscounster' to take up his abode in our midst, his influence may extend to the orders of civilization, for our protection in the enjoyment of affectionate association. If he will but exert his power to protect us from the forced invasions of trade, that would palm upon our weakness noxious devices, which in naught would advantage the invaders, but make us wretched beyond measure, he will insure our eternal gratitude."

Mr. Welson in response said, — The eminent Lobscounster, if insured from increasing merit a continuance of Heraclean favor, he cannot be forced from his allegiance, and in earnest of his intention thankfully accepted the extended privilege of becoming their permanent guest. But would most devoutly beg to decline acceptance of the cognomic title bestowed upon him by the savage embogator; as to the English ear it was euphonious with smack of a descent from an ancient sea cook, and in no way likely to insure reverence among sea-faring men. Indeed, the individual referred to would have strongly suspected collusive substitution if the interpreter had been well versed in the aquatic lore of ocean English.

When the visitors were about leaving, M. Hollydorf announced his intention of entering upon his microscopical investigations on the morrow, reminding Correliana of her promise to render him assistance.

"With life and health I shall most assuredly be present," replied Correliana, "for I have a woman's curiosity to test the wonderful magnifying powers of your instruments, which so far exceed our untutored conceptions of mechanical refinement. As we have some practical knowledge derived from the observations of animalculan life, we hope that our assistance in your department of science may eventuate in relieving your anxiety, occasioned by the delay incurred from the aid you have rendered our people." With this enigmatical proposition, bespoken with the earnest zest of sincerity peculiar to all her variations, Correliana and her parents bade the members of the corps good-night. Long after the departure of their visitors, the members of the corps, puzzled and perplexed by Correliana's seemingly frank intention, commingled with implied reservations, and a knowledge of the world incompatible with the complete isolation to which their people had been subjected for ages, endeavored to unravel the clew to her powers of premonition.

After listening in silence for a long time to the various suggestive expositions of others, the padre suddenly exclaimed, " You may reason and think what you please, but for my own part I know that I have not been myself since *she* first came on board of the *Tortuga;* and if everything was fair and above board, as they would have us believe by their words and actions, they would speak out at once, and not hold anything back to make us feel doubtful of our souls' safety. For by the mouths of a cloud of witnesses, we know that the powers of darkness have wrought from the beginning of the world their designs for the temptation of souls, through the agency of woman's allurements; and for myself I can truly say, that I can't avoid doing as *she* wishes to have me without a word of direction. Besides I am altogether too happy to have it natural or lasting; and the method of educating their children separate from each other, and away from the example of their parents, is barbarous and unnatural."

At the completion of this impulsive padric, Mr. Welson quietly observed, — " If we are to judge from appearances, we could not question the source of your improvement. But as appearances are deceptive, and the evil-disposed seek solitude for indulgence, the cloud of a witness rose from beneath the skirt of your coat, with the odor of tobacco from your suddenly concealed pipe, to confirm your shame in the presence of purity. If your soul has been tempted, it has been from gross indulgence to purity."

The padre abashed relapsed into silence. But Dr. Baāhar, who had for a butterfly consideration furnished him with the means of indulgence, undertook his vindication, which he commenced with the syllogistic proposition: " We will certainly admit that your spasmodic sarcasms are poetical refinements upon fact, but I contend that you are neither scientific or logical in your deductions. If God created man with reasoning instincts, they were undoubtedly intended for invention and indulgence. Again, in depriving children of their natural protectors' care and example, is in open controversion of Divine will. As for me, I do not assume to be more wise in my day, than my ancestors were in theirs. By the assumptions of your theory, founded upon the partial knowledge of these egotistical Heracleans, who have been shut out from a knowledge of the world from time immemorial, we should repudiate the transmitted experience of our ancestors. I shall not be guilty of so gross an act of ingratitude ; my father the counselor, and his progenitors, ate their saur-kraut and sausages, drank their beer, smoked their pipes, and were excellent swordmen and genealogists, and I intend to do honor to the habits they inculcated."

Pettynose the buzz recorder of sound, and Lindenhoff the genealogical curator of sound, with Viscouswitzs the photographic artist, sided with Dr. Baāhar, the latter sensuously remarking: " The women may be accounted puritanically beautiful, but they lack the bouquet of civilization, as well as the natural flavor peculiar to the creole variations ; and as to pleasure, I could derive as much by an association with marble busts in the atélier of a sculptor. There is an air of repulsiveness about them that repels geniality, so that I never feel comfortable in their presence, and but for the encampment of the Vermejo Indians on the lake, I would, with the first opportunity, throw up my engagement and return to the haunts of civilization ; for of all things I abhor pedantry in men and puritanism in women."

"We are as yet novices in the ways of the Heracleans," urged Mr. Dow, "and but imperfectly understand their motives of action or system of self government. To judge them from our partial impressions, which your personal opinions bespeak, is proof positive that the cavils of surmise, peculiar to individuals, originated the prejudices to which you have given voice. To me the addenda to their morning salutation and evening anthem of praise, as rendered by M. Hollydorf, bore advisory reference to the source of their happiness." M. Hollydorf fully endorsed Mr. Dow's views.

CHAPTER XII.

M. HOLLYDORF after morning salutation mustered his assistants for the inauguration of the legitimate duties entailed by his commission; as he had become fully impressed with the necessity of "working up" a sufficient number of experimental proofs for the basis of a preliminary despatch of intention. Selecting a retired portion of the latifundium for his field of operations they commenced their labors in good earnest. Of all the civilized nations of the world, we can claim for the Germans a just preëminence in those departments of science devoted to the investigations of the habits and associations of insect life. In truth, the enthusiasm shown for insect explorations has extended itself to every department of their national existence; from the palace to the cabin particular attention is devoted to hunting, impaling, and preserving their cadavers, arranged in order, genera, and species, in mausoleum cabinets for mummified exhibition as shrines for the enraptured gaze of Teutonic devotees. Even the mediæval Gael of the Scottish Highlands never possessed, in living endowment, an attritive iota of the associate luxurious zest imparted from their joint stock investments, or the Egyptian, of yore, in his necropolitan collections, a source of such vainglorious gratification.

M. Hollydorf's first day's investigations were rewarded with the discovery of old species, familiar to his eye, under new and strange combinations, affording conclusive evidence of exotic transfusion in propagation at some remote period. In semi-meditation, with a disinclination for food and midday rest, he continued his preparatory investigations while his assistants refreshed themselves with their accustomed rations and siesta. Availing themselves of his invitation and leisure, the prætor, Correliana, Mr. Welson, and Dow made their appearance. Apologizing for interrupting his studies, Correliana requested the privilege of subjecting a flower from her garden to the magnifying power of the tympano-microscope? Assuring him, with its presentation, that she felt certain, from its extreme beauty and purity of fragrance, that it would attract a high order of animalculan existence capable of appreciating its rare combinations. After a close examination with his unaided eyes, he declared it to be of an unknown species and as peculiar in its rare beauty, novelty of its perfume, and delicate pungency of its impression, as the Heraclean representatives of woman kind were superior and distinct from their civilized genera in the purity of their habits and customs. With this combined pronunciamento of comparison as a vent to his enthusiastic admiration, he placed the flower in the field receptacle of the tympano-microscope for focal magnifying reflection of its parasitic habituary residents, for inspection and classification in substance and sound. With an exclamation of surprise, compounded of fear and amazement, he started back from the instrument exposing to view the petals and pistils peopled with a multitude of diminutive human beings, who were convulsed with sneezing spasms of laughter, which they tried in vain to suppress with expedients in common use by our kind. The tympanum in sound articulation reverberated their tiny cachinations and sternutatory explosions with such comical effect, that the prætor and Correliana were compelled, notwithstanding all their efforts to avoid the impulsive sympathy of contagion, to join issue with this mirthful introduction of our savans to a kindred animalculan representation of our race. While equally subject to the uncontrollable spasms of mirthful laughter and dumb amazement, the spectators to this scene of apparent conjurement were held speechless.

The leader of the diminutive apparitions at length leaped lightly, as if propelled by a sneeze, upon the stage within the reflecting compass of the tympano-microscope. Then, after a few ineffectual attempts to regain his composure, he finally succeeded in obtaining sufficient control to offer the following apologetic address, which gradually recalled us to our senses; but not in sufficient degree for a realization of their actual existence as human beings, free from the magic attaint of fears conjured from superstitious instinct. He thrice repeated to attract our attention from the stupor of amazement: "Men of science, and deliverers of the Heracleans, our protogean affinities!" Our partial attention secured, he continued. "If through the disability of our Dosch, or chief advisor, our selection as Manatitlan ambassadors to welcome you, in our people's behalf as the preservers of our co-affinities in affection, should prove a source of discredit from our undignified appearance on presentment, it would prove a source of lasting sorrow. But we feel certain that you will extend to us the favor of believing that we are not inclined to untimely mirth, notwithstanding the example we have given to the contrary. With the concerted desire to impress you at a suitable moment with the reality of our existence as a race, Mistress Correliana probably forgot the keen sensitiveness of our schneiderian membranes to pungent odors, and with the intention of giving as much eclat as possible to our introduction, selected from her garden the most beautiful and fragrant flower of its parterres. The novelty of our emprise withheld our attention from the flower until it was placed in your hand for examination, then too late to effect an exchange, we braced ourselves to resist its effects. Hence our humiliating condition when exposed to your view and hearing! Thrown off our guard by the transformation effected in our size and sound of our voices, and above all by the consternation manifested in the expression of your faces, we could not resist the impulse of our naturally mirthful dispositions. That the infection should reach and overpower the more staid humor of our cousins, you will not wonder, when you recall your own and our disordered extremes. If you will control your perturbed emotions for a moment's reflection you will be able to realize the irresistible nature of our impressions under these combined effects. Withal, when our existence and presence in auramentation becomes familiar as a recognized reality, you will find in our joyous dispositions a ready explanation for these ante phases of our first personal introduction."

Upon this hint, Correliana conquered sufficient composure to introduce the speaker as Manito, the Prætor of Maniculæ, the chief city of Manatitla. Then with the accompaniment of a spasmodic inclination to sneeze, as they leaned over the serrated edges of the petals, the tribunes were introduced individually by name. This process was lengthened by occasional suppressed tendencies to mirthful outbreaks, which gave M. Hollydorf and his companions an opportunity for partial recovery from their dazed state of amazement. When sufficiently restored for intelligent comprehension, the flower was changed for one of less pungent odor, and Manito from the rostrum point of a petal continued his address.

" From our diminutive size we willingly subscribe to the designation your nomenclature bestows upon insect animalities which are but partially visible to your unaided eyes. Still we do not disdain our size, for with the Manatitlans it has received the compensating privilege of a perception that enabled them to distinguish the evident object of mankind's intelligent endowment above the instincts of associate animality. " Like individuals of your race, ours vary in size. Some among the Manatitlans have reached in stature a height approximating in a remote degree to your well formed dwarfs of a standard monstrosity in the diminutive extreme sufficient for the excitement of wondering surprise. Our own divisions are expressed in terms rating from the smallest in stature, which are called tits; these form the masses, but with a sensible diminution in numbers from an upward tendency to the second degree of elevation from the majority. The middle class are styled mediums. With every generation this grade has been increased in proportion with the decrease of the tits, and ranks in status with your "well to do" money grade of merchants and speculators. The giantesco enjoys the highest statutory standing in the ranks of size, representing your titled duke commanders, and subalterns of lordly and knightly degree. But these distinctions are only perceptible to the eye, and in no way arbitrary in the assumption of prerogative stature rights above those below. As our scholastic term of education commences with the infant at the age of two years: the first stage that directs and controls the infantile perceptions and cravings of instinct is styled papilage, and is under the supervision of the censor and nurse, who hold the instinctive exaggerations of parental fondness in check from birth. This habilitative stage of matriculation is the most trying for direction, as upon it depends the matriculant's after power of self-control. The second stage of nonage commences at seven, when the self-devising perceptions begin to expand into individuality, that require educated direction, and leading encouragement. At fourteen, or the pubertal stage, the first indications appear for the premonitory inauguration of status rank established for the distinctions of size.

The initiatory discipline of the scholar entering upon his senior term, induces the tractor disposition of the censorial advisor, in association with his juniors; in place of your form system of " bullying " the nether " fag," whose weakness makes submission a virtue, when subject to the classical distinctions of arbitrary power. The seniors become assistant tutors to the censors and teachers from the age of fourteen until the close of their twenty-third year, when they graduate; and after a probationary term of three months' " courtship," with the connubial censors' selection of affiances, are married. This cursory glance will serve for an introductory insight into our natural system of education designed for the direction of our immortal endowment in perceptive flight above the body's ephemeral gratification of instinctive desire.

" Of other matters, pertaining to our actual realization of an enduring happiness, you will be advised by our advisors; as our interview was designed solely for your recognition and realization of our existence as a race in diminuendo alliance with your own. Our associations with your race are of a privileged description, which from the concentrated acuteness of our sensitive perceptions, enables us to divine your thoughts by auramental espionage. If you will give a moment's investigation to the impressions of thought, when free from the turmoil of suspicious doubts, which now assail and render your efforts for reasonable perception void, you will find that they are all distinctly enunciated in the thalmus auditorium, which is the focal centre for maturing sensorial observations. Our size, and practical knowledge of the sensitive departments of your ears, enables our giantescoes to gain the aural sinus without provoking titillation, and its proximity to the vibrating portal, or vellum auditorium, permits our sensitive perceptions of sound to realize your thought articulations before they are matured for rententive comparison, or the vocalized utterances of speech communication. So that in reality, we hold the gigas (the name word we use for the designation of your race in contradistinction to our own) subject to our direction, when free from the ruling habits of instinctive indulgence, which defy control. As the previous knowledge of our advisers has preferred you to their confidence, I will state that our means of direction are through thought substitution, which the giantesco is able to modulate with ventriloquial variations of voice for the receptive nullification of those derived from their own sensoriums. Of course, the effects vary with the intensity of the subject's command over his own sensorium, and the absorbing influence of educated impressions imparted from habits and customs. As an example, I will now state that M. Hollydorf, in his turmoil of doubts, feels that Mistress Correliana has in some way imposed upon his confidence; but my informer says that his impressions are in no wise capable of assuming the power of self control, so that upon our own responsibility we will exonerate Correliana from all deceptive intentions; as she was subject to our control in withholding from you a knowledge of our presence, as the mysterious source of her guiding premonitions, and means of obtaining information of human affairs in the world beyond the inclosing walls of their isolated city. Now, in turn, we ask you to withhold from your companions the result of your day's explorations, that you may observe the influence we are able to exert for their mystification, and the development of the intangible resources of instinct, which subserve for the delusive beguilement of reason from the intelligent direction of creative indications. This much, will prove sufficient for your night's cogitations, but to-morrow the Dosch and his advisors will instruct you in the weightier matters pertaining to our educating system devised for self control. As you are still hovering in the clouds of doubt, we will regale your senses, for composure, with a musical olio. M. Hollydorf, at the period of our first introduction, was considered an excellent judge of music, and at times amused himself with amateur compositions, one of which pleased me, and on my return to Manatitla I presented it to our musical censor, who adopted and incorporated it with our salutations. We will now render it, that you may pass censure or commendation upon the accuracy of our version; for of all the selfish kleptomanias, that of stealing musical compositions, and mutilating them in transposition for an author's reputation founded upon a lie, is the most contemptible within the range of barren instinct. Fortunately, only the younger branches of the mouthpat tribes of our species have ever been guilty of a witless invention base enough to seek gratification from so mean a subterfuge."

With this apologetic prelude Manito marshaled his choristers along the borders of the dependent curves of the petals facing his bewildered auditors and rendered the following stanzas with an effect that revived them from their superstitious fears:—

> "From darkness dread, the dawn appears!
> Mother of day, whose dewy tears,
> Distilled from the labors of the night,
> Greet with joy, the sun birth of light.
>
> "Hail, glorious mother of morn!
> Beautiful type of woman's form,
> When hallowed from instinctive night,
> She hails, at birth, a son of light."

M. Hollydorf recalling the occasion and source of inspiration, glanced at Correliana with a furtive look of anguish. For the prompting source of the stanzas, was a longing desire that woman's beauty should be adorned with more lasting "graces" than those bestowed by the fashionable dressmaker, dancing master, and boarding-school mistress, in hopeful premonition of an immortality with joys exceeding the gossiping allurements of a heaven of sense. The look of sympathy he received in return banished from his thoughts doubts, and suspicions of supernatural agency. Manito, observing the confidence expressed in his glance, and the more ready belief of Mr. Welson and Dow, that the Manatitlans in reality represented a diminutive department of human mortality, said, that as his mission for the day had been fulfilled in degree beyond expectation, they would not prejudice their success by prolonging the interview, but would leave them with a new zest for the transmission of one of their best melodies. He then rearranged his choristers and rendered " Home, sweet home," with an effect that caused them to join in thought sympathy with the affectionate harmony of Manatitlan expression. At the close the prætor and tribunes of Maniculæ bid their first giga audience good-by, and disappeared from view. Correliana then signaled the stoop of her favorite falcon Merlin from his circling wafts above the latifundium; after a short perch of a few moments upon her wrist, he was despatched, as she announced, to Maniculæ, bearing back the prætor, Manito, and tribunes.

Mr. Welson was the first to break silence after their departure, with a long drawn, — " Whew," as a prelude to the exclamation, "Ah, ha! mistress Correliana, we have the secret now to all your mysterious enactments, which inclined those the least superstitiously prejudiced to credit you with an inheritance tinctured with the pretensions of your sibylline ancestry. But our wondering amazement is scarcely less than it would have been under the superstitious impression that you really possessed the power invoked by the ancient sibyl. Still the manifestation of a visible source, however small, is far more agreeable to our perceptions."

Correliana answered, with a pleading smile, " You

will surely forgive, and pardon me for retaining a secret of such importance, in the face of all your kind and confiding acts, now that you have learned that I received it in trust from a source so well qualified with the essentials of prudent direction? The Dosch, however, will more fully state the many causes that rendered its retention desirable. But of this you can rest assured, the Manatitlans are *bona fide* representatives of animalculan humanity; and when I state that we are solely indebted to them for our redemption from the bondage of instinct, you will understand the nature of our trust in their direction."

Beckoning the stoop of a falcon, it alighted upon her wrist. She then exposed, beneath what they had supposed to be an ornamental attachment of designation, a howdah. Then taking from her pocket pouch a reel of filmy thread,—attenuated to a degree that rendered it almost imperceptible to the eye, she wound the free end around Mr. Welson's finger, then asked him to try its strength. With his utmost exertion, tried with many devices for its separation, the thread remained unparted. She then explained that the materials, from which, in perfect combination, it was drawn, were mineralized with flexile and vis inertia substances in adaptation for a great variety of purposes, subserving for the protective furtherance of health, comfort, and personal purity. Also for protective defense, "as it is impenetrable to the swiftest fledged missiles when wrought into textile fabrics." But its most esteemed peculiarities are repulsive resistance to uncleanly cohesion, combined with a nonconducting neutrality in the transmission of cold and heat, causing the refuse execretions of the body to evaporate without obstructing the rejecting orifices of the ducts, when used in its adaptation for raiment. In part, we have been able to imitate this valuable acquisition for the protective preservation of our persons from decomposing agencies, which are constantly in a fermentable and putrefactive state of conceptive action for the production of renewed vitality varied in degenerative series. But of these matters the Manatitlans will advise you in due time. In your present state of perturbation it will but little avail to extend our conversation into details that require for a complete understanding consecutive exposition."

After Correliana and her father had taken their leave of the four favored witnesses of the new grade revelation in the status of humanity, they remained standing in the same position, absorbed with contending emotions of doubt and belief, until aroused by the approach of Dr. Baāhar and the padre. Then, with a forced recovery, M. Hollydorf announced his intention of discontinuing his explorations for the time being; which afforded his assistants a desired relief, for with their few hours' occupation they had discovered in themselves an unwonted dislike for the professional details of their occupation. While on their way to deposit the tympano-microscope in the house designated by Correliana as the one intended for the reception of the Dosch, the four maintained their thoughtful silence until after they had bestowed upon the instrument of revelation a careful disposal. Then M. Hollydorf sententiously remarked, " Although still perplexed, I am confident in the full integrity of Correliana's assurance that these Manatitlans are *bonâ fide* embodiments of humanity, with intelligent capabilities superior to our own! But it is hard to reconcile them with any of the preconceived ideas of our race. They certainly advocate, with practical demonstration, a more direct and reasonable way for the attainment of present and prospective happiness, than that of redemption from sin by saving grace?"

" By all that there is in us, capable of assuming the control of judgment, we cannot avoid their own, Miss Correliana's, and the confirmation of our own senses in attestation of the fact of their real presence," added Mr. Welson.

"For my own part," said Mr. Dow, "there is to me nothing more strange in their discovery, than in that of the Heracleans, now that we have recovered, in a measure, from the first startling effects. It has occured to me frequently, of late, that there must have been some interior creative object in the gradations of instinct, and ultimate alliance of superhuman intelligence with the highest grade? It is certainly impossible for me to reason myself into the belief that we have been endowed with a perception of goodness, and the necessity of purity for its attainment, to have them dispensed with in life for the substitution of the instinctive greed of selfishness, with the accommodating proviso of repurification by an act of saving grace! Neither can we disguise the fact, that we now think and act quite unlike our former selves, with a sensible improvement in happiness, in freedom from the selfish accessories we formerly thought necessary for its assurance."

At this point they were interrupted by the entrance of the prætor with his wife and daughter, who came to inquire if M. Hollydorf wished to suggest any change for the better accommodation of his instrument with regard to light? In the expression of his satisfaction, M. Hollydorf alluded not only to the wonderful preservation of the buildings, but furniture, which appeared, in style, to have been coeval in manufacture with the remnants seen in old Heraclea. In explanation the prætor said that it was much easier to preserve from decay than to restore ruins. But the means of preservation had been bestowed by Giganteo XVI., Dosch of the Manatitlans, as a legacy to the sons of Indegatus, associate prætors of Heraclea, who were the first of our race that became personally acquainted with animalculan humanity. "You will find all of the unoccupied houses of the city in like good condition with this, and equally free for your inspection and occupation."

As the occasion was opportune, M. Hollydorf consulted with those present how he might prepare a statement of the day's developments sufficiently credible for the acceptable belief of the Home Society? The prætor advised him to defer his cause of perplexity to the Dosch, who would resolve it readily, from a personal knowledge of the characteristic peculiarities of the members of the R. H. B. Society. Then Mr. Dow preferred his petition for their united aid in the advancement of his historical compendium of the Heracleans. This all were pleased to accord, as it was through his indomitable perseverance that the discovery was accomplished, before the City of the Falls had been reduced to the tenantless condition of its senior counterpart. As he was referred to me for special aid in compilation, from his lack of knowledge in the constructive use of the Heraclean idiom, — which was to us personally a source of mutual regret, — it will be well to state in anticipation of a similarity in diction of our separate labors, that I have been in no way beholden to him for the style I have adopted in recording the historiographical account of the corps investigations. I trust that this egoistic explanation will prove sufficient in efficacy to redeem me from plagiaristic odium?

CHAPTER XIII.

THE Prætor and his family, including Cleorita and Oviata Arcos, with the Four, awaited, on the morning succeeding the eventful day of Manito's animalculan introduction, the coming of the Dosch of Manatitla in the audience chamber of the house, dedicated by Correliana in aptitude to the developing powers of the tympano-microscope, "the auriculum." After a short delay of expectation, the courier falcon appeared at poise, from which in swift descent it came in downward incline direct to its perch on Correliana's wrist. But a second elapsed before the tympanum reëchoed in cheery tones of salutation the voice of our expected visitor. Our attention attracted to the field of magnifying reflection, discovered a coterie of animalculans, of nearly the same size, grouped about the speaker. With the salutation, "Afferens scientiam errantes gigantes," he addressed us as follows:—

For ages untold, our race have waited in patient expectation for the morning's dawn when they could salute yours face to face, and impart to you a source of happiness that in life realizes communion with immortality. To us has been vouchsafed this coveted privilege, and it shall be our study to improve it to your advantage. Notwithstanding the mal-a-propos accident — casting upon Correliana an arch glance that wrought for her face a scarlet veil — of yesterday, which detracted from the dignity of an introduction so important to the regenerative welfare of your race, we were glad that auspicious mirth was the trophy of the occasion, rather than tears of grief, of which we shall be mindful in adjudging our censure to the cause. Joyous mirth we have esteemed an evidence of goodness, for it declares itself beyond the reach of selfish impediment that breeds evil intention ; even when the foibles of our kind become the subjects of humorous provocation. Mirth is ill timed, when preconcerted with a knowledge that a portion of those present will be unable to appreciate the humorous incentive ; as it opens wide the door of suspicion with your peoples, who have been educated under the partial sway of national habits and customs. Dissimilarity in habits and customs, under national patronage, begets from seeming incongruity a disposition to gibe with missile retorts, fledged and tipped with ironical sarcasms, as rankling in effect as the pointed weapons in the mouth of Mr. Welson's knighted chief. To be frank, if the ludicrous scene of yesterday had occurred with matured acquaintance, I should not have spared the demure, but conscious blushes of the fair medium. Our first acquaintance with you, although not mutual in personal recognition, is of older date than yesterday, and upon it has been founded our predilections, which in train have led to the many concurrent circumstances favoring the happy issue of our more direct scheme, devised for the liberation of your race from the pampering trammels of instinct. It would have been quite easy for our first giantescoes to have obtained an introduction to your race, if they had emulated the desire of being exhibited as an iotian monstrosity for the gratification of giga greed and curiosity. But fortunately for our present hoped-for issue, our system of education, devised for the development of affectionate confidence, encouraged the past generations of our race to wait for an opening free from the entailment of experimental disadvantage. A knowledge of our race for the gratification of your scientific savans' curiosity, would have been as profitless for good, as their sight-seeing acquaintance with the moon and stars. Our Manatitlan sages have from the earliest period recommended extreme caution to prevent the premature introduction of our race to yours. The favorable indications to be watched for in premonition of a successful issue were those of extreme folly, heralding a closing cycle; for the contrast afforded by the result of our happy example would attract kindly imitation of those inclined to affectionate goodness.

Desideratus, one of our most approved prognosticators, deposed that the affections of woman afford the best test of a closing giga cycle. When frivolity and the gossiping comparisons of vanity gain the ascendency over natural affection, inherent as the birthright of woman, then you may know that the symbolic serpent's tail has received its final circle inclination for union with the mouth. This inclination was foreshadowed in the eighteenth century, with invention of power looms ; which with the largely increased acceleration of steam, fabricated in excess of the world's actual requirements for healthy protection and comely adornment. With steam as an inductive aid to civilized progression, the Eugenic era was ushered in, when the frail mortal tenements of women became subject to empirical vanity, and in rivalry, the standard-bearers for cumbersome mechanical products, to the utter perversion of healthy elasticity, comfort, and their special vocation of fostering for immortality affectionate goodness. This derilection of giga women from their manifest duty, has brought in train domestic and dynastic miseries, while from dreary self conviction their hopeless prospect closes with the grave. As we have now adventured the only opportunity that has ever occurred, with a prospect of success, for extending the influence of our happy experience to your race, we will with our introduction premise a description of *Our Country*.

Manatitla is situated in the Andean district of La Plata, with a southern aspect. It occupies a space between the parallels of 20° 40'' and 30° south latitude and 40° 50° west longitude, embracing an area of forty square furlongs, of Manatitlan measurement. Its surface is diversified, combining in well-defined variety mountains, hills, and vales, with their concomitant streams, lakes, and brooks; affording with arable advantages, prospects unrivaled in beauty, which have been enhanced by the grateful labor of its inhabitants in acknowledgment for the benefits bestowed. The climate is salubrious and free from the extremes of heat and cold, having a valley altitude varying but little from six thousand feet above the estuary of the La Plata. The adjacent country is occupied by the giga and animalculan wild hordes. The Minim is the largest river. Its source is derived from Lake Areta, located in the Andean spur of Ultisimma; flowing in a northeasterly direction it finally becomes tributary to the Vermejo. On the northwestern bank is situated our chief city, Maniculæ. Forty of our miles below, on the same bank, is situated the City of Iota, containing twenty thousand inhabitants. Nearly opposite the last named city, is the town of Speck, its inhabitants, in transition, being chiefly occupied in the manufacture of auro-silicate for edificial construction and textile fabrics, rendering them indestructible and repulsive to cumulative adhesion. The entire population of Manatitla is estimated at eighteen millions, with a healthy tendency to a continued rapid decrease in number, from causes which will be described hereafter.

The Traditional History of Manatitla, is coeval with the imaginary date of Mauna Che's advent as a deity from the La Plata into Alta Peru, reaching in your time measurement to eleven thousand years, which probably embraces relics of truth, among others a like origin with the Heracleans; as we are without doubt descended from castaway parasites of gigas from the eastern continent. But as it is a constant repetition of acts of oppression, in kind with your classical written history, we will not shock you with their rehearsal.

The Actual, or Written History of Manatitla, was commenced in the latter portion of the reign of King Primus, from which dates our transition period, or emancipation of our people from the instinctive rule of the stomach and its engendered lusts. But from its resemblance in factional disruptions to your own, culminating in a parallel to their cycle condition, we will only allude to the causes that immediately preceded, and in tendency wrought the changes that finally effected partition from old habits, and the reverenced usages of instinct. Arbitrary, religious, and civil exactions, seconded by compulsory persuasion against all nonconformists, signalized the tendencies of the period, and gave birth to an ultra instinctive race, styled liberal democrats, who claimed the inalienable right of suffragian equality bestowed upon the lower orders of the animal creation, in the exercise of their untrammeled state of field and forest freedom. The regular national church, and king, persecuted the nonconformists and schismatics with dire vengeance, under the patronage of godhead personification, translating the living heretics with tortures, burnings, and repetitions of drowning suffocations by resuscitations from a moribund state, and like admonitory chastenings in transition for the final judgment of their long enduring and merciful godhead. The persecuted schismatics emigrated to distant lands, in order that they might worship their God of reformation in freedom from invidious restriction of rites. When located, they in turn used the same strenuous arguments to subvert the tribal forms of worship. Gaining the ascendency, with destructive agents, they deprived the aboriginals of local option, forcing them to conform, with death and displacement, until they had obliged the remnant descendants of their benefactors to accept a conditional exile on the outskirts of progressive civilization, in transitu for a grave ultimatum. The notable invention of letters signalized the latter portion of the reign of Primus, and to it he laid claim as king rief discoverer; which in the law of entail declares the subject a utensil to be used for the exaltation of kingly prerogative; being identified with everything that pertains to the glory of the throne and its legitimate scionry, his assumptive appropriation was sustained with ministerial affidavits and legal opinions, in attestation of King Primus's great literary and inventive capacity, allied to clemency, justice, and generosity. But after his death, there was found concealed in the hut of a bard, who had disappeared just anterior to the announcement of the king's invention, parchments inscribed with the newly introduced characters, which set forth the bard's adverse claims in these terms: —

> With symbolic signs, I have found,
> The art of representing sound.
> On distant business one can send,
> Or with them greet a distant friend.

From this scrap of post circumstantial testimony, it is evident that he either intended to filch from the king, or that the king did obtain his reputation for literary invention from the fior's or bard's genius. The latter presumption receives probable confirmation from our auramention of similar pretentions to authorship advanced by giga potentates of the past and present age.

The rule of King Primus was of the most despotic description ever enforced by an arbitrary will over the weak subserviency of plodding human instinct, which in kindred affinity with the dogs, is content to give vent to a growling yelp when the freedom of its tail is ground by the heel of the oppressor. Whenever these constitutional growls foreboded an insurrectionary show of teeth, the gregarious spirit of commune revolt was allayed by the grant of a new charter of rights, but if this precedental sop failed to lay the retaliative spirit engendered by oppression, the current of their wrath was turned against their neighbors, with arbitrary conjurations as the provocations, of war. As an infallible test of his infallibilty death displaced him to make room for a successor. The people put on sackcloth, and rolled in the 'dust of humiliation, in mournful semblance of grief for the loss of their demi-god, whose dealings had been grievous and past finding out.

After public eulogistic exaltations, funereal orations and lamentations had subsided, his only son was proclaimed successor with jubilant rejoicings. But Justinatus, the son, resolutely announced his determination to reject the succession, recommending the people to select from the wise men of the nation a council to decide upon a form of goverment best suited in adaptation for the requirements of the people; but they with their faces and thoughts turned to the rear, in reverence for past usage, clamored for a king. But they found in Justinatus a man as determined for the enforcement of right, as his father had been for wrong. He commanded them to turn their faces to the future, and act according to his direction, not for themselves or their generation alone, but for those who were to succeed them. Submissive to the letter of his direction, but in conformity with precedental creed, they elected eight men by ballot, and instructed them to proclaim Justinatus king. With this evidence of their precedental stupidity he assumed the power of directing them for their own good, selecting four men of as well approved wisdom as his judgment could discover, he placed at their head his early instructor as chief advisor, with the titled designation of Dosch. After this inauguration of an advisorial system, Justinatus, as a pupil, received from them instruction; combining, with his advance in knowledge, his aid in promoting the practical development of means for insuring equality in thought and judgment, necessary for the promotion of the common welfare.

In consideration of the fluctuating variations incident to common usage, their first endeavors were directed for the devisement of a method that would insure exampled conformity in act. The difficulty of effecting uniform compatibility, in the then present habits of the people, soon became apparent. As a dernier of preparation, a division of labor was enforced, according with the personal healthy capacity of each individual. Under this system of equalized industry for community support, the drones were soon discovered, and subjected to the taskmaster supervision of those capable of exercising self control for the common good. Of course the outcry of slavery and oppression became rampant with the ill disposed and vicious; but compelled association with the good soon wrought a happy change; but not before many revolutionary schemes of revolt had been planned by the democratic majority, and nipped in the bud. The great bar to the full success of the renovating process, was the all absorbing lust for selfish gratification, procured from the sacrifice of others' welfare. Exhortations and demonstrations of the evil effects and instability of pleasures having a material dependency upon the appetites and passions of the body were of no avail. Stimulating provocations, for the production of inordinate appetites, had held an increasing sway from time immemorial, and the infatuation still continued to subvert the efforts of the Doschate of advisors for the establishment of a rational source of happiness, that should extend its blessings for the reciprocal appreciation of all. Laws and penal restrictions proved of easy evasion, and the local option of individuals native to Manatitla, having a desire to establish in perpetuity the happiness of their people, as a beacon light of example, were openly defied by aliens. To restrict emigration, which was claimed as a privileged right ordained as an inherent instinct of animality, they did not dare! as it was declared by the majority an assumption that would directly controvert the rights of septs and nationalities guaranteed by deity. The civilized progenitors of the races represented by tribes and small nationalities occupying the country adjacent to Manatitla, had undoubtedly been parasitical attaches to giga castaways like those of the Manatitlans. This stumbling block of perversion, continued from generation to generation for centuries; until the advent of the Dosch Desiderata, who with the aid of his advisors, turned the tide anarchy by the adoption of foreigners as guests, withholding the privilege of citizenship for bestowal upon their children's children of the third generation. This inaugurated an era memorable for the change of precedental precept, based upon warlike achievements, into a source of abhorrence with the increasing minority. Thoughtful consideration bestowed upon example for the transmitted improvement of future generatons in goodness, produced a wonderful effect upon the actors of the then present generation by the induction of harmony from reciprocal goodwill. Through his wise deductions, that clearly demonstrated the necessity of self government in association with others, woman threw off her shroud of vanity, and labored earnestly for the renewal of her lost prestige of trust, bestowed for the transmission of purity and goodness. The incipient struggles of the minority, under the direction of Desiderata, were short and decisive; but for the time being evoked with groveling bitterness fierce invective from the majority. A memorial address of remonstrance, from the democratic majority, against the abrogation of the rights of citizenship, in the first and second degree of alien residence, set forth, that God had created all men free and equal without respect to color or habits, with the command that they should work out their own way of salvation, and that each individual was guaranteed an inalienable right to participate in the government of his fellow man. "And that, whereas, as hereinafter stated," the citizens of Manatitla represent different nationalities, it was but just and right that they should have a voice in the council of advisors, in order that they might guard and protect their own liberties and safety. With this preamble, imitated from giga precept, the contest can be realized without repeating the stale platitudes of democratic subterfuge. The promulgated reply was as follows:—

"The Dosch of Manatitla and his advisors, to the alien guests (heretofore, in acceptation, adopted citizens) of their people and country, greeting! We have received your petition, and have reviewed with care the requests you have proffered. Our answer is set forth in the subjoined proclamation.

"'To the residents of Manatitla of foreign birth! As it is our matured desire to emancipate the people native to our country from their own degrading habits, and the deleterious example of those derived from extraneous source, we herewith announce the corrective enactments we have devised for the collective well being and happiness of all within our advisorial control. As it is manifest, from the conclusive evidence of creative design, that mankind are in bodily and functional alliance with all the different grades of animality, through the representative agency of omnivorous instinct; it is also as clearly evident that his endowed superiority resides in his privileged capacity for self control, with an ultimate intention equally apparent. For a rational realization of this saving clause, an easy estimate can be made of all the tangible sources of happiness held by human kind independent of the body's instincts. As upon these depend our hopes of happiness in life, in premonition of immortality, it is imperative with all to hold them in reciprocal cultivation for the confluent control — in subjection — of the passions inherent with the vital functions of animality. As woman, the endowed source and mother of our race, when free from the attaint of man's selfish invention, expresses a natural repugnance to everything opposed to purity and goodness, and in the full fruition of her endowment is reverenced as the germ ideal of immortality; we have through her a direct indication of the immortal source of happiness bestowed with creative intention for the local option of mankind. In negative assurance, that purity and goodness is the endowed source of happiness; woman when lost to their sustenance, becomes hopelessly degraded; sinking with loathsome taint below the vilest brute, and utterly lost to the instinctive ties of affection, will not hesitate to sacrifice mother, husband, sister, and child to the poisonous lust of her reptile selfishness.

"'That the cause of this ferocious degeneration, which has the power of transforming woman from the glorious ideal of immortality, into an object too repulsive for her destroyer to find in his vocabulary words of beastly vulgarity sufficiently strong in the odor of putrefactive designation for expressing in comparison the foulness of his scorn, is derived from man's insatiate devisement, cannot be denied! For the exampled amendment of this woful cause of degeneration, we have provided family censors, and nurses, in sufficient number for present requirement, whose duty it will be to hold in check parental indiscretions, and mutations incited from the instinctive variations of fondness and petulance. With the close of the second year, the provisorial charge of the family censor and nurse will be transferred with the infant to the national school of the department in which they reside, their guardian duties continuing until the seventh year, at the commencement of which the child is matriculated as a pupil, with full scholastic adoption by the censors and teachers. For the additional furtherance of our system, subserving for the vindication of creative indications for the elimination of our immortal endowment, we have separated the sexes that in the process of educational attainment they may remain free from the natural temptations inherent with instinct.

"'The benefits conferred by our national system of education you have realized in the peaceful confidence and unity imparted in after association; also the sequent inseparable unity of our marriage conjunctions. In truth, they are happily too apparent to be gainsaid. So that in accepting our hospitality as guests, you cannot avoid, in courtesy, a willing recognition of our rights of freehold preëmption, for preserving our habits and customs of purity and goodness, intact from the infringements of foreign attaint; or question the justice of our privilege of enforcing their observance; or in default, question our corrective enactments devised for the culprit's realization of practical liabilities incurred by the transgressor. These will be strictly enforced. But that there may be no cavilings, with the hue and cry of barbarous excess in punishment, we have provided accommodations adapted to the specie degradation of the lower orders of animal instinct, of sufficient capacity for associate occupation by human emulators of bestiality in kind; through all the gradations from the *omnium gatherum* 'swine,' blood thirsty 'tiger,' down to the reptile conservators of poison. For the correction of women who have lapsed from their vocation of conservators

THE MANATITLANS. 141

of purity and goodness, into the incipient stages of gadding and gossiping detraction, we have provided cage apartments for their mutual accommodation with birds representing their kind, in the hopping vent of thoughtless words. We have provided for initiatory correction pavonicas (animalculan peacocks) for the exemplar admonition of the vain-glorious; and jabboracidas (jackdaws and magpies) for the likeness of gossipping repeaters; for the loud mouthed and strepitant clackitas (parrots and cockatoos), and for the 'fashionable' imitators, simia curios (female monkeys). These, as occasion may require, will extend invitations to their 'likes' of the human sex to attend their levées, which will be subject to the auditorial outside inspection of the public, if morbid curiosity should prompt witnesses to the ordeal of misery. These provisos and corrective conceptions have been devised for, and proved to be of universal benefit, with the evidence of well attested experience; and we desire your coöperation, as guests, for the perfection of our system designed for the advancement of purity and goodness. But shall strenuously insist that your children shall become participants in the privileges conferred by our system of education.'"

As you will readily conceive, this proclamation of Desiderata and his associates caused the fulmination of bitter invectives and threats of vengeance, which served to vindicate the wisdom of the predicated precautions. But the writers of the period state that in a few generations the influence for good extended to savage tribes, who petitioned for admission of their children into our national schools. The improvement was so marked in its demonstration of a realizing source of happiness, that but three centuries elapsed from the period of organization, before the foreign nationalities were peacefully absorbed, their subjects becoming educated citizens of Manatitla.

With the illustrative sketch that I have given of

Manatitla's transition period of extension, you, and the readers of the historiographer's transcript, will readily understand the inceptive source and stages that premised the establishment of our practical system of education. But owing to the limited number of words and terms for the expression of purity and goodness, with their practical variations, in your languages, we are of necessity obliged to use them in frequent repetition.

CHAPTER XIV.

AT this stage, M. Hollydorf interrupted the Dosch, with the assurance that he was fully convinced not only of the actual existence of animalculan humanity, but of the tangible wisdom of Manatitlan providence, shown in their inauguration of rational system for educational discernment, necessary for the fulfillment in life of happy intention. "But the difficulty of making the home society realize by letter the multiplying wonders in the course of our discoveries, puzzles my invention for a credible method of imparting the information without subjecting my sanity and integrity to impeachment. If you can, in your wisdom, resolve me how I may absolve myself with credit in my official correspondence, I shall certainly feel grateful."

The Dosch smilingly assured him that he had no occasion for fear, as the sensational novelty of truthful record, with a little auramental aid rendered by the Manatitlans in the substitution of thought, would suffice for the ready adoption and belief of his report, as a marvelous indication of the age, in evidence of rapid progression under German lead. With this closing advisorial suggestion the Dosch and his companions departed for Maniculæ.

The abstracted mood, fitful and irrelevant conversation, with the daily convocations of the four conservators of the Manatitlan secret, in the house under the northern temple's eastern wall, did not fail to attract the wondering curiosity of their associates. But as M. Hollydorf had emancipated the members of the corps from field duty, they found no lack of pleasing occupation in rendering useful aid to the Heracleans. Doctor Baāhar had enlisted the padre, for a quid pro quo, in the pursuit of butterflies; the two curators of sound engaged in herding and woodland pursuits; Jack and Bill, under Heraclean and Kyronese instruction, engaged in "navigating" a small garden plot in the latifundium, with amusing success, while Viscouswitzs, the artist, wooed the Indian maids of the Vermejo tribe.

The Dosch, in continuation of his historical sketch of the Manatitlans, passed to the period noted as the Heraclean epoch.

The third century of your Christian era was well advanced before they were aware that there was a race of white gigas occupying a city not far remote from Maniculæ. At that period distance was measured by the time occupied in conveyance by the insects then in use for transportation; but as the vitality of their bodies was subject to deciduous tenure, travelers were obliged to confine their researches within the limits of populated districts, between which adventitious paths were well defined. The defective means of communication with remote Manatitlan provinces had ever been a source of sincere regret. Still the lack of advancement in the art of locomotion had never interfered with the actual realization of happiness. The wood roach and beetle were used as insects of draught in the preparation of the soil for cultivation, and the flea for equisaltation, it being the favorite mount for distant journeys and pleasure excursions. The first innovation upon these time honored extra locomotive adjuncts, was effected by the persevering ingenuity and daring courage of a medium named Bussee. He had from an early age devoted his thoughts to natural history with the practical intention of improving the native stock which was too diminutive to be made available for transportation. As a boy he had been noted for a quick practical judgment, displayed in his ability to eke out from scant means the fulfillment of a desired end. Many of his improved domestic utensils are still in use, in evidence of an inventive genius in advance of his age. His habits were erratic, showing an impatience that disdained restraint within the bounds of precedental usage. Still his affectionate desire to confer public benefits attracted a grateful solicitude whenever his absence was unusually prolonged. But as he rarely returned without some valuable acquisition, confidence in his ability for self-protection waived anxiety. At length an absence of two months without communicating with his family, aroused public sympathy to such a pitch that a search was decided upon. In preparation the citizens of Maniculæ had collected in the anthemique to consult upon the most feasible means of conducting the search.

When the direction was decided upon, and they were issuing forth for its prosecution, they were startled by the gyrations of an apis isolata (solitary bee) in close proximity to their heads. After a few eccentric evolutions which excited a commensurate degree of alarm, their fears were relieved by a shout of laughter in the jovial tones of the absentee, who, by a skillful direction, caused the bee to alight in their midst. When sufficiently assured of freedom from danger, his parents and the Dosch approached near enough to obtain a view of his mechanical appliances for guiding his prize. Between the wings of the bee, upon his back, a net with latticed films, supported in dome shape by stiff fibres, was attached. This turret was retained in place by filaments, which passed beneath his body, in the articulation between the body and thorax, so that his movements and winged action were not impeded. To the antennæ, on either side,

7

were attached filamental guides, or reins, for directing his course, the proximal extremities being coupled within the pilot cone. When assured of the strength and security of the attachments, the Dosch and parents of Buzzee ventured on a short experimental flight. As the insect circled, in company with his mate, to gain a bee line, the daring volantaph caused him to execute a variety of intricate evolutions, which at first alarmed his passengers, who expostulated with him in reproof for his temerity. But when he explained his wish to show them how completely the movements of the bee were under his control, they no longer offered objections, their fears being turned to admiration. When satisfied that air flights could be conducted with more ease, safety, and swiftness, beyond the most sanguine expectations of ancient or modern Manatitlan prognosticators, his enterprise was highly commended. When landed the Dosch and advisors expressed a desire, in behalf of the people, to listen to a relation of his adventures in the anthemique, as it would be the means of avoiding rehearsals from hearsay, with the defects that of necessity were attendant upon individual versions. An hour before evening song the anthemique was thronged with the citizens of Maniculæ anxious to hear Buzzee's relation of the expedients used for a capture so important in its prospective bearings to the people of Manatitla. To enhance the clearness of his demonstration, and at the same time show the dazed docility of the bee, Buzzee directed his flight to the cantilor's rostrum, and after he had settled addressed the assemblage from the pilot cone, in substance, as follows: —

"Although no stranger to your manifestations of affection, I am well aware that in appearance I have been remiss in rendering you suitable returns; but am certain that your confidence in the integrity of my intentions will exculpate me from meditated indifference. I am now happy in being able to bring you tangible proof that my wanderings were not prompted from motives of selfishness or disdain. From my childhood I have listened in silence to the oft repeated regrets that our extraneous means of locomotion were limited to insects so lacking in the instincts of intelligence necessary for successful direction. Those available for locomotion were too ephemeral in their term of existence to be trusted for conveyance far beyond the habitable limits of our country, which from the illimitable firmament seemed to be but a mere speck upon the earth's surface. With a curious desire to learn the wonders of creation overshadowed by the starry canopy, my earliest thoughts were directed to the acquirement of the means necessary for safe transportation above the earth's surface. My thoughts were at first naturally directed to artificial wings as the indicated means of progressive transposition from earth to atmospheric space, without giving thought to the consideration of ponderable adaptability. Human mortality, which requires omnivorous support, declares itself ponderable in the vis inertia of earth, in contrast with the airy attenuations that bespeak adaptative intention in creating the tenants of space. In addition, with the successful achievement of working wings, there would be inevitable friction with the uncertainties of wear and derangement in flight, with awkward position of ponderable suspension in space for repairs. So my inventive genius was fain to hold itself convinced of the futility of subverting the order of elementary adaptation, designed by the Creator for the perfection of His intentions. Self convicted with the foolish audacity of my labors to safely suspend with motion, and locomote with facility ponderable humanity in space, I bethought myself of man's privilege of making subordinate organic vitality, with legitimate kindly motive, subservient to his desired facilitations. For the elucidation of my thought suggestions I directed my investigations to insects of flight, to select from their varied species one suited to our requirements. The primary qualifications necessary were sagacity, supporting wing expanse, strength, longevity, and equal motion in flight, with instinctive perception of individuality sufficient for submission to our kindly direction. The efficient qualities indicated for the selection of a winged conveyance, were first, size, with an adaptation for control, in combination with a supporting buoyancy in excess of its individual requirements. In the second degree intelligence, with a longevity sufficient for compensative training, and memory capable of retaining the imposed impressions, subject to the recognition of personal direction foreign to their own volition. Added to these essential qualifications, it was desirable that the insect should be naturally inclined to sustain a long and swift flight. Bees had early attracted my attention, but there were many objections to their adoption that seemed insurmountable. Multitudinous in association, and individually aggressive, were primary defects in disposition; while in industrious habits and vocation they were subject to routine enactments, which together with the tenacious nature and method of collecting and disposing of their food threatened to end my ambitious projects, in trial with them, in death from suffocation, or waxed adhesion to their bodies or cells. The fear of being stalled and borne to their cells for living incorporation, raised an insuperable dread, that prevented me from coveting an experimental acquaintance with the working orders of their kind. Often in my wanderings I have passed beyond the boundaries of Manatitla in search of a locomotive desideratum, which I had supposed necessary for the welfare of our race, as well as a gratuitous vehicle for the gratification of my covetous desire to rise into the realms of space, to survey beneath our terrestrial place of abode. A month since I was returning homeward sad and dispirited with continued disappointments, when at the close of day, while the glowing tints of the setting sun still lingered in the glory of their parting adornment to foliage and flowers, I was attracted by the swift whirr of strange insect wings. In a moment my attention was drawn with intent desire toward a pair of insects bearing a hybrid resemblance to the bee family. After a careful reconnoitering inspection, seemingly directed, first, to the quality of the flowers of a tropical honeysuckle, and secondly, to see if they contained insect occupants, they alighted upon the petals of the fairest. Unlike the hoarding selfish instincts of their congeners of the bee kind, they premised their labors with playful dalliance, partly upon wing and with sprightly pedal evolutions, while darting in chase and counter chase in and out from the petaled cups of the flowers. In a few minutes their playful antics and fondlings ceased, then the male with an autocratic appearance of gallantry assisted his spouse to load herself with the sweets and waxy exudations of the flowers, this accomplished he sent her unescorted away, evidently to unload in their store house. During her absence he devoted his time to a general inspection of the flowers, with the evident intention of selecting the best. In one he found a belated droniva (a tropical representative of the bumble-bee family) who was ejected without ceremony, although double the size of the audacious usurper. His activity, independence, and cleanly regard for his own person, disposed me to excuse his cavalier exaction of service drudgery from his mate, as the duty seemed to afford her pleasure. In fact the pair impressed me so favorably, that I determined to avail myself of the opportunity to secure a permanent attachment.

" Years anterior, as you are aware, I perfected a harness in anticipation of the fulfillment of my hopes of being able to make a capture suited for our locomotive requirements. This I had carried with me in all my excursions, and while my coveted prize was engaged in his erratic flights, I placed myself in ambush in the fairest flower of his selection, and had the gratification of securing him in leash before the return of his mate. He soon became aware of unusual restraint, and curious to learn its cause made experimental flights which gave me an opportunity to test the success of my invention, and I was delighted to find that I could direct his course with ease. Seemingly puzzled at the loss of his voluntary power of direction, he made every available effort to learn the cause of his sudden bereavement, and was pursuing his investigations when his spouse returned. With man dibulations he quickly communicated to her the restraint that had been placed upon his movements during her absence. With evident anxiety she commenced a search for the impedimental cause. In a few moments she discovered the filamental guides that I had attached to his antennæ beneath the carapace, but failing in her attempts to remove them, after a short consultation, they rose in flight from the flower to the bee line of their home with a marked show of anxiety, which made me feel a glow of regret that my selfishness had been the cause of their disquietude. Once only, in homeward flight, did I attempt to subject him to a variation in course, but it caused such a trepidation in his mate that it was with difficulty that she recovered the balance movement of her wings. Reaching their cell, which was in a fissured ledge of basaltic formation, they held another consultation and investigation, during which my turret cone was subjected to a close examination, but the tough silicothed filaments were too strong for removal by her feeble efforts. Finding his strange investment inevitable, and attended with but slight inconvenience, he, at last, with cheerful philosophy, soothed the anxiety of his spouse with endearments, abated of their autocratic patronizing air of superiority. This show of appreciation for his mate's solicitude, at once bespoke a high degree of sympathetic intelligence attained by a union of instinctive equality. In contrasted proof of the evident assumption, I will adduce the ants, and our neighbors of the human species, who live in a state of concubinage, to show that sexual gregation begets a condition of brutal selfishness in the males, causing them to use physical strength for the reduction of their females to serve as bond slaves of passion and labor for multitudinous production in kind. From their continued dalliance after nightfall, I was pleased to learn that their habits were semi-nocturnal in perceptive activity. When they finally retired for the night to the shelter within their cell, I suffered retributive spasms from the powerful mellific odors that pervaded the cell, which caused protracted coughing and general relaxation, so that in my extremity I was prompted to make my escape into the open air, but the intense darkness and my weakness prevented me. As my air passages became accustomed to the acrid irritation, I in sequence suffered from mellific narcotism, and fell into a stuporic medium between waking impressions and fantastic visions of instinct that precede the waking dawn from sleep. These variations continued until the bees' emergence into the open air, in the morning, revived me. After their matutinal salutations they rose in flight circles to their bee line, but winged their course in an opposite direction from the honeysuckle plot so memorable in their previous day's experience, probably attributing the cause of restraint to some inherent property of the flowers.

"My elevation and swift passage through the air, reminded me, with its bracing effect, that I had not taken food for a donsenack, so feeling at ease I unstrapped my script and made a hearty meal, with a zest that the words of our language will fail to express. Shortly after I had closed my morning meal, the bees commenced their circlings in downward descent, and ere long I discovered below, on the rocky declivity of a hillside a growth of honeysuckles, the goal of their attraction. In the circling support of their buzzing wings they remained suspended over the flowers for some time, until their safety had been tested by dronivas and humming birds, then with caution they ventured to settle upon the petals, and after some hesitation, the female was loaded and dispatched with her first cargo to the cell. The male, as on the previous day, employed his time during her absence in an investigation of the floral resources of the hillside, with an occasional essay of his belligerent propensities directed against humming-birds and other collectors of sweets. This disposition, which seemed to have received an aggravated accession, in the vigorous temerity of daring assaults, from the restraints I had imposed, was treated with a gentle admonition to test their directing efficacy.

" The first essay provoked a display of resistance, but without avail in thwarting the changes I meditated, except for the production of a marked degree of discomfort, as the tension of the filamental bonds from opposing obstinacy caused a spasmodic action of the wings from axillary compression. Disconcerted, after frequent trials of his voluntary powers in opposition to my guiding mechanical appliances, he settled upon a petal for reflection. Then, seemingly, after mature consideration, an instinctive impulse would cause him to dart away in flight as if to test anew his strength in controlling volition, but only to be turned back before reaching the object of his destination. When successful, after frequent failures, he seemed to be quite as much disturbed as with the contrary results of his trials. I soon found in these practical essays, that my studied calculations for his direction fell far short of the absolute requirements of necessity and safety. In his short flights I discovered a power of resistance that baffled my attempts to direct his rise and descent, which was evidently independent of head and wings. Looking backward, when making a short tack, the resisting part was made manifest by the movements of the cartilaginous rings of the body. In studying the changing results in controlling direction, I found that the body acted as a rudder in flight for upward and downward inclination, and until I could obtain its concerted action with head and wings, instinctive volition would oppose my usurpation of its natural rights. With the view of effecting temporary control I rove a ring with a line attached to the four terminal quarters of its circumference, to act, when adjusted, as a tip to the body. This I confined in place without much difficulty, and passing the lines through corresponding guides to the carapace reflected them through pulleys back to the cone. These additions to my managing devices, met with no decided opposition, but the victim kept my movements under the watchful supervision of his eyes, but more in curiosity than in fear or anger. On the return of his mate, an antennæ inspection was improvised for tracing the new additions, but as their labors were quickly resumed, I interpreted their quiet resignation as an act of submission. After the departure of his spouse on her second homeward trip, he engaged in a flight trial to learn the extent of the new vetoes that I had placed upon his volition correspondence with members of his body corporate. His diminished lack of self-control begot a vengeful desire to retrieve compensation by inflicting retributive discomfort and stings upon the innocent. After his test trials had convinced him, that in movement he was no longer capable of commanding himself, but subject to a mysterious power, he fought two rounds with dronivas, the odds being four to one in favor of his opponents, each exceeding his weight by two thirds; after sustaining his preëmption right to the sole occupation of the flowers with them, he matched his vengeful speed and tactics against a score of humming-birds, proving himself equal to his undertaking. Besides these emprises of valor and speed, I subjected him to a test of my guiding improvements to which he not only submitted in freedom from irritability, but seemed to recognize the new sensations and eccentric effects as a pleasing supplement to his involuntary powers, superseding in part the necessity of volition. Desiring that he might become accustomed to my guiding presence, and familiar with my person, I exposed myself as often as possible to his own and consort's eyes, and on their return to the cell at nightfall, I felt certain that they had accepted me as an attached presage for good. As in oft repeated subjection to deleterious influences, the narcotic effluvia of the cell was far less offensive than on the previous night.

"With the dawn of the third morning the bees rose to their line and settled in descent upon the flowers in bloom on the vines subject to the previous day's levy. As if in anticipation of my intention, the usual four cargoes were dispatched in less than an hour, then both circled upward to the line for homeward flight, when, to their astonishment, I turned the lead of the male to the westward. This deviation from ancestral custom, and sequent habits, aroused the most obstinate resistance, which after several pseudo starts succeeded in baffling my intention, and but for the fortunate discovery of the cause, which was the fouling of the sinister guiding line, I should have been obliged to succumb to the instinctive obstinacy transmitted for the preservation of formalistic routine. With hazardous determination I succeeded in righting it, notwithstanding the increased velocity of their homeward flight, accelerated by the instinctive impetus from the imparted zest of their return to the line of old habits. When again subject to my control, the course of the male was changed to a northwesterly direction, but the female coaxingly endeavored to turn him back with the voice of her wings, as she kept abreast in equal flight. Finding it impossible, she reluctantly resumed her station in the rear, yielding protestingly to his lead. At first a natural feeling deterred me from casting a look below through fear of being surprised with giddiness, but gradually this passed away under the exhilarating elasticity of the air, which appeared to raise my spirits to an equality with my ambitious aspirations. Presumptious mortality even ventured to cast a scornful glance upon things mundane; when lo! in advance, rising to the bee line I discovered a meroptic bee-eater which dissolved in fear my exultations.

"Luckily my naturalistic studies enabled me to disappoint him of his premeditated tid-bit gratification, as I should have been included in his bill of fare, with a vale for the improved means of locomotion I had obtained for my people. Grateful for the presence of mind which in great emergencies baffles instinct, I abruptly changed the course of the bee northward. But the pursuing merop was not to be disappointed without an extra effort to secure his prize, for he immediately tried his chances in chase; but as he was soon distanced he gave up pursuit, still soaring above the trees to intercept those which he expected in train, but for once, at least, he was foiled in following the transmitted impressions of ancestral instinct. The curved flight of the merop, even with the advantage of superior swiftness, would have been quickly distanced by the undeviating line of the bees' air trail when once in advance of their pursuer, unless retarded by the greed of an overload; of this fact the instinct of the bird is apprised, but hunger sets at variance all rules, and if he fails in intercepting, he often pursues. The apiaster, after his first capture, if his prize proves to be a honey bee, builds his nest beneath the line, for the purpose of surprising the homeward flight of the workers when loaded with his coveted sweets. My escape from sudden death, although easily avoided from seasonable discovery of the danger, served as a timely warning, which kept me in careful watch for unknown perils.

"Only a short space of time had elapsed from the start, when in advance I discovered a beautiful and highly cultivated valley. The giga laborers were Indians who were under white taskmasters. Passing over the valley, which extended to the northwest as far as the eye could reach, I changed our course to the northeast; rising to a line above the mountains two cities opened to my view, both inclosed with walls for protection. The largest city was built in the basin of an amphitheatre of surrounding hills, with an opening, and corresponding gate of the city, looking out upon the valley we had overflown. The second city was beautifully located upon the summit of a hill, overshadowed by the spray of a large waterfall that flowed over the brink of a precipice, which extended its barrier for miles north and south, its perpendicular descent being only broken by a zigzag roadway cut in its face for communication between the two cities. The wall of circumvallation around the City of the Falls was not fully completed, for thousands of workmen were still engaged upon the portion inclosing the large plain that sloped from the summit in broad expanse to the limits of the walled stream that flowed from the basin of the falls, without the foundation of the walls, to unite again without the cinctus gates. From the circling lash in the hands of the taskmasters, it was easy to comprehend that the laborers were bondsmen, their color indicating aboriginal birth.

"In homeward flight the bees were allowed to take their own course, which, from the accelerated rapidity in the motion of their wings, declared a nostalgic haste to enjoy the hospitalities of their sweet home. After the morning's labor of the bees was completed on the succeeding day, I directed their flight over Maniculæ to discover whether you were over anxious on my account, but as my family appeared to be free from disquiet I again turned my bees westward for new explorations. Alighting at midday on an island in a lake, south of the valley discoveries of the previous day, I found it un-settled with an animalculan race of tits, whose sole occupation seemed to be devoted to sociable potations of a fluid that excited amicable quarrels, in which the families engaged with wild enthusiasm, without respect to age or infirmities. The domestic amusements were varied with cockroach racings, worshiping, drinking, dancing, fighting, and hunting pediculas, in the latter sport the women and children engaged with peculiar zest. In verification of our sages' demonstrations of instinctive cause and effect, when subject to gregarious association in folds, in freedom from the directing intelligence of Creative endowment, their bodies gave sure indication of reactive bestiality. Disgusted with the extremes they exhibited of wailings and vociferous jabberings, as the product of instinct bewrayed with madness, I was glad with grateful relief when my cleanly transport bore me again into the pure atmosphere beyond the sound reach of the reviling pretexts of these ape libels upon Creative intention. Assuming the privilege of a sub-lunary discoverer, I named the island Greenpat, from the emerald beauty of its tints, and the inhabitants Mouthpats, from their unthinking volubility, bespeaking the unkempt scragginess of their natures.

"Having tested the ready sagacity of my transport acquisition, and applicability for quick conveyance, I now propose to make use of it to obtain others of its kind, with a view to propagation, as I feel certain that it can be domesticated for mutual advantage, as both male and female evince an increasing confidence in the controlling influence of my presence; and of the enduring longevity of the species I feel equally certain."

The assemblage, at the conclusion of the narration, enthusiastically congratulated Buzzee upon the result of his successful perseverance, saluting him as a public benefactor, with the title of apiamaster. In the course of a century there was not a family in Manatitla without a pair or more of the apis isolatas, which became known in common usage as the bee phaeton. Their introduction as locomotive facilitations contributed greatly to extended sociability, as they were able to bear with ease twenty giantescoes, forty mediums, or their equivalents in tits, and we have evident reason to believe that they instinctively enjoy their domestication with us better than in a wild state, for in our pleasure parties they harmonize the voice vibrations of their wings with our songs. To Buzzee's inventive skill we are also indebted for the imperishable combination used in building, and the preparation of our texile fabrics.

After the discovery of the Heraclean cities, with the increase of our people's means of communication, they were visited daily for the purpose of influencing the citizens to bestow more kindly treatment upon their aboriginal benefactors. Evoce (quick perception), a giantesco, had gained the ear of a cruel taskmaster, for the purpose of using his voice in expostulation, when to his surprise, he distinctly heard vengeful denunciations without the utterance of words of speech from the mouth of the brutal auramentee. Satisfied after frequent experimental repetitions that the enunciations were vocalized impressions heralding audited words of speech that could be suppressed or spoken, he made known the nature of his discovery to the Dosch and his advisers. The coincident impression of their own thought enunciations, confirmed Evoce's suggestion that thought enunciation, and also

THE MANATITLANS. 159

instinctive mental impressions, were vocalized by an enunciator in proximity with the ear, and in communication with the combined organs of sense. Upon these suggestive conclusions was founded an experimental course of investigation, which resulted, not only in the full verification of their deductive anticipations, but with the development of the power of substituting extraneous impressions for adoption by the giga auramentee, through the modulated induction of the giantesco voice to an accord with the mood of the subject. Great care was required in the ventriloquial modulations of the auramentor's voice for exact correspondence with the characteristic peculiarities of the auramentee's; for any remarkable deviation was sure to alarm their superstitious fears. For the acquirement of facility in the substitution of ideas and thoughts, it was necessary to obtain humoristic ease in the detail expression of idiomatic phrase peculiar to the auramentee's use of language. With the naturally good, we were soon able, with the mutual incitement of novelty, to evoke and cultivate the germ of pity, while with the instinctively bad our efforts served to arouse superstitious fears for the negative advancement of our object, through retributive apprehensions of vengeance in return for their cruelties. These, with strange inconsistency, caused sacrificial oblations, with deputized prayers, to be offered in commutation for the continued gratification of their evil habits and passions. Yet, with all the perversity of ruling instinct we have been able to accomplish much good through the means of thought substitution with your race.

CHAPTER XV.

HAVING given you, by quotation from our chroniclers, a synoptical view of two important discoveries which facilitated our communication with your race, I will now, continued the Dosch, refer you to your own impressions, and the eccentricities of the uninitiated from thought substitution, for the clear demonstration of our auramental powers. Or if, in review, you can recall examples of instinctive spiritual manifestations, you will be able to judge of our method in dealing with the instinctively stupid, partly with hopes of reflecting the extremes of absurdity, and in sub-part for our humorous gratification in tracing the commotional hubbub of selfish instinct in its search for the means of saving grace to rescue folly from its own attaint. You will soon be able to judge of the limits that we are confined to in auramentation. With the instinctively evil, our efforts excite fear and ritualistic prayers for propitiation, and exorcism of supposed inimical agencies foreign to self. But with the good we are able to impart happy encouragement. Selfish excess, in all of its forms, bespeaks a material agency and end, and as this is the god of realization with the gregarious democracy of the gigas, the influence of our auramental efforts — if their source was known — would be denounced with bell and book, as heretically pedantic and puritanical. But goodness imparts an animus joy that affords in life tangible impressions of immortality.

We now will pass to our fourth important epoch,

noted for the personal introduction of the Dosch Giganteo to Indegatus, Prætor of the present City of the Falls. In the process of rehearsal we shall allude to the third or falcon era.

Indegatus, Prætor of New Heraclea or more properly Heraclea of the Falls, was a man of indefatigable energy, and at the period of Giganteo's introduction had just rescued the city from great peril. The peril from the besiegers was in fact less dangerous than the factious dissensions of the populace within the city walls. Aware that idleness was the mother of envy and turmoil, he had caused the latifundium to be divided into garden plots apportioned to the size of each family, for the cultivation of edible roots and cereals. While engaged with his two sons, Unipho and Gnipho, in the cultivation of their land, a bee alighted on the father's shoulder, attracting his attention from the singularity of its appearance and fearless confidence. Apparently satisfied with the attention it had received it flew to a neighboring flower occupied by a companion. Shortly after he felt a sharp sting on the back of his hand; a quick glance discovered a speck variegated with dark and shining particles, which he was about to brush away, supposing it to be an insect; when something peculiar in its movements attracted a more minute inspection, this resulted in the recognition of a little body possessing the dressed outline of the human form. Startled with superstitious fear from an apparition so manifestly supernal, he called his sons that their stronger eyes might confirm or dispel the impression of his more attenuated sight. After an inspection of a few seconds they burst into a merry peal of laughter, exclaiming in a breath, "It's a little man in Heraclean armor and sagum, flourishing his sword and spear as if he wished us to understand his signs!"

"My sons," urged the father with anxious fears, "give more reverend heed! He appears in a guise that betokens admonition from the regions of the nether world. Give earnest attention to his direction that the import of his visit may be revealed."

Gnipho. " The little stranger points to one of my ears as if he wished to be admitted to a hearing ? "

Indegatus. " My son, must I admonish you a second time to be more reverend in speech when addressing a being bearing tidings, you know not from whom, or from whence ? "

Gnipho. " You have advised us, father, to follow the example of our superiors, and this stranger phantom appears to be in no serious mood, for he laughs at your fears. But I will admit him to an audience, that he may declare the object of his visit ! "

Indegatus. " Presume not to take advantage of his levity, for as you are well instructed, and know that when I advised, it was for your dealings with mortality ? "

Gnipho. " Now he laughs outright, and my ear resounds with his mirth, as if filled with the infantile chirping of a joyous cricket. But now he speaks ! "

Indegatus. " Listen ? "

Gnipho. " Yes father. He asks if I can hear him distinctly."

Indegatus. " Then in virtue of my office as prætor and augur, I will address him. Speak Nuntius : What tidings bear you from the spirit world ? and from whose realm do you come in this disguise ? "

Gnipho. " Again I hear his small voice in the chuckling check of merriment, as if he would fain speak in reply."

Indegatus. " Then listen, my child, to the message he bears ? It surely cannot presage ill if he is in merry mood ! "

Gnipho. (Listening.) " He says he is not a spirit, but of mortal birth, like ourselves. But I will repeat his own words. ' Say to your father, that I have been long acquainted with his goodness, and desire to relieve his anxiety for the self-imposed misery of his people. Also to render him other efficient aid in a small way!'"

Indegatus. " Ask him, with grateful thanks, his name, and from whence he came?"

Gnipho. (Laughing.) " He says his name is Giganteo, the Dosch or patriarch of the Manatitlans, a race of animalculans whose country lies six hundred stadia to the southeast of the deserted city of Heraclea."

Indegatus. " Ask him how he proposes to help us?"

Gnipho. " He says by adding to your knowledge, in a privileged way that enables the small to help the great! He expresses a wish for us to retire with him to the parapet steps of the northern wall, where we shall be comparatively free from the shrill vibrations of the cicada's winged notes."

Indegatus. " We will move as he directs."

Gnipho. " Now that we have complied with his request, he charges me to listen, and treasure all that I hear, that I may repeat it to you."

Indegatus. " We will keep silence that your attention may not be distracted."

After an hour's close attention, Gnipho rehearsed to his father and brother, The Admonitory Request of Giganteo, Dosch of Manatitla. " Your ancestors of old Heraclea trained falcons for hunting, and through their borrowed use the Manatitlans obtained a knowledge of giga and animalculan nations beyond the ocean. We wished to recompense the service by imparting the source of our happiness to the people of Heraclea in return. But tyrannous ingratitude had so blunted affectionate sympathy, that your immediate ancestors alone listened to our warnings. But even they would have shared the common fate, if we had not found among the slaves those capable of judging between the good and evil. The majority of the enslaved were as relentless, as the doomed were blind to their impending fate. They had determined that none of their hated oppressors of either city should be spared; but the Manatitlans through the same means that I propose to offer for your aid, foiled the deadly intention of the slaves. While the old Heracleans were reveling in the height of their prosperity, falconry, as with all the pontine races coeval with their transatlantic progenitors, was their favorite pastime. Aeriolus, a worthy successor of Buzzee, visited the mews of old Heraclea, and with equally well devised skill in preparation, conceived the idea of utilizing the swift flight of the falcons and powers of abstinent endurance for crossing the ocean, the shores of which he had visited with the limited powers of the bee volant. Adventuring, with associate volantaphs, trials for their control in hunting, he soon perfected guiding attachments as efficient for directing their movements in flight as those devised for the bee. Selecting the swiftest and strongest he gradually accustomed them to long sustained flights over the ocean, insuring their welcome back to the mews by increased docility — under direction — to the will and lures of the falconer, when in the field. The anticipated difficulties from opposing wind currents, and means of obtaining food sustenance, and disposing of it while in flight, had been successfully overcome by prolonged observations verified with tests. Food was obtained by directing the falcons' attention to flying fish as objects of prey, which, with parachute aid, they were able, after a little practice under the stimulus of hunger, to devour in mid air.

"In memorial of his success Aeriolus gives in testimony the transcribed after observation. 'The transition from meat to fish, for a " fasting " flight of instinct, was adopted with far greater avidity than in human acts of ritualistic conformity to mythical injunctions, which we have seen practiced by the sectarian devotees to creeds, as negative compensations for over indulgence of the carnal affections.'

"When the arrangements of Aeriolus were fully perfected, he and thirty associates, with their wives, bade a hopeful farewell to the people of Manatitla, and started from the lochia (plaza) of Maniculæ upon their adventurous air voyage of discovery, with a leading falcon and three followers. Studying to aid the falcons by every possible means they, to their joyful surprise, discovered land on the morning of the fourth day from the start, and, at an early hour thereafter, alighted upon the lofty peak of an island mountain, since known as the Corcovado, a mountain summit of Corvo, one of the Azorean Isles. After regaling the falcons in relief from their lenten diet, they, of their own accord, continued their flight to the mainland. Our joy was much depressed, while passing over the beautiful land scenes, by the fierce cries of giga hosts engaged in battle encounter. In our course eastward, to a country of colored races resembling the aboriginals of our own, not a day passed without our forced observation of a battle scene, with fields and smoking ruins that bespoke the devastation of warful rage.

"Sick and despairing from the constant recurrence of murderous acts of despoliation, we at last reached, in returning, a cluster of islands in the western ocean to the northward of our point of arrival. On the largest island we found a hardy species of falcon, and, with the lure of our own, obtained four. After a few days' training of our transport addition we returned to the island where we first landed. Anxious to return to our people, and the cheering welcome of loving affection, we only tarried upon the island a sufficient time to accustom our newly acquired birds to devour their food while sustained by the parachute and their wings in mid-rest. Starting, homeward bound, on the morning of the sixtieth sun from the date of our departure from Maniculæ, we reached it again on the third day with the first notes of the evening anthem of thanksgiving, in which we gratefully joined in our descent to perch.

"Some days were occupied in the public rehearsal of the events and discoveries transpiring in the progress of our voyage; the resulting issue proving a source of congratulation, nathless, our disappointment from the unfavorable prospect afforded for an affectionate reception by the animalculan residents of the many countries over which we passed, from the effect of giga example. After many repetitions of the voyage, it was decided that colonization in the chief cities of Europe and Asia offered the only means for the effectual regeneration of the animalculan races for a happy appreciation of our exampled resources of loving affection. When the proposition for colonistic volunteers was proclaimed it received such a general sympathetic prompting of affectionate obligation, that every Manatitlan held himself and family in readiness for the service. As it was necessary, for self-defense, to have a majority of giantescoes and mediums, to overawe treacherous designs, the required number for colonizing Rome, Constantinople, and Jerusalem, were obtained by lot. When, on the eve of departure, the Dosch advised them to live apart from the natives of the cities, and in self dependence upon their own exertions for support; but to receive all healthy children of the required age, placed at their disposal for the Manatitlan term of education. The emigrants numbered among their volunteers representatives of all the mechanical branches of artisan labor, and especially those well instructed in the departments of indestructible house building and defensible vestments. So that little fear was entertained for their safety, as they could with ease repel the largest armies of tits that could be mustered. But their chief reliance was upon a sturdy adherence to their native habits and customs, yet ever open for the reciprocation of affectionate goodness. They were also admonished to make all possible application for the speedy acquisition of lingual idioms spoken in the different countries of their sojourn for future availment. Also, whenever favorable opportunity offered for the cultivation of giga goodness, to use their privilege of auramentation and thought substitution for encouragement and fruition. These general directions were improvised more for encouragement than from actual knowledge of the process best adapted for the controversion of habits and customs opposed to affectionate association and self-government.

" The first appearance of the Manatitlans in the cities of their destination attracted universal awe and curiosity on the part of the resident animalculan tits; for but few of the natives reached in stature the medium size. Their sudden and mysterious advent, gigantic size, quiet demeanor, and the great affection that they manifested towards each other, and in all the rela-tions of life, proved a source of emulous wonder and admiration with the good, and a rankling, envious source of disdain to the evil minded.

" Selecting their sites for residence from the advantages of inaccessibility to giga approach, arable soil, and capabilities for irrigation, their habitations were quickly constructed, with a cleanly elegance of adornment that added a new element of wonder to the lazy imaginations of instinct, in the superstitious belief that they were visitors under the patronage of divine agency. With these introductory advantages, which the colonists disclaimed to be other than those within the reach of all grades of mortality, which their appearance and vocation were intended to impart, the schools were organized, and flooded to overflowing with applications for admission. The monthly visitations of the children's parents confirmed the belief that mission was under the special direction of the Godhead.

" The unselfish warmth of their children's affection

opened to their view a source of realized happiness that truly bespoke the impressions of immortality, from the continued joy imparted in anticipation of renewal. The reputation of our advent, from the representative example of our neophytes, soon extended our influence to remote animalculan dependencies, so that the extension of our colonistic schools well nigh drained Manatitla of its effective resources of vitality. In the course of a few centuries, dating from the Manatitlan colonistic advent in the countries of the eastern continent, the system of education introduced had been generally adopted, under our supervision, by the animalculan races; notwithstanding the instinctive opposition of the chivalric portion, who followed the ancestral prestige derived from the preferred imitation of giga military school organizations, designed for the classical attraction of the senses with tinsel display, and 'glorious' din of martial music.

" Meanwhile your Heraclean ancestors had completed their third wall of circumvallation, and had extended their predatory excursions to the nether ocean beyond the dominions of the Yunka Machicas (Alta Peruvians) into those under the rule of Mauna Chusoes (children of the sun), whose women were esteemed very beautiful, being compared by a Heraclean chronicler of the period, 'to all that was lovely in person, with a complexion that blended upon a surface of white, the reflected rays of burnished copper and gold.' This comparison conveys the rich expression of metallic voluptuousness so much coveted by your Roman ancestry. These ravishing toys of passion heralded the end. Your immediate ancestors had obtained an asylum in your present City of the Falls. 'The end came and with it our hopes of communication with our colonies beyond the ocean.'

" Although aware of the approaching catastrophe, we had supposed the falcons would be spared, but the Indians included everything living in the sum of their hatred that had contributed in any way to the oppressive pleasures of their taskmasters, unfortunately including all but eight of the falcons in the massacre, sacrificing with them some of our best volantaphs. The eight were employed in transporting our Mouthpat neighbors to Rome, Gaul, and Iberian Asturias; but it was hard to keep in advance of their reproductive tendencies with so small a fleet.

" There was, however, a slight improvement in getting rid of the old stock, but the Dosch little thought of the possible injurious effect the Mouthpats were likely to exert in retarding the progressive prosperity of our colonists. But when, in the course of a few years they were deprived of their last falcon, the Manatitlans were led by the troublesome dispositions of their neighbors to reflect upon the evil ingraft they had imposed upon the labors of their people abroad. This source of anxiety has been so greatly magnified in the course of centuries which have passed since the loss of our last falcon, that, in our distress, we now appeal to your suffering sympathies for aid in reclaiming those of the descendants of our carrier breed who have, from ancestral habits of association with your race, made their eryemews within the circuit of your cinctus walls. In like return, when your sons and daughters have redeemed for us the means of more safe and long sustained flight, we shall be better able to render you service against your enemies."

Indegatus. "I have listened to my son's transmission of your request, and we will thankfully comply with your desire!"

Giganteo. ". That you may be enabled to effect the good I contemplate, it will be necessary for you to restrict your confidence to those of your family who have arrived at the understanding age of discretion. For with your people's knowledge of our existence and communication with you, our efforts would be rendered void."

Indegatus. "We can readily understand the many ways in which its publicity would compromise your endeavors to render us aid, and ycu can rely upon our watchful discretion and submission to your direction. But I would wish to be resolved upon a subject all important for the fulfillment of our higher responsibility? Your discursive narration of events in your locomotive attainments, has implied a reliance upon a higher source of aid than our gods. It would appear that you claim for creation a sole Creator, who has bestowed upon mankind a duality, compounded of instinct for the support and prompting of material manifestations of the body, with an affectionate guide in readiness for an alliance to perfect individuality for a happy earthly initiation of the animus into the blissful current of immortality? This has reflected through the darkness of our customary usages a path of light, most cheering in prospect of immortality! Do you deny the existence of Gods whose favors are to be propitiated with acceptable prayers and sacrifices?"

Giganteo. "We have within ourselves all sufficient evidence of a supreme Creator, who has created mankind with a privileged superiority from an alliance with affectionate purity and goodness. A knowledge of this optional endowment we have derived from its practical observance in exampled association, founded upon an educated preference above, and for the affectionate direction of our bodies self-sustaining instincts. Of our method of education, which adapts the body's instincts for the allied entertainment of animus purity and goodness for affectionate anticipations of immortality, we will practically instruct you in season for adoption."

Indegatus. "Then you not only deny the existence of our Gods, but erect an altar within the body for the sacrifice of animal passions, in purification for the reception of a proffered alliance with affectionate goodness?"

Giganteo. "Your quick comprehension surprises me! It will, however, lead you to a ready appreciation of our system of education for insuring allied reciprocation."

Indegatus. "The cause of my augur sight is that my parents offered with example a happy impression of attainments in kind with those you describe. But as the hour of reflection approaches I will ask you to join us, that you may be refreshed, for the continuation of your suggestive history, with its application to our needs under the direction of your people!"

Giganteo. "Gratias, for your kind proffer! But I must not allow my appetite to act the parasite in your famishing need. My wife occupies the howdah of the phaeton, and has brought at least a month's provision, so that in our plenty we are better able to share with you, and I should, at least in the form of courtesy, have asked you to test Leoptilea's skill in the culinary art; for I can assure you, she has an excellent reputation in the art of appetizing food combinations."

Gnipho and his brother, with all their restraining efforts, could not refrain from a hearty outburst of merriment at this courteous sally of the Dosch, whose commissary stores for a month's supply for himself, wife, companions, and volantaphs, were the scarcely perceptible burden of a bee. Indegatus catching the infection, the trio startled the hereditary silence of the latifundium with the unusual echoes of jocund laughter, causing the distant laborers to suspend their occupations in wondering surprise at the vent of emotions which had been so long suppressed with the rule of discontent and anxiety. The cause of this ebullition lent his mitey chirrup to swell the chorus, and incite its continuance with Gnipho. Changing to the ear of Indegatus after the more urgent emotions had subsided, the Dosch complimented him for his well preserved sympathetic mirthful tones of voice, expressing in commendation his surprise that the long disuse of mirth provocatives had not caused the resonance of his intonations to become dry and wheezy. Then, in continuation, he said, " Now that I have gained your kindly appreciation seasoned with the genial sympathy of a hearty laugh, I will rejoin my family while you are absent with yours during the heat of noontide."

CHAPTER XVI.

WHILE Gnipho was rehearsing the wonders of his marvelous interview with the Dosch of Manatitla, to his mother and sisters, as he was about closing he became suddenly silent, with his eyes drawn attentively to his right ear. The strabismic impulse startled his mother and sisters, but a bright smile on Gnipho's face relieved their fears. In a few seconds he held out his hand, and presented, with an introduction, the Dosch and his wife, with their companions. When female curiosity had subsided the Doschessa intimated her desire to hold an auramental interview with the eldest daughter! Her compliance was accompanied with evidences of trepidation, but after a few minutes these subsided giving place to an expression of vivacious interest, indicating a discussion of matters pertaining to female economy. The Dosch observing these symptoms of female confluency, reminded the prætor that the confidence of the sex was formalistic, and never free in the presence of males! Acting upon this hint Indegatus with the Dosch, Gnipho, and brother withdrew to the thalmus auditorium. The renewed interview of the Dosch with Indegatus and sons was opened by Gnipho, who petitioned his father, "May I question the Dosch to obtain further knowledge of this power of self control? It appears so natural and free from the delusions of our worship, in which we are constantly supplicating for what we neglect to obtain from our own endowed resources, it must insure happy contentment in life, and as he says, a pro-realizing foretaste of immortality. Indeed, father, I have before felt that there was within my control a peaceful joy that would serve as a shield from self-deception and the wiles of hypocrisy, which, in grateful thankfulness, I have wished to impart for others' benefit."

Indegatus. " My son, I have looked back with reverence upon the ceremonial forms of worship practiced by our ancestors, relying upon their efficacy without questioning the authenticity of their divine origin. Even on the appearance of the Dosch, as a stranger, in a form so questionable, and in accordance with my preconceived ideas of disembodied spirits, I did not doubt but that he was a nuncio of some special admonition, in answer to my supplications for aid in controlling the disaffected Heracleans, who have so greatly increased the misery of our position. But since he has convinced us that he is in reality a diminutive impersonation of our mortality, and has spoken so directly to our understanding, my eyes have been opened to the profane delusions of our long practiced ritualistic rites, addressed from and to an instinctive void, in evasion of our privileged endowment of goodness, which should direct our grateful thanksgivings to the Supreme Creator. We are surrounded with sad realities, which require self reliance for their correction, and from the source of goodness we can alone hope for directing aid. The discourse of the Dosch harmonizes with a host of new thoughts, which convict me, from past admonitions, of willful infatuation and stupidity in avoiding the animus impressions of my better nature. In the sincerity of truthful surprise, we can now look forward, my sons, with the confident hope of inaugurating for future generations a source of happiness that will reflect the current rays of immortality. But we should address our grateful emotions to the Dosch, who has interested himself for the redemption of our race from selfish infatuation."

Giganteo. "I must again express my astonishment for the apt perception you have shown in discovering the means premised for rendering the Heracleans amenable to self control, and offer grateful acknowledgments to ancestral auramentors for the presage they bestowed for the easy enhancement of my success. The volantaphs will now describe to Gnipho and his brother the position where the nest mews of the falcons can be found, and when transferred to those in the house adjoining the one you occupy, you will receive the necessary instruction for rendering them serviceable."

After a few quotations from the old volantaphs with reference to the treatment and training of eyasses they returned to the apartment where the Doschessa and her companions were entertained by the family of the prætor. The mother and daughters were so deeply engaged in curious inquiries that the return of her husband and sons remained for some time unnoticed. When listening had become tedious to the Dosch, he requested Gnipho to congratulate his mother and sisters upon their freedom from awe in conversing with disembodied spirits, as their hoarseness gave evidence of a busy, if not a clear, occupation; which, from the sequel they had been privileged to hear, seemed to be devoted to the worship of a doubtful divinity. The mother replied, that they had been taught that appearances were deceptive, and he could have but little reason to wonder, from his own and people's extraordinary size, if first impressions seemed to verify the adage in a most remarkable way. Especially when they reflected that the whole race of Manatitlans, consolidated, would but little exceed in size a single Heraclean. "But the moment we became accustomed to the pipeleo voices of your wives, and could understand what they said, why bless you, we knew at once they were mortal women, for every word and accent of their tongues bespoke the nature of our sex, and we acknowledged without thinking the reality of their minute personalities. But then, they expressed themselves so wisely in the unity of their affection, that we again doubted; for it appeared so far beyond the reach of mortal attainment, in the power and reciprocation of individual control, that we felt within ourselves the impossibility of a near approach to their sympathies in genial merit. But quickly perceiving our new source of regretful dismay, they described to us how they had been educated, and what was proposed for the benefit of our people; so we were consoled that the difference in attainment was only in the degree of perfection, which we should realize in progressive ratio from the grateful reciprocation of future generations. Our children appreciate the advantages, and are determined to act in consonance with your directions, but we cannot hope for a near approach to a love as disinterested as your people's, who have never known the misery entailed from the ranklings of envious detraction. If the impression you wish to make on those of our citizens who are hardened in their conspirations for misrule, prove successful, although, for the time being, it ministers for good through the superstitious vagaries of their perverse blindness, we will bless you in their behalf for the legacy of affection that will return to them through their children's dutiful love. Indeed, they will be ignorantly grateful, that your people made them subserve as bridges for the safe passage of their children over the slough of accumulations that flow from the sewerage gratifications of sensuality. Yesterday, we worshiped, with them, gods of man's creation, bearing the kindred impress of decay, and, with our authority, would have punished with death those subject to a defection like our own of to-day. Yet, we have often been led, from the exampled enactments of our parents, to question the happiness of a heaven where the aggravating fluctuations of our earthly associations would be continued. For, with even less faith than my husband, I could not realize the wisdom of a divine economy that designed, in defiance of original intention, to elevate brute mortality, in human shape, to the privileges of purity self-refused by earthly election. To our great relief, you have resolved this trying source of perplexity conformably with our wished-for reverence, sanctioned from an endowment of purity. Thankful to the source of our enlightened preservation, we can now clearly discern, through Creative indications, the path to immortality, purified from the adventurous impositions of superstitious instinct. Grateful that the realizing perfection, in the increase of attainment, will be reflected back from generation to generation, in recompense for the interest of our indebtedness to you, we now proffer it, with the involving title it confers of reducing past and future to present embodiment."

Giganteo. (In whispered enthusiasm to Gnipho.) "Your mother is an oracle of giga understanding, and the wisdom of her responses has proved an heirloom to her children which should cause you to be ever grateful with thankful manifestations in songs of praise to the Supreme Source of all good."

Gnipho. (In enunciated thought) "We are truly grateful, and love her beyond expression."

Here Gnipho raised his hand impulsively to his ear, before thought, from loving engrossment, could check the movement from the impression of cause.

Giganteo. "Pardon me! In my nervous desire to reach the tragus, that I might witness the expression of your mother's face as my wife imparted your testimony for the increase of her joy, I trusted my whole weight, with the impetus of a catch, to one of the vibbrillæ in a tender portion of your ear. The twinge of pain I caused was well repaid with the glance I caught of the radiant joy that suffused her face."

Gnipho laughingly explained the cause of his sudden grimace, cautioning his mother to be more guarded in exciting the admiration of her guests while they were tenants of others' ears! Before she could reply a number of the leading conspirators, with others of the disaffected, called with terror-stricken faces, imploring Gnipho, who received them in the audience chamber, to intercede with his father for their forgiveness. Gnipho from auramental dictation replied: "My father will receive your acknowledgments of treasonable designs against your own happy preservation in the temple fora when overshadowed in the sun's decline from the brink of the falls!" With "repentant" fear and its prompted "sorrow" they humbled themselves with submissive servility, beseeching the son to present to his father their humble duty, with the hope that he would forgive their past transgressions. Gnipho promised that he would deliver their message to his father, with the assurance of his forgiveness, if in token of their sincerity they would endeavor to controvert the injury they had inflicted upon the community by casting a suspicion upon the integrity of his family, when, as they were well aware, his family had been devoted to the public welfare. They departed, upon receiving this admonition, giving voice to those abject terms of submission, common to democratic expression when detected in acts of base ingratitude. This interview, with others that followed in quick succesion, gave evidence that the promised leaven of Manatitlan aid was working, which caused the prætor's family to express in the warmest terms their grateful admiration. After a few weeks, employment of Manatitlan talent in the revived art of auramental thought-substitution, with the addenda revelation of secrets in embryo, from presumed miraculous intervention of divine power exercised through the prætor, the citizens, without exception, were brought into subjection to his direction.

This led to the immediate inauguration of the Manatitlan system of education. The transformation of the temples of the foræ, for the reception of the children, inspired a feeling of instinctive awe from the audacity of the undertaking, which was heightened by the humorous devisements of auramentation practiced by the volantaphs engaged in directing the education of the brothers and sisters in the art of falconry. With the completion of the temples for the reception of the children, the families of the prætor's nearest relatives and adherents supplied the schools with teachers and censors, and in a few years all the citizens became warm supporters of the new system, fully impressed with its manifold benefactions from an increase in affectionate confidence. The children of Indegatus soon became proficients in the successful training of falcons, and were then able to place a large fleet of birds at the disposal of their benefactors. Since the time of Indegatus, the daughters of the prætors have assumed the charge of the mews as an hereditary heirloom.

At this stage of the historical relation M. Hollydorf, with the suggestive aid of the Dosch, completed his summary of the events that had transpired from the commencement of their river explorations to date, which was addressed to the secretary of the R. H. B. Society. Afterwards, Mr. Welson, with the same aid, directed letters of inquiry to his "friend" M. Baudois, a French scientific gentleman, resident correspondent of the R. H. B. A. of Paris, at Montevideo, who employed his time in fishing, for the classification of the inhabitants of the La Plata estuary, with the intention of comparing them with the fishes of the Mediterranean Sea, to determine the migratory tendencies effected by variations in the current monsoon, to and from the Strait of Gibraltar. He had also traced the glacial indications of the neighborhood, in search of transition tracks of rocks in the diluvial currents of the prehistoric periods of the earth's immersion, before its surface extension regulated with its axis movements, the winds, and tides. He wrote a second letter, of like import, to Don Pedro Garcia of Buenos Ayres, an antiquarian of note, expressing a desire for his coöperation with M. Baudois for elucidating the probable origin of the Kyronese ; and in the collection of all available collateral evidence for substantiating the approximate period of the Heracleans' advent upon the Mauna Luna shore (American);- urging] him to separate and classify his proofs so that there might be no Mandevillian interweaving of facts with traditions and conjecture, as they were intended for Mr. Dow's use in his elaboration of Heraclean history. With the desire expressed for their aid in behalf of Mr. Dow's undertaking, he did not forget to advise them of the essential advantages he had derived from the discovery of the representative remnants of humanity descended from castaway exiles of the eastern continent. In illustration of the effect produced he described, for the benefit of Don Pedro's family, the impression of Correliana Adinope's presence upon the wife of one the Vermejo chiefs, who was of Spanish birth, having been kidnapped in girlhood from the settlement of Amelcoy.

"You, and yours, will become more perfectly impressed with the comparative effect produced upon me from intercourse with the Heracleans, under Manatitlan direction, by repeating, in your own language, the testimony of a mother of your own race who has been subjected to the wifely use of a savage chief of the Vermejo tribe since her abduction at the age of twelve years. ' Ay moi ! ' she exclaimed, after a visit from the Heraclean maiden we rescued. ' When Correliana comes there is something new and good in my body that comes forth to meet her, for I feel no longer like myself, I am so happy. Then I talk to her in a way quite unknown to myself; ay me, how placid my heart grows with the light of her presence, and love, which makes me feel and forget how much I have lost. But when she goes away the darkness returns, and I am a beast again; then my children ask wonderingly. " Mamma what makes you so good when she comes, and then scold so badly when the men come back?" I try to tell them of the light that comes with her, and the darkness that my people bring to put out the light of my love; for when the hombres talk the good leaves me, and feel that I am lodo again. If she could always be with me, what a source of joy I could be for my children. Yet, she says, that my children, who are nurslings, will be permitted, by their fathers, to attend the Heraclean schools, to learn how to comfort me when I am old, with a love like hers. If this should come to pass, what love there will be in store for me? But we are not like you!'

"With my comforting assurance, that her children, if intrusted to the charge of the Heracleans before they became accustomed to the ways of her people, would be taught by exampled association the same soothing sympathy that had proved so grateful from its influence imparted by Correliana, she anxiously asked, after a few moments of thoughtful meditation, whether her children would not love their teachers better than their mother, for their goodness was constant in its brightness, and prefer to live with them to her neglect? When I was able to make her understand that the object of the school was to encourage an undying love in children for all that was good in their parents, so that its brightness would extend with increased strength beyond the present life, her mind became enraptured with the thought of increasing her own worth to merit the fulfillment of my promise."

CHAPTER XVII.

DURING the interim of letter writing in readiness for the anticipated opening of courier communication with St. Lucia and Anelcoy, Captain Greenwood had advised Correliana of his wish that the padre and sailors Jack and Bill should meet the steamer at the latter place. His despatch urged haste, as he was about to leave the gold spit, which they were then working; its deposits had become nearly exhausted. When she made the wishes of the captain known, Abdul Nycaster, the son of the mayorong, volunteered to act as courier under the conduct of a party of lower river Indians. These Indians, called by the Mestizoes, Vermojotes, ranked next to those of the upper valleys in trustworthy intelligence, so that no fears were entertained for the safety of those intrusted to their care.

A few mornings after the despatch of the courier and his party, the Dosch resumed the historical thread of his narration.

You can well imagine, the chronicler exclaims, the enthusiastic admiration of the Prætor Indegatus's children for the Manatitlans, when they saw the anxious expression of their parents' faces give place to an unspeakable joy, which imparted its radiance alike to his former adherents and foes. In evidence of their grateful sincerity they were unremitting in their endeavors to perfect themselves for the duties of censors and teachers, as well as in the more direct returns of material aid to their benefactors' affections, from their success in raising and training falcons, which promised the means for the speedy accomplishment of a reunion with colonistic correlatives. A year and a half had scarcely passed before the volantaphs were able to extend their flights a day and a third's distance in stretch over the ocean, for their own instruction in the management of the birds free from exhausting irritation, as well as to accustom them to devour their food while sustained with parachute and outstretched wings. The volantaphs, while disengaged from the active duties of their profession, kept the democratic instincts of the Heracleans in mindful dread of harboring thoughts of disaffection, held in legacy from hypocrisy, the progenitorial mother of hatred and misrule, as they had been taught, with lessons of chagrin, that their thoughts were no longer their own. The result of this knowledge enforced sincerity, which begat cheerful confidence in association, an effect that soon became manifest to the besiegers. Elasticity of thought, unprejudiced by suspicion, soon imparted its health-giving impression to the movements of the body, and action of the senses, directing them to the cultivation of useful occupations devoted to the common welfare. This freedom, in surcease from the treacherous enactments of suspicion, produced symptoms of reviving alacrity in the unanimity of action, which the savages detected from their perch on the brink of the falls' precipice, with puzzled surprise, evinced by the changing increase of numbers, and curious gaze of the watchers. The first practical use made of the falcons had been devoted to watching the Indians to learn their projected intentions, with the purpose of defeating them by anticipation without loss to the Heracleans. The unaccountable improvement in the condition and cordiality of the citizens made the savages more wary and watchful. The river savages, suspicious of the valley Indians, kept a large body of their number constantly before the gates to prevent treachery. From couriers, which had been sent to the most distant of the river tribes, it was evident that some new and more energetic scheme was in progress to bring the siege to a close. While the valley harvesting was in progress, the volantaphs had observed long trains of loaded llamas proceeding up the Lepula and Vermejo valleys, and their destination was traced to a cave in the basaltic continuation of the falls' precipice, about a mile to the north of the city. Giganteo explored the cave, and found that it contained extensive stores of dried fish, squillated meats, (hardened by the combined action of heat, pressure, and smoke), corn and maize parched, ready for grinding, in preparation for their favorite murmiel, also dried fruits in abundance. As the extent of the hoard foreboded large auxilliary accessions he was alarmed, and only thought of adding to the defenses of the city. While in flight around the city to examine if there was in the walls an accessible foothold for the savages, an accidental discovery suggested the idea of appropriating the stores of the cave for the benefit of the Heracleans. Satisfied that the moats and walls were free from adventitious aids of encouragement for savage emprise, his attention was attracted by a jetty of basaltic rock that projected into the nothern basin of the falls from the outward shore. Measuring its distance from the terminus of the wall and base of the precipice, he found that the space would admit of the circuit swing of a bridge sufficiently long for secure lodgment against the jetty. His brother, an engineer of ability, had a model of a bridge with the required measurements prepared for the prætor with a descriptive statement of its object. Great was the joy of the prætor's family when this projected source of relief was explained with the assurance of its working practicability, which promised to render nugatory the designs of the leaguers, by depriving them of their ready means of subsistence, thereby provoking suspicion of treachery, with the probable result of disruption and dispersion. The prætor immediately paid a visit to this loophole of promised good fortune, accompanied by the most skillful Heraclean artisans, who declared, after consulting the measurements, that with the floating material the plan was not only feasible, but the bridge could be quickly constructed. The Dosch recommended that the northern crematorial temple should be dismantled, as its timbers were well suited in length and seasoned lightness for the purpose. But this proposed act of desecration created a momentary impression of dismay in the mind of Indegatus, to which was added his fears of reviving the citizens' superstitious prejudices, as it would be held as an open defiance of the avenging gods. The Dosch appeased his misgivings, with the promise of anticipating religious objections. This was accomplished, but it required skillful substitution of thought, notwithstanding the prospect of plenty offered in the event of success. The labor imposed, in the quick execution of the work, aided in subduing the conjurations of danger, while the veil of mist rising from the spray of the waterfall effectually screened from the eyes of the Indian sentinels the work in progress.

By the time the valley Indians had gathered and garnered their crops, adding their quota to the stores of the cave, the bridge was launched for trial, and from the buoyancy of the timber was found to be portably light and strong, so that in reversed movement against the current it could be easily managed. Gnipho was its sole occupant in trial essay, guiding with a rope the safe lodgment of its distal extremity against the jetty. When well tested in all of its working movements it was drawn back with comparative ease; but not before the adventurous Gnipho had reached and reconnoitered the entrance of the cave. For the prætor's reassurance of the favorable acception of the enterprise by the citizens, the leaders of those who were formerly disaffected made a public acknowledgment of their transgressions, at the same time tendering their full submission to his direction. Although greatly shocked with the atrocity of their meditated treachery he forgave them without reproach.

The river savages, feeling secure against surprise from the watchful care of the guards before the cinctus gates, and sentinels upon the brink of the precipice overlooking the city, left but few of their number to guard the cave. When sufficiently dark to screen their movements, on the night set for the fruition of their enterprise, the men, women, and children of Heraclea were astir, and ready to use the utmost of their strength for the success of their foraging expedition. When the Indian camp before the gates had become quiet the party selected to surprise the keepers of the cave started and without difficulty effected their purpose. The prætor leading the surprise party had ordered that the Indian guards should be secured without the loss of life, if it could be effected without endangering the success of the undertaking. But their savage desperation in using their teeth rendered the destruction of life necessary. This was effected by suffocation, advised by the Dosch that marks of violence might be avoided, hoping thereby to involve with mysterious dread the cause of death; as the river savages were known to hold as strong a belief in the agency of evil spirits, as the Heracleans in the vengeful ire of their gods. While the bodies of the savage guards were being placed in imposing attitudes to excite the awe of their companions when discovered, the work of transporting the stores of the cave had already commenced, great care being taken that no vestige should be dropped by the way to indicate the course, or from whence, in identification, the despoilers came. The llamas, after the transportation of the stores was accomplished, were stabled in the southern crematorial temple, under the screen of the cloud mist of the falls, which had formerly subserved, under the ritualistic ceremonies of priestcraft, to mystify the superstitions of the Heracleans.

With the first gleam of the sun on the dial brink of the falls, on the succeeding morning, the Heracleans offered their first pæan song of thanksgiving, before the open portals of their houses, in gratitude for the inauguration of an era of plenty; the first in the provisionary record of centuries. After their morning meal they engaged in their usual avocations with wonted composure; at least in as much as the savages could detect from their perch on the brink of the precipice; but to the close observer there was an elation in the expression of their faces that gave sure indication of recent event of extraordinary import, proclaiming a happy emancipation from anxiety. At the suggestion of the Dosch the volantaphs watched the movements of the different tribes to observe the effect produced by the discovery of their loss. Until after the meridian hour had passed the vicinity of the grotto granary gave no indications of life, then a heavy rain set in, that served to still further delay visits of inquiry prompted by the non-appearance of the store guards. But early on the following morning the wildest commotion prevailed, the tribes in scattered bands were seen hastening from every quarter toward the cave. For the first time panic fear, in attraction, made them forget the objects of their undying vengeance, for the camps were deserted on every side, leaving the city to its own guard. This opportunity offered for a second sally was not neglected. A large amount of forage was collected from the deserted camps; but from the signaled report of a retrogade movement, still more disordered in the haste of fright, the Heracleans were about to abandon it to the flames, when they fortunately recollected that they held in possession the arms of the savages, so they easily turned the current aside and garnered their second trove safely within the city gates.

A bee phaeton had been held in requisition by the Dosch to observe the effect of Gnipho's ghastly array of dead savages, in pantomimic postures, with eyes distended, and outspread hands as if to guard them from a sight of horror, while their backs were half turned as if deprived of life in the act of escaping. When the Dosch arrived the savages were collected around the mouth of the cave, none having the courage to enter, but all, in act, were seemingly desirous of obtaining the intervention of his neighbor's body as a shield of protection from apprehended danger. But at last the luskols (Indian priests) were forced into the cave, the lesser grades being used as a wall of protection for the higher. The Dosch described the scene as horribly ludicrous when viewed from their interior position. "The fearful contortions of the diviners, as they were pushed forward with their unwilling features half exposed to the light from the mouth of the cave, in contrast with the dead dramatis personæ, furnished a study that we had no desire to prolong with the concurrent evidence of fear derived from auramentation. It was a hideous sight to behold these otherwise untamable brutes in human form, so abjectly appalled by the dead bodies of their late companions, simply from an arrangement in posture at variance with their traditional ideas of cadaverous propriety. Bewildered with the first glance, they became, of themselves, immovable with fright, until the reactive alarm from the cavernous sound reëchoing the breathing catches and grunts of the struggling mass behind, caused a frantic effort of wild desperation to regain the freedom of the open air. This contagious spasm of fear relaxed the energies of those obstructing the entrance so that they were held intact, a helpless mass immovably impacted over which those from the interior made their way in scrambling disregard to the means used in effecting their liberation. Paralyzed in voice utterance, the only sounds heard were shuffling struggles accompanied by a succession of ucks from the by no means gentle action of elbows upon opposing bodies. When at length the blockading mass crawled forth, bruised to the necessity of quadrumanal progression, their luskols had disappeared in flight. After the cave was cleared we took a high bee-line, from which we were able to see at a glance the many curious scenes enacted in their flight from the self pursuit of fear, which in variation kept us constantly convulsed with laughter." The day was well advanced before all the stragglers regained their despoiled camps; then, without apparent regard for their loss, they commenced a second exodus to a grove under the precipice to the south of the city, with the evident intention of being as far from the reach of the cave's scroul influence as possible. With the certainty that fear would prevent the savages from trespassing within the prescribed boundaries to the north of the city, Giganteo assured the prætor that the citizens need have no fear of using the pasture and arable land, accessible by the basin bridge, in the night time, if they would only take the precaution to dress in white, as that was esteemed by the superstitious of all tribes and nations as the favorite color of spirits blest and damned. The besiegers soon became aware of a marked improvement in the physical condition of the besieged. This they attributed to the unfitness of their luskols, who were deposed and sacrificed upon their own altars.

The Dosch, relator, here remarked that instinctive fear excited from variations in natural cause and effect from accustomed routine was alike common to all the grades of animality. The dogs howl in dread from the sun's eclipse; the cattle of the plain, and swine, the omnivorous congeners of mankind, will pass unheeded the dead of their species when the cadavers are natural in position, but when suspended from the branches of trees they become affected with the impulse of dismay, and like the savages endeavor to escape from the scene without the motive power of direction. The birds of the air are paralyzed with the same impulse of instinctive terror when warned of the earthquake's shock by the herald hush of preternatural silence. Taking advantage of these controlling fears of instinct, that prey upon themselves in retributive reprisal, " philosophic " and designing mythologists have conjured creed distinctions of imaginary attributes, which in combination are unitized under the style of soul. Upon this mythical assumption of attributive materialization, they have founded a system of compensation for its salvation, in a mazy labyrinthine series with a graceless cordon of conditional graces under the signs manual of saving, efficacious, sufficient, and redeeming privilege. In contradistinction to this undefinable process of instinctive soul elaboration in the scale of rewards, follows the retributive punishments, but so inextricably intermixed in chaotic confusion that ritualistic words of lunatic designation are used in substitution for the intelligent expression of thought. The priesthood of the sects, or herds, that become adherents to the formalistic use of words and material rites administered for instinctive regeneration, talk in public discourse to distract attention from the peaceful meditation of goodness. Notwithstanding the multiplication of these most daring and glaring inconsistencies, which have banished with truth, sincerity, and confiding affection, the masses of humanity are still held in blind subserviency to the fantastic rules and rulers of instinct in kind. You will scarcely wonder at the slight impression that we have made with auramental thought substitution, while the instincts of your race are constantly distracted with the bellowing exhortations of sectarian recruiting preachers, in combination with inebriate oaths of the passers-by in derisive profanation of the worshiper's selfish deity. Perversion and prostitution have so degraded the legitimate powers of perception, that the pleasures of instinct have become a source of misery from nauseating excess in over-indulgence. Indeed, from sheer disgust, we have been inclined to discontinue auramentation altogether; for your pretentious civilization, and enlightened progression, is in fact nothing more than savagery refined with art inventions for the morbid gratification of instinctive sensuality; which in recurring product have given birth to toil and turmoil, greedy vexations, strife, hatred, and kindred passionate distempers cultivated in infatuated expectation that they will yield in reversion, after death, instinctive soul purification and a heaven of peaceful rest.

The Betongese, although accounted savages, would disdain to acknowledge ancestors who had tried in ecclesiastical courts of luskol diviners, dogs and swine for murder and witchcraft, with the farcial appointment of civil pleaders of the legal fraternity for their defense! notwithstanding the special qualifications of the latter for clearing the defendants. From your ready appreciation of the higher dispensations of purity and goodness, in exampled enactment, the Heracleans can scarcely realize that you ever participated in the ruling delusions of your race. Your physical comfort, and freedom from insect plagues, in Heraclea, are derived in legacy bestowal from ancestral purity, devised in healthy enactment by the Heracleans to fulfill present attainment. By following these corrective indications your race would forefend their kind from the imposed penalties of curative professional plagues, who flourish from maladies bred in the flesh from over indulgence, in reckless regard of the certain recurrence of like from like. Our falcons in their three days passage across the ocean emit the osprey's fishy odor; and with assimilation the English, French, and Germans exhibit in national crudities the instinctive effects of diet. These are inevitable facts that admit of no palliative variation in deterioration in the process of hereditary transmission; this an observer of a single generation cannot fail to discover if possessed with ordinary powers for comparative discernment. With this deviation from our historical path, designed for Mr. Welson's analytical aid in the classification of evidences pertaining to the gradation developments of instinct, we will now continue our relation by quoting from the chronicler Titview's record of the 2d Falcon Era.

CHAPTER XVIII.

THE sons and daughters of Indegatus had become so well instructed in the art of propagating and training falcons, and withal so much interested under the direction of the volantaphs, that little danger was apprehended of another interruption in the supply. When all the preliminaries required for the voyage across the Atlantic had been well matured, Soartus, with a fleet of five well equipped falcons, and fifty giantesco companions with their families started from Maniculæ on their adventurous flight, and on the evening of the third day arrived at Corvo without acci dent, and were overjoyed, in their descent to perch on the Corcovado, to observe signals of welcome as if their coming had been anticipated by premonition. In explanation, after a joyful welcome, the Corcovadians said, that a watch had been kept constantly on the alert in expectation of their reappearance, during the interval of the many centuries which had elapsed since the last falcon departure. From whatever cause the delay, their confidence had grown in strength, that it would be overcome in time by the enterprise of their parentcedors.

This reunion of the Manatitlans with their Corcovadian colonistic outposts occurred on or about the 17th day of August, 1071 of your era. The voyage had been well sustained by the novice falcons, notwithstanding their recently acquired art of taking flying fish, and feeding while beating support with their wings under the favoring aid of the parachute. But hunger is an apt teacher of method with available means for its appeasement. The first warning the aeronauts received of their near approach to the land of their destination was the invasion from the windward of a suffocating odor, of the most disgusting taint, that pervaded the howdahs and assailed their olfactories, causing a violent retching, that made them apprehend pending calamity. But the pilots, when sufficiently recovered from the sudden invasion, consulted their charts of odor currents "laid down" by old navigators, and found that the nauseating cause of alarm proceeded from the confluent waft of Celtic and Congo exhalations from humanity, with the conjunctive loom of garlic odor in eructation from the inhabitants of Portugal, Spain, and south of France conveyed seaward by the evening land breeze, marked *Fœdisima allium exhalata ab homine*. After giving vent to humorous instinctive comparisons referring to the gross habits necessary for the production of odors so foul in their distant waft, the old peak of rendezvous on the island of Corvo saluted their glad vision.

Our reception by the Corcovadians was affectionate beyond comparison, fully enlisting our utmost resources in reciprocation. When the exuberance of our mingled congratulations had subsided into the calm current of sympathetic inquiry, we soon became aware of their loving troth to Manatitlan habits and customs. In evidence of the lealty of their alliance they had diminished a third in numbers, as they averred, with a marked improvement in all the essentials of affectionate purity and goodness, manifest alike in the conservation of physical and thoughtful development. In answer to our solicitous inquiries with regard to the welfare of the Animalculan and Giga population of the islands and mainland, they answered that the inhabitants of the islands had become more barbarously enlightened and destructive in tendency than they were in our ancestors' time; our

Mouthpat deportations having added fuel to the degenerative tendency of the islanders. Still the good example of the colonists had attracted a desire on the part of parents that their children should be educated under their instruction.

After a week's sojourn with the Corcovadians, Soartus continued his flight to Rome, taking with him as many of the islanders as the howdahs could accommodate. "Soaring to our first poise above the peaks of the Asturian mountains, we hailed with matin song of praise the broad disc of the sun as it reflected, in ascendency from the horizon, the inequalities of the countries beneath in panoramic light and shadow. As it rose in the full splendor of its mellowed morning beams, dispelling with sparkling reflection the dewy mists of night, a scene of surpassing beauty greeted our vision. From north to south, on the western confines of the Iberian peninsula, a lofty range of intervaled mountains formed an inviting attraction for the mist clouds, which dispensed their moisture as an ever replenishing source of rivers and streams, that in descent received tributary contributions rendering with their water supply the valleys fertile. In continuation from the extreme south, a sea-coast range of lesser height formed an interior basin, by circle inclosure to the east and north. Within this, cities and hamlets were scattered with seeming indications of peaceful repose. Our eyes were held entranced with the beautiful scene; and we wondered how man, gifted with rational powers of discernment, could fail to discover in the lovely blendings Creative indications designed for his direction in the paths of peace and purity. The falcons, left to the guidance of their own pleasurable instincts, just cleared the topmost sprays of the trees in their gliding circuits, but in soaring above villages and cities the volantaphs raised their flight beyond the reach of missile weapons. While passing over the city of Leon from north to south, we saw men, women, and children, flocking in crowds to the western gate. Curious to learn the cause of this early commotion, so unwonted from the descriptions we had read of Iberian habits, the falcons were directed over the point of attraction. Clearing in circling descent the spires and towers of the cathedral church, our ears were saluted with the mingled bellowings of a seemingly enraged animal, and loud shouts arising from a multitudinous collection of voices pitched in range from the shrill stridency of childhood, through the medium grades of maturity, to the vacuous piping tones of senility. Over-reaching the gate towers, we beheld collected in an amphitheatre within wooden barriers, a large concourse of people, intently gazing with boisterous plaudits upon an encounter between a horned animal of the quadruped species and armed men with garments covered with tinsel. The sides of the enraged beast were dabbled and trickling with gore, from the many wounds inflicted by the knives and spears of the "sportsmen." In quick apprehension of the effect of this cruel pastime, which had caused our wives to give utterance to a cry of pitying horror, the volantaphs suddenly depressed the falcon's tail from the "bishop's run to the pope's nose" causing it to fan-spread, intercepting the view of objects beneath from the howdah; then soaring to a poise, soon left the revolting scene and arbiters of cruel instinct beyond the compass of eyes and ears. One of our Corcovadian correlatives then explained the nature and origin of the amusement. He said that it was styled by the Spanish a taurista, and when it commenced at sunrise it was styled "taurista hiquete para almuerzo." The practice was said to have been derived originally from a race inhabiting an island in the northern ocean, who utilized the flesh of the slain brutes, and as a sequence assimilated their pugnacious characteristics.

Here the volantaph directed our attention to the

cultivated beauties of the Apuljarras of the Sierra Constantina, glowing in the morning sun with the brightest tints of verdure. Wooded and vine-covered slopes in ever varying contrast with the colored transitions of Moslem taste in the adornment of their dwellings, offered the strongest evidence of peaceful desire, notwithstanding minarets indicating fanatical worship abounded in cities, villages, and hamlets. As we neared these scenes, citizens and busy cultivators were seen engaged in their varied occupations, their forms reduced by distance from giga size to tits, reminded us of our own happy Manatitlan homes. The swift flight of the falcons conveyed the impression, from change in perspective, of gliding transitions of the same persons into varied employments, as if endowed with illimitable versatility. These pleasing illusions, which in slower flight, and nearer approach, would have truthfully depicted the miserable realities of servile selfishness, so much dreaded in foreboded encounter with the Giga races of Europe and Asia, caused Uffea, the wife of Soartus, to exclaim, with a long drawn sigh — " If they could only see and know us, I am sure they could not resist our happy example that would make this scene a reality? They surely would not refuse a joyous boon that would make these blooming valleys and verdant hills echo with songs of gladness raised in morning and evening praise? What joy for the future Manatitlan voyagers while floating over these lovely creations, if our people could be made the means of making these groves and hill-sides resound with songs of praise to unite in their fullness, with our peoples, in mid air responses? Our hopeful sighs united with the desire that her vision might prove prophetic. But alas, our falcons had scarce attained their poise for descent to the Valentinian shore, when fierce human cries, and the loud clangor and blasts from cymbals and trumpets, resounded from the plain below. The volantaphs reverse in the direction of the falcons for descent, brought into view two hosts engaged in battle encounter, each in defiant utterance shouting their war cries, one, " Dios y Santiago," and the opposed " Allah ó Profeta." No persuasion could induce our wives to venture a glance toward the fearful scene, and at their request the volantaphs changed the falcon's course to shut out the view.

Our Corcovadian correlative in defining the casus belli, said, that those fighting under the standard that bore the device of a grotesque head covering, supported by crossed keys, were styled Christians, and the words, God and Saint James, which they ejaculated with their blows, were the names of the alleged author and favorite supporter of their creed, which theoretically inculcated peace and goodwill among men. While those arrayed under the crescent banner, who were as vehemently calling upon their god and prophet, were enjoined by their creed to destroy all who denied the divinity of the prophet. Both alike, in practice, upheld the absurd inconsistency of their creeds with the destructive fanaticism of instinctive passion. " As you will in Rome be forced to give heed to the brutal enactments practiced under the style of religion, and opposing variations in Constantinople and Jerusalem, it will be well to advise you of the texts that are quoted in train by the Christian sect, who are exhorted, in the battle cry, to strike for the God of Israel, and Saint James, — kill and spare not! The grateful teeth of the saws read in this wise, Do good to those who despitefully treat thee! He that smiteth thee on one cheek turn to him the other also! He that gives to the poor, lendeth to the Lord. Cast your bread upon the waters, and after many days it will return to you increased an hundred fold. These renderings of implied selfishness in mild vapory language, you have seen exemplified in enactments this morning; but you have still to undergo the painful infliction of sympathy for physical torments that priestcraft imposes, with the anticipation of 'hellish' power, for the punishment of dissenting heretics. The Moslems, who are fighting under the crescent standard, advocate sensual lunacy, with the prescribed belief in a heavenly haram peopled with houri, — a name that they bestow upon sensuously beatific women, who in Christian terms are styled angels, — these are used as lures for inciting the lusts of the faithful in blind subserviency to the commands of their leaders. Like their Christian opponents, the masses depend solely upon priestly interpretation for the reconciliation of contradictory passages in their creed, openly bestowing their reverential fealty upon the sensually mad, who are called saints or santons, esteeming the touch of their lewd filthiness as a vise for heavenly reward. Their priests, who are styled dervishes, wear the same hermaphrodite vestments in ceremonial enactments that distinguish their Christian counterparts, deriving oracular inspiration from dizziness invoked by rapid whirling gyrations. But as I see that you are oppressed with sorrowful disgust, in view of the rank stupidity you are about to encounter, we will allow you, with these premonitions, to verify the construction we give of the ancient giga Roman senator's apothegm, which should have been, 'whom the gods wish to destroy, they first make sensually mad.' With the instinctive example of destructive hatred inculcated as a morning lesson for appetizing the kindred germ of children's passions, in the bloody struggle between inhuman art aids and brute strength, exhibited in the bull's blind rage, we can easily fill the intervening space in life with occupations in qualification for the battle enactments of the meridian."

With the volantaph's announcement that he was about to bring the falcon to a poise which would afford the occupants of the howdah an equidistant view of the peninsulas of Spain and Italy, a retrospective glance was cast backward. The keen Manatitlan sight, intensified with ardent admiration for the glowing beauties of the Iberian landscape, soon became absorbed in tracing the rare combinations of mountain and valley verdure, merging and varying in blending tints from the sun's declining cast of light and shadow, until startled with Uffea's sudden call, Alew! (look!) Recalled by her startled cry, our attention was attracted by her steadfast look, to the Valentian shore, and there beheld the victorious Christians in pursuit of the vanquished Moslems, vindicating the cry of their priests, "Kill and spare not." "Alas," tearfully sighed Uffea, "is it not enough that they yield the victory? of what avail the bodies of the flying, living or dead, that the victors still ruthlessly pursue and slay?"

Anxious to escape from even a distant view of the carnage, the volantaph brought the falcon to a poise and in descent opened to view the peninsula of Italy. In anxious search for the abiding place of their colonistic correlatives, their attention was soon attracted to the largest collection of buildings, and as the falcon's circuits narrowed in near approach, their eyes sought for signals of recognition. These, from a "bird's eye" view chart of the prominent buildings in the vicinage of Rome, soon became visible, not only from the coliseum, the chief settlement of their colonies, but from every town and hamlet within the reach of vision. With a near approach to the arcades every available place in the Ionic range was filled to overflowing with beaming faces and outstretched arms in token of joyful welcome. But a few moments elapsed before the falcons were safely moored in the old "New Port," and the howdahs thronged with forms and faces that required no introduction by speech for the test of nátionality. Without words, joyful tears, sighs, kisses, and embraces were not alone conceded as the special privilege of womanly affection, but the interchange of these instinctive tokens under the kindly promptings of gladness became general. If there was hesitation indicating speech, those on the outside pressed forward to interrupt the useless waste of words. For at least a full half hour this voiceless scene continued unabated, then the prætor of Coliseo parleyed. "Citizens of Romelia, forbear? What have these, of our kin, done in the lapse of ages, that you suffocate them with kisses and embraces? Are they, in the fullness of our joy, to be denied the viaticum of welcome words? or do you intend to despatch them with the silent interdiction of tidings we have waited and longed for while yet unborn to the world of mortality? Make way for your prætor! Fie upon you, Oluissandra! that you, the wife of the chief magister of Coliseo, should fail to use your tongue in speech, when its words would be welcome!"

With this laughing admonition a tall active giantesco sprang into the howdah, and seizing a sprightly medium woman by the shoulders, as she was about to embrace Uffea, turned her briskly round, with the exclamation, "Now for some system? I am ashamed of you, Olui! Where is your boasted presence of mind and pity? and voice, so easily aroused in gratuitous sympathy for your Giga auramentees? Do you suppose that this little handful of women can withstand the battery of Coliseo's thousands, softly placable as they are, without having their lips and faces flayed? Now Signorina Manatitliana, this is my wife Oluissandra Peasiffea, of the twelfth generation in descent from the Peasiffeas of Maniculæ,— that is, I am, and she shares my loving pedigree, with a worthy merit that exceeds my own."

Before the last clause, Uffea had embraced the wife of the Coliseo prætor with cousinly warmth. The husband laughingly thanked her for honoring his spouse with such affectionate returns upon the strength of relationship, but hoped that the slight deviation from the "giga lineal" would not prove a bar to the full expression of loving confidence. This diversion set the women's tongues in motion, with trills and fugues, which were followed by the men's deeper tones, in more measured accents of curious inquiry. With the sun's decline all united in a hymn of thanksgiving. As the gathering shades of twilight deepened, Oluissandra reminded her husband that the Manatitlan howdahs were poorly adapted for the reception of the prætor of Coliseo's guests upon an occasion so extraordinary? "If the public records are to be trusted," she said, "they will bear testimony that it has been customary for the chief magisters to offer their guests the hospitality of their houses. It will hardly consort with precedent, if the scribe should in record addenda state that the prætor Peasiffea entertained his Manatitlan relatives in their own howdahs, at the reunion of their peoples, after centuries of separation. I am sorry," she continued, "that courtesy obliges me to give you this public reproof, but your 'head and heart' should have been on the alert to give a suitable welcome, as the long delay might conjure sensitive doubts questioning the sincerity of your joy."

The prætor gave his wife a look of quizzical gratitude at this rejoinder couched in Giga style, reminding the chief and his wife of the Dosch's humorous sallies. In like manner they were constantly reminded of home faces and scenes by revived similitude in impression. Among the Manatitlans, and their Corcovadian correlatives, the questioning query would pass, "Who does this or that person remind you of?" or "How familiar that voice sounds." When the resemblance was mentioned, the likeness was found to have been transmitted by collateral branches. The prætor, acting upon his wife's suggestion, our falcons received the attention they required from Coliseo volantaphs, the mews having been kept in readiness for their reception from generation to generation, in constant expectation of their reappearance. Our own apartments, which adjoined those of the prætor, were in the foliated cyma of the capital surmounting the second arch of the Corinthian Arcade. On our way the prætor pointed out the improvements devised and executed by his predecessors, regretting that their comparatively indestructible works should be ingrafted upon one of the perishable follies of the Roman empire for its more extended time durability.

At dawn, on the morning succeeding our arrival, anxious throngs were awaiting to greet us with salutations of joyful welcome. Many of these had come from distant districts to participate in the rejoicings of an occasion so auspicious for the united welfare of the colonies. Among the visitors, there came from the St. Angelo department the Dosch of Romalia. He had started with the first announcement of the falcons, and traversed the city of the gigas during the height of one of their saturnalian feasts of flesh, which precedes the lighter indulgences of fasting upon a fish and lentil diet. After the morning salutations, the falcons were employed in excursions for the practical instruction of the native volantaphs. On the third day after our arrival, the Dosch and prætor consulted the chief of the Manatitlans upon a subject that had been a source of disquietude to the colonists.

" We have deferred the unpleasant subject we are about to introduce, to the latest possible moment consistent with the responsibility of our charge," said the prætor of Coliseo, " that the impression of our welcome might remain cheerful until you had fully compared the extent of our worthiness with your expectations. The sooner anxiety is dispelled with a knowledge of impending evil the better. The Mouthpat seed our mutual ancestors sowed, just before the close of the first falcon era, in taking root assimilated with the native Animalculan races, and in process imparted their own deleterious habits and prejudices to their entertainers. With a wider scope for the gratification of their sensual instincts, they soon united with the democratic rabble of Rome in opposing what they termed our pedantic puritanism. With their coadjutive stimulation our allies in the exampled practice of purity and goodness were subjected to annoyances and persecutions and were denied the right of local option, as natives, in selecting for themselves a choice of education for their children. As we could not extend our protection to the good throughout the broad expanse of Rome, we established self-supporting colonies in the country as asylums of resort for the oppressed, so that in consolidated association they might receive our more effective support and aid. Still, in defiance of intimidation and actual injuries, we had more applications for the admission of Roman children to our schools than we could accommodate. The deported Mouthpats, from the first, became adherent imitators of Giga habits and customs, and fanatically zealous in support of the Catholic dogmas. Before their advent, the Animalculans of Rome had been content that the Giga priests should enact their parts in ceremonial worship. But the Mouthpats urged that the practice of vicarious worship through a race barred from direct communication by mouth interlocution was a subterfuge of the most damnable tendency. Their labors for the reconstruction of the ritualistic tenets, and regeneration of the Animalculans from proxied dependency upon Giga religious administration, was finally rewarded with the election of a pope of Mouthpat descent. This pope, Innocent the First, in Mouthpat designation, now occupies the silicoth residences relinquished to the first cargo of his ancestors, by ours on their removal to the Coliseum. In imitation of Giga example, although he claims higher merit from his strict administration of the ordinances of the church, he has established a court of inquisition for the trial and punishment of heretics.

" One of the first acts of the court was the proclamation of an interdict prohibiting the Romans from holding communication with the Coliseos under the penalty of excommunication from the only true and holy church. We well understood the term excommunication in context, although the court was wary in using threats of torture and death, against the parents who had entrusted their children to our care, until they had tested our disposition by overt acts of intimidation. For this purpose they have arrested the parents of our pupils on their return to Rome after paying their monthly visits to their children. To-morrow they celebrate a saint's day, in the calendar of the Mouthpat Church, and are erecting in the gutterleads of the church of San Lorenzo, lists with barriers and the usual requisites necessary for the accommodation of spectators, and actors that engage in the barbarous follies of a tournament. But the real object is the inauguration of an inquisitorial chapter. The pomp and ceremonials only serving as an introductory blind, that will hold pity in check by arousing the passions with chivalric show in brutal enactment as a placebo for the final catastrophe. In order that the pomp might equal that of their Giga exemplars, who are engaged in preparing for a like celebration, they in anticipation sent challenges to the most celebrated and valorous representatives of Animalculan knighthood throughout the courts of Europe, subject to a like defiance from their Giga contemporaries ; whose heralds acted as the locomotive beasts of burden for the transportation of their parasitic knights errant. To prevent imposition, the field kings of all the countries in Christendom, subject to Animalculan sway, were requested to add their attest to the order and standing of the knight applicants, also to the service reputation of esquires emulous of achieving the honorable distinction of wearing golden spurs. The requirements of those desirous of contending for love and honor in the lists, on roachback with spear and sword, were — to be of pure lineage, of not less than three generations, in affirmed descent, free from the attaint of mesalliance. The squires were to be second sons of a parentage alike eligible for the distinctions of knighthood. That the trains of each knight should be well satisfied with their allotment, the third day was set apart for their contention with arms suited to their stations in life; ample means were to be furnished-for the eating and drinking entertainment of all comers. The knights summoned had already arrived in train with the Gigas who had been cited for the tournament in preparation for the first crusade designed for the redemption of the holy sepulchre from the possession of the infidel Saracens, in which the Animalculans will also engage. The first course, in this tournament of human instinct, will be inaugurated after a grand high mass, to be held in conjunction with the Gigas, the priests of both races joining in the ceremonial administration of the rites. After the grand 'entre' the first course will be run between Count Marceroni, the Roman champion, and any knight bold enough to accept his challenge. The first encounters are to be on roachback with spears, in support of the affirmed superiority, in beauty, of the contestant's countrywomen, or "ladies," in the style of the challenge. The proclamation sets forth that the encounter will be conducted in freedom from the slights of incantation, or other surreptitious advantages, in fair and open battle, The first unroached will be declared vanquished, yielding to the victor the right of heralding the supremacy of his countrywomen's beauty, with the privilege of selecting from them a queen to reign during the continuance of the jousts, as the empress paramount of love, honor, and beauty. The second day's joust will be a contention with axes, maces, and thorn sticks; the victor will be awarded the privilege of selecting a lady to preside over the distribution of prizes to the successful in the melée, or herd encounters of third day's strife.

"This sketch of the announced proceedings, will give you an idea of the amusements patronized by the Gigas under the 'angelic' supervision of women, with the sounding style of chivalric. But with both races, the preliminary amusements are devised as placebos to invite in transition awe, rather than indignation and horror for the final tragedy of human sacrifice. Your opportune arrival will, with falcon aid, render our service effectual in baffling their intended murderous enactment, if our emprise meets with your approval."

The chief and his associates warmly approved of the course proposed, offering to undertake alone the hazard of its successful issue, that the reproof might be in effect more significant of intention in the event of future transgressions. The scenes enacted at the tournament, the Dosch said he would relate in quotation from the chronicler Titview's letter to Giganteo.

CHAPTER XIX.

Coliseo, departmental colonial settlement, or chief city of Romalia, on *the fifth of Ratu, of the one hundred and eighth century of the Manatitlan era, decio multiplex.*

REVERED ADVISOR:— In accordance with your request, I send you an account of a visit, which in company with the chief and our associates, was made to the temporal dominions of the Mouthpat pope, Innocent First, for the rescue of Roman tits, who had been kidnapped on their return from the monthly visitation of the Coliseo schools. The object of the pope, prompting this human theft, was to effect intimidation by offering them as a public sacrifice, or burnt offering to the god of their worship, after the manner of the Tenockitlans, Manchees, and other civilized nations of Mauna Luna. The vatican, or stronghold of this Animalculan sect, styled Christians, is situated on the leads of the Giga church San Lorenzo, formerly occupied by our colonists, and by them transferred to the Mouthpats as a leasehold to be secured in perpetuation for the consideration of reciprocal good will. The church of San Lorenzo, as it is now called, was remodeled by the present Giga Christian dynasty, from the temple of Faustina, by Antoninus, surnamed the Pious, as a distinctive memorial of regard for his wife. In the process of renovation required for the new form of mythological worship, the outer walls, with their architectural ornaments, were left unmolested above the entablature of the cornice, which upon the inner face, looking out upon the leads, contained the sacred dove-cotes or cells. These the Manatitlan colonists had appropriated by ejecting the unbaptized usurpers of the pagan oracles, which in turn they relinquished to the Mouthpat sympathizers of the new ritualists succeeding to the interior of the edifice. The roof was accessible to the Animalculans by a jutting rear X buttress erected for the support of the inner incline of the lateral walls. For a long time the cells were jointly occupied by the Manatitlans and dove descendants of the legitimate Giga divinity of the church, who found themselves controlled in flight by a mysterious power adverse to their natural inclinations and gregarious flock associations. "These could have been made excellent substitutes for falcons, but like all of the animal species that congregate in flocks and herds, they were subject to unclean parasites, rendering them obnoxious to purity, which caused their ejection by a process as mysterious to their comprehension, as that which had previously controlled their flight. Ten months previous to our arrival the Giga pope had announced a tourney, or 'passage-at-arms,' upon a scale of unprecedented magnificence, for the advancement of a crusade against a race of Moslems, who exacted tribute from Christian pilgrims on their way to the holy sepulcre of their creed. The chief honor to be awarded the victor, was the command of the forces levied for the holy war. But each of the contestants was obliged to offer for the acceptance of the king herald, irrefragable proof of his blood nobility, and faith in the immaculate conception, pope's infallibility, and Catholic efficacy of saving grace. We arrived in Rome four days previous to the one designated for the grand ceremonial opening of the Animalculan court of valor, which, as a special distinction, was to be honored with the pope's presence and arbitration. As the descendants of the Mouthpats have not improved in the industrial habits of self-devisement in the economy of time for useful purposes, we were able to secure our position on the roof balance of the portico without being observed. Directly opposite is the vatican, or oracular palace of the Pope Innocent. This is one of the larger dove-cotes which had been finished in silicoth by the Manatitlan colonists as an anthemeque, or place of public assemblage, and is now occupied by his "holiness" as the dove successor divinity of the roof. An hour or more after our arrival, the familiars with their working aids commenced the labors of preparation, and from the character of their employments it is evident that there will be no cooing notes of sympathy expended in behalf of suffering victims.

In the northern gutter of the leads the lists are erected. The galleries rise in backward ascent from the arena. The terraced gradations are adapted to rank founded upon material possessions; the lowest are intended for the nobility, or landed proprietors; the next grade is alloted to the rich in gold and silver, or materials in trade, these are styled merchant commoners; the highest or gallery, the tribune informs me, is destined for the gods, as the rabble are facetiously called. If the terms of expression interest you, their generic source can be traced in the Chinese and Babylonish manuscripts of Animalculans deposited by our old travelers in the archives of Maniculæ; from them you will be able to judge of the progressive attainments of the Romans. Our position will enable us to see and hear all that passes.

The canopy of the ladies gallery is of richly woven material, contributed by a kabulistanee convert to the papal creed; as you will not find an equivalent for the word "lady" and its co-appelative "gentleman" in any of our indigenous writings, I will give you the Coliseo tribune's version. "A lady in Giga acceptation, is a woman who employes, not only her own, but the time of her servants in the adornment of her body with cumbersome material, greatly in excess of her requirements for comfort and comeliness; in addition she does not hesitate to apply pigments to her face. These aids are only limited by the metallic means of supply; and, as your judgment will decide, they detract not alone from personal purity, but render the persons of the Giga females, in fact, repulsive. Still auramentation, with the confirmation of the senses in support of our labors, has proved utterly powerless for the successful stay of the fantastic follies that follow in train from the gratification of woman's envious rivalry. The term chevalier, or gentleman, is still more vague in acceptation, and application with the Gigas. Like virtue, conscience, morality, and saving grace, the meaning depends upon arbitrary intonations of the voice in application, without true intrinsic value for the expression of means or substance realization. If you, or at least a tit, should ascend to the tier that will be occupied by the gods, and suddenly accost one of the meanest and most blasphemous of the noisy rabble, with the words, "You are no chevalier, or gentleman," you would be saluted in return with a blow or volley of vile epithets too horrible for exampled utterance."

He was about to give farther proofs of the illusive beguilements of Giga usage in word conversation, but was interrupted by the blasts of sackbuts, and rumble of kettle-drums, from the tents of the challengers, which were partially concealed by a flowery bosque, which had its source from soil and seeds deposited by the doves. These signals quickened the movements of some blackamoors who were decorating the canopied galleries set apart for the ladies, the pope and his apostolic-cardinals. Soon a living stream of Animalculans began to deploy in descent from the buttress turrets of the parapets, and when within hail commenced a series of pantomimic imitations of the blacks' deformities, accompanied with gibing words, as if in cruel preparation for the scenes of the day. Notwithstanding the serious nature of our mission, enhanced with the fear of after self-censure, we could not withhold our instinctive sympathy when the Ethiopians discomfited their democratic Mouthpat assailants, who showed themselves in every respect inferior to their swarthy opponents. The tilting touches of the blackamoors' tongues, often caused the Mouthpats to wince and brandish their thorn sticks with the well known hereditary twirl, as if they desired to relieve the smart with customary material arguments. The arrival of the higher orders of Animalculan life, mounted on gayly comparisoned blatidean roaches, beetles, and ants, created a diversion with less freedom in expression, yet the democratic tendency of crowds to turbulence was still apparent, for the knights in their turn became subject to the natural flow of depreciation, but in tones subdued in measured prominence to a comparison within and without the reach of the subject's lance. Finally the ladies were passed in review with rabberly freedom, but ever mindful of the length of the knight attendants' spears. And lastly, after the beauty, dress, and palfrey management of the ants by the ladies, had been freely discussed, the squires and grooms in livery, received the dredged overplus of scurrilous scoffs and taunts. When the flourish of trumpets and beadle cries reminded them of the more important calls of selfishness for the rival displacement of early comers, we had the disagreeable opportunity of witnessing a scene of uproar that baffles description. During the struggle for places the bailiffs' staffs were used with a freedom that denoted the lowest degree of servile subjection on the part of the herd.

While forced to listen to the vile language of the " plebs," we were constrained to hold in the balance of thought the questionable advantage of speech for the advancement of purity and goodness, for as yet, an affectionate word, or one of sympathy had not been spoken within my hearing. The Coliseo tribune, who acted as my mentor, informed me that the Giga children of Rome lisped oaths with the first impressions of speech, and with the adventure of sentences used them for anathematizing imprecations in demands for selfish gratification. The pursuivants' calls, "Aller laisser!" Gallic Latin, of imperative command for space, again attracted our attention to the lists, and from the commotion, it was easy to perceive that the curiosity of the vast assemblage was on the tiptoe of expectation. The prolonged flourish of trumpets, with the clashing rattle and rumble of cymbals and kettle-drums, prepared us for the heralded announcement of the approach of the accredited knights who were to lead as challengers and challenged in the jousts for the awards of honor. While passing in review before the ladies' and popes' pavilions, we had an excellent opportunity for comparing the relative intelligence of those representing the different nationalities and septs.

The first group were Austro-Germans; these were large in size and heavy in feature and form, with a pervading vis inertia in their sluggish movements. But to my surprise, the tribune informed me that in ritualistic gymnastry, spear, and sword exercise, they were esteemed the leading nation, although rarely sucessful in their encounters, as they were too slow and methodical in delivering their blows, and as a general result, they were unroached while deciding upon their point of attack. Calling my attention to their movements, I could not fail to observe the sluggish halo that seemed to invest both riders and cockroach steeds; the latter, he said, grew to an enormous size in Germany, as in omnivorous habits they assimilated with their national contemporaries of the human species, whose intelligence was an instinctive reflection from stomach distention to the brain. Notwithstanding the swinish cast of their eyes, and sodden dullness of facial expression, there was outshadowed a nucleus reflection, through the gross embargo of fungous flesh, that bespoke with its oppressed rays the still existence of the animus of goodness; but it was so faintly luminous that it failed to make its source self-apparent. From the remarks of the tribune, it appeared that the composite peculiarities of the German septs had been derived from a dietetic source, dependent upon the digestive energy of the stomach, and powers of distension for the disposal of sectional ingesta, selected for the expression of divisional patriotism. The mental effect produced was graduated from the wrangling source of irritation, to the philosophic effect of over-distension, conjoined with an owlish expression of vacuity in the region of the eyes. Although in outward expression ritualistically Christian, their mouths and stomachs were infidel in observance on fast-days, strenuously advocating the future resurrection of the body in the stomach. The females belonging to the family of the Count Palatine Von Lushmywitzs, possessed the elements of physical beauty, but there was a subdued expression of the eyes, with a pervading depression of deportment, exhibiting the guardian effects of discipline administered in their male sponsors' philosophic moods; also in combination, a settled, disgustful despondency, emanating from an association with their lords after they had become filled to repletion with philosophic wisdom. It was easy, however, to perceive the struggling elasticity of purity and goodness in outreaching desire for kindred association, and hopeful deliverance from an entombed death of corruption. Pointing out several knights and ladies of more compact physical attainments, and vivacious movements in the expression of thought correspondence with features, the tribune stated that their improved appearance was solely dependent upon an admixture of foreign blood, opposed to swilling barley mead, and coarse gluttony. Many of these had sent their infants to Manatitlan colonistic schools, and, by the after adoption of their children's example, had raised themselves to a comparative appreciation of the privileges bestowed for the elevation of humanity above the coarser instincts of its suballiance with the lower orders of animality.

The next train, in the knightly roachalvacade, embraced in the retinues of its leaders the representative extremes of humanity in the British Isles. They had bluff apetital features, with but a slight remnant predisposition to the philosophical swinishness that enveloped with fatty folds the direct descendants of their Saxo-German antecedents. In the place of jowled lethargy, they presented bold taural fronts, with a canine expression of tenacity about their mouths, that would cause you to involuntarily shrink from a collision with their heads and teeth. Indeed, their general aspect was stout and still in manifestation, indicating strong bodily self-reliance. The women exhibited a robust air of independence, with the pleasing accompaniment of rosy complexions, in strong contrast with the pale, inanimative features of their German cousins, which declared at sight their freedom from servility, with the exhibition of a co-existent power that could turn or tame the obstinate fronts of the males, when within the circle of their domestic domain.

Following close in the rear of English lead, came the Gallic French, who with chattering volubility and grimace, more than supplied the paucity of words used by those in advance. In personal characteristics they were in every respect foreign to their preceding neighbors. Without listening or replying they all talked in medley, which impressed me with the conviction that they had no definite ideas beyond the present evidences of their existence, or hopes of a future free from the predominating influence of their bodies' frivolous selfishness. This impression was confirmed by the tribune, who said that sympathy with, and confidence in their kind, were alike ignored, each holding that the gratification afforded by the passing moment, was the only happiness and real hope, life afforded.

"In accordance with these assumptions, which they verify in act, we have found them loquaciously factious, and quite as unsettled as their Giga countrymen and exemplars; expressing in a breath the strongest terms of wordy affection and hatred for the same person, and in their constantly recurring civil feuds, destroying their rulers and relatives with less compunction than individuals of an opposing race. In verity from frivolous habits in thought and act, they have become despicable examples of the disorganizing effect of infidelity to Creative indications designed for our self-control. The French women, as you observe in the boldness of their public demeanor, are as vivaciously apt for the illustration of the opposite gender's instinctive ideas of happiness in domestic association, as the German are submissive."

The Spanish knightly representatives next passed in review. As roach cavaliers, they were far above their predecessors in graceful bearing and dignity of deportment; but their thoughts were inwardly disposed to the sole devotion of self-contemplation, looking down from the lofty pinnacle with supercilious complacency and condescension on all alike, expecting from others the deferential consideration they paid to themselves. The women were veiled and demure in bearing, but the instinctive sparkle of their eyes defied the gauzy fabric's concealment.

The Italians came next, yielding, as entertainers, precedence to the four leading nations. The Mouthpat descendants of Italian nativity, as is the wont of the stocks derived from their audacious ingraft, assumed the lead. Closing the roachavalcade were knight adventurers from lesser nationalities, lacking in numbers sufficient for separate display. Alas, for my own and companion's vaunted power of self-control over our instinctive passions, when exposed from inexperience to human acts of cruelty committed from the wantonness of power. As these human novelties passed under our curious inspection, panoplied and strangely garbed in glittering armour, one of the sturdy knights attached to the rear, in mere wantonness for the exhibition of his prowess without the risk of rebuttal, with the pretext of a gibing inquiry with regard to the health of his maternal parent, suspended an unoffending tit dwarf by thrusting his spear through his jerkin, holding him aloft, while the blood from a flesh wound trickled over his hose. It was well for the successful issue of our undertaking that a kindly disposed herald not only relieved him from his uncomfortable plight but gave him the wherewith for the purchase of another garment with the largesse he had received from the other knights, and in retribution ejecting the offender with disgrace from the lists. This justly merited punishment pacified our irate desire to call the aggressor to an account for his wanton cruelty. The warmth of our sympathetic pity caused the Coliseos, with a smile, to warn us that if we attempted to redress all the wanton and cruel aggressions that would be forced upon our notice, with the infliction of bodily punishment, we should be obliged to forego the benefits of peaceful example altogether. " We are powerless to stay the barbarous cruelties they practice among themselves, and our interference would only serve to aggravate the spirit of retaliation. From the beginning, the course pursued by our people has been beset with many temptations provoking arbitrary interference for the redress of wrongs. But they have managed to escape free from serious difficulty up to the present time, and our presence here to-day was compelled by affectionate sympathy to save the lives of well disposed parents whose children have been intrusted to our charge. With your aid we can now make a lasting impression that will prevent future aggressions, for they have a hereditary dread of the actual presence of the unadulterated Manatitlans."

The trumpets sounding a parley for the announcement of the regulations of the day, and terms of the challengers, caused us to give hasty thanks to the tribune for his timely admonitions, with the promise that we would try and hold our feelings aloof from passionate excitement in behalf of all except those in whose aid we had enlisted. The lull that succeeded the herald's call permitted us to hear the alheu of the pursuivant, but the special conditions of the challengers were drowned by the renewed buzz of voices. Gathering strength from personal retorts, they gave birth to sounds ranging from the high nasal to the low guttural of the German dialects, accommodating themselves to fragmentary enunciations of words, and the most dissonant syllabic combinations possible in conjuration from the rumbling intonations of indigestion. These crudities were jargonized with the curt unconsolidated English, the fustian croak of the Dutch, Gallic flippancy, slipshod Irish, and the nasal bagpipe drone or burr of the Scotch; to which was added the quavering Italian, the soft flow of the Spanish, with cadenced harmony of intonation, and the horrible rasping cartilaginous utterances of the Jewish nose, so repulsively selfish that endurance shudders in memory of the infliction. The tumult of tongues intercepted and confused the herald's announcement, so that we were dependent upon the pantomimic acts of the stewards and beadles for an interpretation of the decrees regulating the combat. The Coliseos were in no way anxious to witness the medlarious enactments, so we were preferred to the most eligible positions for the gratification of our newly fledged curiosity, which I am ashamed to confess was over eager, and for the occasion must have lowered us in the estimation of our cousins. Our position exposed to us, in side view, the entire multitude in its caste gradations, from the canopied pope and cardinals and their created nobility, with the "fair ladyes" occupying the pavilion opposite, upward through the terraced ascent to the "strident gods," who worried each other with true mythological zest when disengaged from the supervision of those below. When eager expectation was at its height, attention was diverted from the lists by solemn long drawn blasts of sackbuts accompanied with the roll of drums. This new phase closed the mouths of yelping instinct, and with the awed hush, the object cause of the doleful braying made its appearance from the gates of a dovecot abbey. This was a priestly procession moving with slow cadenced steps to the sound of an anthemed dirge. The tribune, who seemed to be intuitively apprized of all the movements in the routine phases of priestcraft, informed me that the procession preceded by the palled bier indicated a judicial combat, which was instituted for the final decision by the judgment of God of right and wrong in argument, as well as the guilt and innocence of persons accused of crimes. Farther exposition was anticipated by the entrance of a herald into the arena, who proclaimed, with an alheu,— "Whereas, it has occurred,— to the extreme mortification of his holiness,— that certain of the consistorial incumbents,— to wit, the holy brothers Bonefacio and Buenaventura, have differed in their estimate concerning the preferred essentials of sufficient and efficacious grace ; giving rise to the implied accusations of perjury, with taunts, that in the warmth of expression exceeded the bounds of Christian forbearance,— it has been determined by our holy father, that they shall have extended to them the privilege of deciding a question so important to their respective personalities, and the general welfare of souls, by the ordeal of battle. In referring the detection of the perjured to the judgment of God, the doom pronounced upon the convicted, — in vindication of the just and holy laws instituted by the gracious clemency of our holy father, — it hath been adjudged that the respondents be awarded the benefit of knight champions, who for the love and sustenance of truth, in devotion to the holy mother church, shall offer, in willing submission, their bodies with arms for the decision of the question at issue, using their utmost exertion in skillful encounter, that the perjured may suffer merited punishment."

The instinctive elements of humanity, like those which hold irruptive sway from excessive accumulation in the earth's interior, involve anticipatory emotions of coming events. As, with the vacuum lull of the forest that heralds the tornado; or, in the ocean calm, the bubbling ripple crests which warn the volantaph of an approaching tempest; and the dread silence that precedes the earthquake's convulsive throes, — the congregated multidude in counterpart had anticipated from the dirge the startling premonition of some horrible gratification. From suppressed respiratory gaze, unbroken except from an occasional Jewish croak, or belching eructations from uncircumcised barbarians unaccustomed to the suppressed control of reverential awe, the multitude, when the cause for the sombre prelude was announced, burst the bonds of thoughtless silence with deafening shouts for champions. The knights no less prompt offered their service in mass, so that it became necessary for the interdicted prelates to make selection of champions; in aid the heralds sounded their call for a parade procession. As the squires brought in the knightly equipments for their masters, the grooms led in the panoplied war-roaches. When mounted the knights passed in review before the consistorial pavilion. Exposed to the searching glances of spiritual and temporal criticism they exerted all their dexterity in the management of the roach-steeds and in the changing exercise of sword and spear to win the favor of an election. Cardinal Bonefacio had with other accomplishments acquired the reputation of being an expert roach equitator, as well as a skillful artist in the use of arms, in his temporal days. Buenaventura, his polemical adversary, aware of these advantages, and his own defective judgment, sought by the chicanery of his eyes to detect Cardinal Bonefacio's preference, as in casting lots he had won the first choice, which would enable him to secure the champion of his opponent's election. Among the knights there was one from Tipperary, a county shire in the Island of Hibernation, a dependency of Albion. This knight bore the title and name of Sir O'Ham Ill Tong, of Scythio-Mongolian extraction. His brogue, and quaint peculiarities of speaking in reversion, had attracted the tribune's attention, who was familiar with the derivation of his sept and lingual idiom. He informed me that the literal meaning of his name was bad pork, as it was customary with the Mongolian tribes to name those outlawed from the hereditary cognomen of Kan Avan, or John, the son of the tribe, with the cause of defection; which in his case might have been inherited from ancestral resemblance, exchanging a bad article for good, or stealing from his own tribe, as each of these crimes were punished with bestowal of a name referring to the cause of attaint. His squire still bore the tribal name, and for economy enacted the part of roach-groom. This unique pair had attracted mirthful attention, and were the especial favorites of the gods; the knight for correspondence with name, and Kan Avan, his squire, for successful apeish imitation of his master's traits; but both were so blinded with boastful vanity that even the scoffing plaudits of the hinds ministered to its inflation. Cardinal Bonefacio, being in disposition humorously inclined, could not divert his eyes from the combined comicalities of their forms, pretensions, and clownish movements. This marked interest at once decided the choice of Cardinal Buenaventura. The wisdom of his selection was confirmed when the generosity of Bonefacio suggested a more prudent election. Finding that his colleague's determination could not be changed, Bonefacio made Don Bacalao his proxy, a Spanish knight who depended upon his dignity and purity of lineage for success.

The cardinals having selected their champions, the lists were cleared for the encounter, the beadles remaining to enforce the ritualistic regulations. When the champions had assumed their positions, the barrier gates were opened to admit the judicial brotherhood, who while chanting their dirge proceeded to the centre of the arena and there deposited the bier with its palled coffin, upon which had been placed a skull and crossed bones of the human leg; over these as a dividing barrier the two champions would be required to thrust their lances in closing career. With this sable wall interposed, gravely premonitory with its relict escutcheon, of sepulchral entertainment, Sir O'Ham's face became blanched, while the point of his spear directed heavenward described in trembling movements a geometrical medley of circles with acute triangles, as if engaged in calculating his chances of being elected a tenant. The face of the Spaniard, more intelligent in expression, but with deeper set fanatical lines, became overshadowed with fitful gloom as he intently watched the ominous proceedings of the cortege. Even the "gods" became silent with awe, in the presence of these foreboding evidences of the end of mortality, involuntarily crossing their foreheads and breasts with their fingers, as if to exorcise inevitable fate. With the sound of the trumpets for the charge, the perturbation of Sir O'Ham Ill Tong caused him to lose all control over himself and spear, the latter in its fall to rest encountered the beadle's head. The blow was as free from the attaint of misdirection as it would have been in the most skillful hands with the intention of producing the prostrate result which it accomplished with the beadle. Although reduced to an attitude of supplication, the victim of this mishap gave voice to language widely at variance with the formulistic words used in Catholic prayer. The beadle's uncontrollable anger, and the increased confusion of the champion, were of that humorous cast that renders mirth irresistible, even under the impression of imposing solemnities, and the restraining presence of those high in worshipful authority. In the present instance, the pope was constrained to turn his head aside, but his jowls shook and vibrated with jellied throbs that absolved even the ladies, in the opposite pavilion, from the painful effort that would have been required to conceal their mirthful emotions. The contagious effect of boisterous laughter would have made irresistible headway with the democratic elements of the godhead, but for the timely forethought of the heralds, and the mettle some impatience of the roaches, which seconded with spirit the trumpet's sound to charge. This the disgraced beadle, with aching head, encouraged with a smart kick applied to a sensitive part of O'Ham's steed, causing it to start in its career with an impetus that nearly unseated the worthy knight. Losing his stirrups at the start the champion of Cardinal Buenaventura was obliged to exercise all the presence of mind within his reach to restrain his fiery roach from the dreaded encounter over the sable barrier; but fed at the table of the Giga pontiff, restraint was impossible. Notwithstanding the mad career of the highfed roach, the knight with instinctive bravery clutched his spear, and would have directed it with his usual skill to the barret bars of Don Bacalao's visor (recorded from his after confession) had not the doomed (by after intention) knight anticipated delayed design by the then present achievement of the same purpose with successful effect; his spear's point catching in the slits of O'Ham's visor he was hurled from his saddle a perjured man. His dangling spear having wounded one of the judicial brotherhood that was added to the sum of his day's disgrace. Hisses, and contemptuous words of scorn saluted the knight's highly sensitive instincts from the mouths of high and low degree when raised from his grovelings in the dust. To his discomfiture was added the pains and penalties of knightly disgrace that condemned his memory to infamy. Stripped of his armor by the beadles and grooms, his spurs were hacked from the heels of his brogans. The tribune informs me, that the name of the shoe is derived from the resemblance of its creak to the Hibernian's brogue. His roach steed, and squire, participating in their master's disgrace, were stripped of their housings, but the former, apparently less sensitive in the appreciation of his perjured condition, shook his blattidial wings with relieved satisfaction, in lively contrast with the crooning wirra of his companions. Then the three unfortunates were reduced as nearly to a state of nature as the regulations of modesty permitted; the knight and squire were haltered and placed upon the roach in a reversed position and led from the arena amid the jeers of the multitude. The heartless lack of pitying sympathy shown to the knight and his companions in misfortune, aroused in onr breasts emotions akin to indignation, but we were again warned that we must not be prodigal with our kindly instincts unless we wished to bankrupt them in utter disgust. Finding ourselves constantly astray with the kindly yearnings of our inexperience in the ways of worldly human instinct, we resolved to set aside the integrity of our home impressions and abide by the direction of the Coliseo's example.

When the judicial ordeal closed, his holiness and consistorials left the lists, that the people might enjoy their more profane amusements free from the embarrassment imposed by their sacred presence, Cardinal Buenaventura showing especial chagrin in being obliged to acknowledge that effectual grace was not sufficient. The only representatives of the priesthood remaining in the lists, who were not in the actual charge of souls, were abbots and priors, with their canons of inferior calibre, none of the Episcopal diocesans venturing to remain, although many cast longing and critical glances to the appointments and bearings of the knights through whose arrayed ranks they filed out of the lists on their way to the vatican. The judicial brotherhood with their dismal bar to joviality, and prisoners, disappeared within the gates of their sanctuary.

The relief from the combined influence of the departed became immediately manifest in the bantering freedom of all the gradations of the assemblage; the clergy leading in the display of wit and gallantry, but in a manner that frequently trenched upon the laity's interested sense of propriety. *The Joust*, from the delay occasioned by the judicial combat, was now hastened, and the two processions were but shortly housed before the first two lances were splintered. Count San Pietro Marceroni, Captain of the Papal Guard, was the Italian's Mouthpat champion, and had succeeded in gaining a slight advantage in bearing off the cockscomb that surmounted the helmet of Count Saint Poll de Parrote, a French knight of great renown, as an adept in the chivalric accomplishments of the age, as well as in the gastronomic art, having been elected by the Pope chief of cuisine, esteemed the most important office within his gift. Between the two knights there had been a warm rivalry for the favors of Princessa Idolisima Canonica, a niece of the pope, after his election to the Papal chair. As upon the result of the encounter the pope's favor depended, the friends of Count Parrote, confident in his successful prowess, had presented him the helmet he wore as a presage of victory. This was surmounted with a cock's head, neck, wings, tail, and comb, all in exultant elevation, as if sounding the highest octave in the clarion notes of victory. His countrymen, with still greater assurance, had caused the artist to execute a wreath composed of hearts, spear-heads, and trefoil, worked in gold; this circled and was attached to the cock's crest. When this was borne off by the well directed aim of Count Marceroni's spear point, I will acknowledge that I felt an instinctive thrill of elation. Still if his signal skill had resulted otherwise than in a bloodless victory, notwithstanding the flippant presumption of Count Parrote, I am assured that I should have felt a tremor of dismayed horror. In sustaining the reputation he had won Count Marceroni successively overthrew the stalwart son of Baron Biermywitzs, a young knight with an excellent German reputation for capacity. Baron Brainoff, a Moscovite with a heavy animal cast of countenance, theoretically accomplished in the Tartar tactics of thrust, run, and come again, which he displayed; but the arena was too contracted for their successful evolution, for he was overtaken and overthrown, his roach participating in his fall. His last encounter was with Baron von Wolfenstein, an Austrian knight with a powerful phlegmatic physique of vis inertia. At the start he placed his lance in rest over the carapace of his roach with the hopeful expectation of finding his opponent impaled thereon after the course had been run. When, from a defect in his calculation of his foe's condescension, he found himself a prostrate companion with his steed, he did not exhibit in his movements the least sympton of chagrin, although saluted with the derisive jests of the democratic plebs, who appeared to enjoy with intense satisfaction the privilege of jousting their superiors with revilings, when temporarily reduced to their own groveling level. After a leisurely survey of his situation he slowly, by easy stages, regained his upright position without the proffered aid of the beadles; then with sluggish movements unlaced his beaver, and raising his visor, showed a face wreathed in torpid smiles that seemed to have had their rise from the recognition of the scoffing taunts as plaudits for the execution of some knightly achievement that had escaped his memory. This innocent act of mazy stupidity was taken by the spectators as a witty assumption in burlesque of victory, and vividly impressed with his supposed humorous aptness they made the welkin echo with their shouts of applause, until he had bowed himself out of the lists. Count Marceroni, by this democratic misapprehension of cause and effect, was completely robbed of the merited zest due to his adroit exhibition of skill in the precise use of weapons, and the Austrian Baron, the least worthy of his antagonists, was established as the prime favorite of the worshipful gods.

Turning to the tribune a look of inquiry, I observed with surprise that his face, for the first time, was suffused with a pleased expression. In explanation, he said that the scene I had just witnesssed was a truthful exposition of the source from which a majority of the Gigas and Animalculans derived their reputations for wisdom, wit, and invention; in truth, he continued, musingly, you can take it as a fair demonstration of the substance of their living realities. Embracing the opportunity, he again advised us not to let our Manatitlan natures interpose their sensitive perceptions for the gratuitous bestowal of praise or pity, as they would serve to mar the relish of comical effects afforded by the hap-hazard novelties in store from the heterogeneous imitations of Giga habits and customs.

While the "fun" provoked by the suppositious humor of the Austrian baron, Wolfenstein, was at its height, the heralds announced the melée. The arena, in this promiscuous scene of antagonism, was so completely filled with knights and their roach steeds that the space between the challengers and challenged only afforded a limited movement, short spears having been substituted for the longer and more cumbersome couch-lance, but the favorite weapons were clubs, swords, and battle-axes. As the combatants were so nearly in contact with each other when unroached, they discarded all weapons save the short dagger, and it was fearful to behold the fierce blows dealt with it, for instead of men the combatants appeared like enraged beasts who used knives, and other weapons, in the place of claws. But as the horrible scene progressed, we were often obliged to turn our faces aside from dizzy faintness. The roaches imitating the fierce example of their riders became infuriate, seizing with their mandibles opposing legs and antennæ, so that many were disabled by mutilating attacks both in front and rear, for in the melée they showed as little regard for honorable usage of the parts exposed as their riders. It was gratifying to see that a majority of the ladies' heads were bowed down, with their hands tightly pressed upon their ears, and it was long after all the evidences of the bloody fray had been removed that they again assumed an upright position. This evidence of sensitive repugnance, on the part of women personally known to them, produced a congratulatory revulsion with the Coliseos, who hailed it as a favorable omen, with the hope that its influence had turned aside the fanatical intentions that meditated the sacrifice of the kidnapped tits upon the altars of superstition. In the enthusiasm of the moment, my mentor exclaimed: "I firmly believe that there is not a Giga or Animalculan mother in Italy, who, in freedom from bias, would refuse the privilege of having their children educated in accordance with the Manatitlan system!"

After the sand had been renewed in the lists to cover the unsightly stains, a light and joyous rondalen was sounded, and the commingled challengers and challenged were seen issuing from their tents in the wooded bosque, and mounting fresh roaches, newly caparisoned with gay trappings, they again entered the lists. When the roachavalcade reached the barrier vestibule, the heralds proclaimed: "Count San Pietro Marceroni, the elected king champion of the lists, in whom is vested absolute authority, and honors that will remain inviolate, until in an approved tourney a worthy successor can be found. Salute your champion king!" The trumpets then sounded the crowning tan-tarra, while the count, assisted by twelve knights-of-honor, dismounted and was invested with the kingly robes of love and honor, then kneeling upon a richly wrought carpet he was crowned by the master of ceremonies with the consecrated chaplet of Mars. As he rose the vast assemblage saluted him with the exclamations: " Long live the noble Count San Pietro Marceroni, king champion of the lists!" the mouthed concussions of boisterous adulation causing the canopies to vibrate with the echoing sound of noisy and noisome breath.

Again the braying of trumpets and clash of cymbals overpowered the tumultuous shouts of the multitude, and when sufficiently hushed for a single voice to be heard, the grand master gave utterance in proclamation to the maxim: "It is not good for man to rule alone! And, in obedience to well approved custom, and the chivalric gallantry of brave hearts, it hath been adjudged that after his own installation the king champion shall, from his own choice, elect a queen of love and beauty to preside in public at feast and festival during the term of his regality. In accordance with this imperative decree our salutations await his choice!"

Mounting his richly caparisoned blattidean roach with a graceful vault, Count Marceroni caused him to execute a variety of changes in pace of the most difficult posing gaits imaginable; and, while passing from a demi-amble to a demi-cavolt, raised from its cushion in the hands of his squire the chaplet designed for his queen upon the point of his lance, allowing it to rest in loop upon its cross-check. This adroit feat was greeted with shouts of applause, which stimulated him to enact a great variety of euphuistic attitudes, — a species of pantomimic male coquetry then much in vogue with the Giga nobility, — which the tribune said would express, with the scenes of the day, how ritualistically void in enactment, the exemplar and imitator were for the real attainment of happy impressions. There was certainly for the instinctive worshipper, much to admire in the studied adroitness of the count's acquirements, as they bespoke disciplined perseverance; but when I reflected that his manly capabilities had been expended upon frivolous expertness to grace a brutal pastime, I found cause for self-reprobation for the sympathy of my admiration. Of the count's predilection all seemed to be aware, but, withal, held their breathing subdued with expectation to gather force for a new outburst of applause when his acceptance was confirmed. Arriving in front of the Princessa Idolisima's pavilion, by the longest circuit, with doublings, that the effect of his euphuistic mimicries might be fully appreciated, he with an obeisance that enveloped the caput-shield of his roach with the plumes of his casque, petitioned with his voice; "Will the most noble Princessa Idolisima Canonica deign to share the brief rule of her humble subject, Count San Pietro Marceroni, as the queen of love, beauty, and harmony?"

With a blush of equivocal import, she nodded her acceptance. Then, while her eyes were yet concealed under the latticed veil of their long, silken lashes, the count, with infinite grace, conveyed the coronet with a "*coup de lance*" to its proper position on her head without disturbing a ringlet with its steel. Plaudits in salvos rose in deafening succession with the successful accomplishment of this dexterous feat, which bespoke long and patient practice with a living model. The princessa represents through ———, her uncle, the pope, one of the original Mouthpat families of Greenpat, with blood variations from European ingraft, which had produced a remarkable improvement, by blending and mellowing with distance the animal traits into an expression, that might with propriety be termed instinctive idealization. Her eyes are black with a shadowy impression of azure softening the snaky brilliancy of the Italian into a sympathetic hue, free from the luric reflection of vengeful passions. When free from exciting emotions, they beam forth an expression of calm dignity, above the sway of lust, vanity, and envy, that gave birth to the cruel frivolities of the day, which she evidently patronizes from compulsion. The composition, in outline, of her other features, corresponds in expression with her eyes, bespeaking the struggling emergence of her thoughts from sensual control, although, from relative association, obliged to conform to the usages that hold ruling sway. But, withal, there is apparent to our eyes an underlying consciousness of purity and goodness, with her responsibility for their cultivation and preservation for transmission. We were so strongly impressed with her longing desire for more substantial attainments than the illusive vanities with which she was surrounded, that we could not refrain from giving voice to our thoughts, which brought a succession of grateful flushes to the tribune's face, foretokening an interest of deeper import than our observation had previously dectected. Frankly acknowledging the predisposition his emotions indicated he begged to be excused from an explanation while subject to so many counter enactments. A request that delicacy conceded without voiced assurance.

While the northern portion of the lists was in the process of transformation into a sectional amphitheatre for the exhibition of an ant and tarantula fight, — which unfortunately our position overlooked, — Count Marceroni led his queen with her maids of honor into a pavilion erected for her reception at the extreme southern portion of the arena, which in construction had received his special supervision. When this, after occupation, was exposed to view, the splendor of its appointments in tinsel adornment received the prayerful adoration of the assemblage. Seated in front of the pavilion on its extended platform, just without the shadow of its dais, were two celebrated improvisorial troubadours of Provence, a small kingdom in the south of France. These were to contend in verse and song for prizes to be awarded by the queen of beauty, love, and harmony, to the adjudged successful superiority of the competitors in their varied styles of adventurous composition. A laurel wreath, or chaplet, had been prepared for the queen's crowning disposal, with the title of laureate; the recipient holding with the emblematic token the conferred privilege of supremacy in his vocation through all the Animalculan courts of Christendom, until a greater star should appear in the musical firmament, for his eclipse, under the like sanction of infallible approval. When the congregation were fully separated for the suitable enjoyment of their tastes in the contrasted extremities of the lists, — the northern having the relative preponderance of eight to one of the southern audience, — a screen was drawn to intercept the boisterous freedom of the borealian sphere.

When the arrangements were fully completed, a maiden herald pronounced in silvery tones, "Alheu!" Then Penny Song with tuneful rebeck essayed his overture with reverent eyes upturned in conceptive supplication to the ruling goddess of love and harmony. To the symphony of murmuring strings in soft prelude to a rippling vision from the fountain of musical poesy, he began his lay. Gradually, with the accession of tributary streams, the current tones increased from the smooth still voiced streamlet, until it merged into the rapid flow of the pebbly brook, combining in sound cascade variations with rippling flow, over shallow inequalities into the bubbling depths of a pool, and from thence onward with accumulating force into the deep volume of the river's tide. His voice in harmonious movement, tremulous with emotions reinvoked from passionate recitative thought, rehearsed in song the Songs of Solomon. Floating with the gilded wings of sensual taste, he fluttered over the luscious lisp of lips united in their sip of virgin nectar from the fount of instinctive love; at first chaste and free from guile in the sparkling depths of pure desire,—but, anon, overflowing in contemplation of the wide expanse of flowery meads, until satiated with its profligate deviation from the legitimate boundaries of its course, he discovers with lamentation the muddy defilement that exceeds his hopes of purification. The song of the bard was wildly descriptive of the longings of instinctive desire, until the hero had embraced with his wisdom the fairest and loveliest daughters of Israel; then imbecile with sensual gratification, he utters in his senility the wailing cry: "Vanity, all is vanity!" and is gathered to his father's dust. The Princessa Idolisima, when the theme became expressive in sensual development, perused attentively the countenances of all within reach of her eyes, but seeing how eagerly engrossed they were with the voluptuous portrayals of the poet bard, she with her selected companions bowed their heads upon their hands with a sorrowful blush of shame. This mood the bard observing, with the cause, introduced into his lamentations strong reproof of himself for the selection of a subject so replete with infamous suggestion. His sarcasm eloquent in self-condemnation aroused the lascivious from their enraptured fancies, causing them, from shame, to break the spell; and when they saw the princessa and her maids with their heads bowed down, many from self-conviction gave evidence that within themselves they felt the cause. Among these was Count San Pietro Marceroni. The bard approved himself to be in the possession of extraordinary imitative powers, for in his recitative picture he held his hearers, of subject kind, entranced and as quickly aroused them with the pungency of his addenda to the dirged lamentations of instinctive wisdom, which would have startled the old hero to a more energetic renunciation of his follies than the vapid exclamation, that his experience was vanity and vexation of spirit.

The prelude of his competitor, Long Bow, was far more primitive in its characteristic portrayal, being timed to irregular cadences of barbaric import. The plaintive melody evoked by his touch from a stringed gourd was in consonance with the strains of his song-theme, which depicted the wild erratic love of savage life, but with far less instinctive merit in recitative description than the love ditties of our Betongese neighbors. At times the savage devotion of the lovers seemed to rise above the animal instincts of sense, but only to sink deeper into the mythical regions of hopeful doubt and despair. The butt at which Long Bow aimed was a rhythmic measure adapted to savage nomenclature and the improvised harmony of his gourd.

At the conclusion of his recitative the herald

maiden called: "Give ear, lovers of the gay and joyous science,—attend! while the queen of love, beauty, and harmony approves with her commendation and award the happy victor in the joust of improvised song?" The count champion offered his hand to assist Idolisima in her descent to the balcony erected for the bestowal of awards, but his bearing was abashed with the conscious self-reflection of unworthiness. Idolisima, the queen, without assistance stepped self-reliantly forward to her position, and then, without ceremonious prelude, beckoned Long Bow to advance. Approaching he knelt in accordance with the prescribed euphuistic forms of chivalric obeisance in the presence of royalty. Placing the chaplet upon his brow she bid him rise. Then calling Penny Song forward she presented him with a zithern, the prize for the victor in song. The assemblage, disappointed in her award of the laureate's chaplet, were appeased by the more substantial gift of the zithern to their favorite, recognizing in its material worth a preference for their choice. This material appeal to their senses opened the only avenue to their appreciation, but their partial murmurs of approval were stilled by a look and disdainful wave of the hand. When subjected to silence, she addressed Long Bow in terms adapted to the usage of chivalry, although manifestly at variance with expectation in expression:—

"Sir minstrel, we have esteemed your claims to the title of laureate as best sustained in the court of fair ladies; inasmuch as you have been pleased to acknowledge with our presence a comparative respect for purity, by which, in fact, you have honored the worth of your own, as well as the common name of mother. By endeavoring to exalt with your muse the instinctive purity and fidelity of your savage heroine, the fair High Water, you have admitted that it is our duty and privilege to excel her untutored example. Still you have paid to ourselves and maternity of our race an equivocal compliment for worth in your far-fetched search for a heroine example of constancy in love and purity. Notwithstanding the implied lack of living examples worthy of imitation for the inspiration of your songful muse, we have preferred your savage precedent to your competitor's glossing embellishment of his hero's vices. With the hope that you may discover, during your adventurous residence in Rome, an Animalculan or Giga representative that will honor your mother's sex with worth sufficient to inspire your commendation in song, we would urge the undertaking as one deserving your earnest attention. In the event of success it will save your muse many weary pilgrimages into deserts and the waste places of earth for the achievement of consolation from savage example. That the emprise may not lack direction I will recommend you to visit the Manatitlan colony of Coliseo. For with them, when convinced of your sincerity, you will find in woman an exaltation of loving affection that will by far exceed your highest conceptions."

She would have extended her reproof, but as she was in the act of addressing Penny Song, her commendation of the Manatitlan colonists brought such an overwhelming "answering sibillation" in denouncement that her brothers, in fear of an outbreak, hurried her away. The minstrels during the excitement approached near enough for whispered communication, the purport of which brought a smile of gratification to her face. They then as a diversion commenced a roundelay of spirited movement, uniting their voices in bluff concert to the sound of the zithernas. As with enraged wild beasts when surprised with the sound of music, the factious instincts of the spectators were stilled. The Coliseos were rejoiced at the steadfast adherence of Idolisima to the inculcations of her Manatitlan education, for the tribune now informed us that she had served her full term in the Coliseo school, promising to relate at a convenient season the cause of her apparent defection.

Although fearfully repugnant to our feelings, we turned our attention to the democratic amusements in the northern sectional arena, within the vestibule of the barriers; as you cautioned us of the necessity of seeing for ourselves all the enactments that we recorded for advisorial judgment. The "pit," as it was aptly termed, was a miniature imitation of the large arena of the lists; but the contestants for the honors of mutilation were ants of the white, red, and black species, and their antagonists were tarantulas distinguished by the same order and variation of colors. But with naturally a higher object in view than their human Animalculan and Giga exemplars, and controllers of the lists, they fought without the artificial weapons that man's superior intelligence has devised for kindred destruction, in direct controversion of the manifest designs of Creative intention. The instinctive impressions of a blind man could have detected without the aid of thought, the tangible distinctions that had furnished the motive attractions for the northern and southern divisions of divertisement predilections. The ants and the spiders engaged in mutilating encounter, were to us, in reality, more attractive objects to contemplate in their active ferocity, than their human spectator instigators, who combined with controlling intelligence the visual evidences of the most abject passions within sphere of animal and its reptile grade of expression. The patrons of the southern pavilion had been held in dalliance with the soft pleasures of amorous sensuality, which unopposed called forth the highest expression of instinctive refinement; but the current tendencies from multifarious indulgence were to the fall and deep whirlpool abyss of passion, hate, and deadly revenge.

The colored specie distinctions of the ants and spiders appeared to be as great a source of inveterate hatred between its caste representatives, as the antagonisms of race in form. This necessitated the separation of the white from the red, and the red from the black ants, with a like disposal of the tarantulas from specie representatives of opposing colors. The first encounter that we witnessed was between white and red ants; the former being universally successful with the odds of nine to one against them, and with the blacks twenty to one. The cause of this concentration of multiplied ability in the white, to cope with red and black antagonism, became soon apparent from their studied aforethought for the serried separation and disposal of their foes before they could effect united opposition. The deadly devices of the whites to lure their foes, with the stupid blindness of the red and black ants in accepting the proffered baits without consideration, afforded a relative study for detecting the cause of the instinctive superiority that rules with white humanity. When the whites in an encounter against overwhelming odds had gained decisive advantage, the cathedral bells of Saint Peter sounded from the chief dove-cot beneath our place of concealment; its funeral peals announced a new phase in the strange enactments of the day, which had appeared to us so perverse as sources of amusement that we were at no loss to discover the Giga creed origin of hell. Fearful that the long-waited-for crisis was approaching, our eyes were upon the alert.

As the yellow tints from the sun's fading rays grew hazy with the twilight heraldings of darkness, the folding doors of a large building of ominous appearance opened, and from its portals issued a procession of Dominican friars. In the midst of these inquisitorial representatives of the human sacrificial order, were the unfortunate tits supported by familiars of the Holy Office, with their limbs dragging helplessly upon the pavement, and their heads half revolving on their shoulders without the power of muscular control. The sight aroused the indignation of the tribune and his companions, who with difficulty restrained us from extending to them our immediate aid and sympathy. This was no slight undertaking on their part, for we were unused to a knowledge of the existence of such heartlessness, and would have adventured their deliverance if the odds had been an hundred fold greater; but we finally yielded to the wisdom of their discretion.

On the arrival of the sad cortege at the knight's encampment they joined the procession escort of the doomed, among which were Sir O'Ham Ill Tong, and his squire, Kan Avan, who had been evidently condemned to the stake. As the gates of the northern barrier were obstructed by the occupation of the vestibule arena, in which the ants and spiders were engaged in an instinctive joust, the inquisitors in deference to the multitude awaited the decision of the contest; the latter, with reverence, giving place for them to witness the result of the sportive entertainment, as a prelude to their premeditated enactments. At the conclusion of the melée fight of the ants, in which the white champions had proved victorious, a large, whitish gray tarantula spider had been "pitted" against five mottled white ants, both parties having been subjected to a training abstinence from food to increase their ferocity. For a time the battle had been equally sustained, but a bold venture cost two of the ants their lives; in the feat the tarantula was maimed with the partial rupture of a carapace and the loss of the distal joints of two legs. Two of the remaining ants having a firm mandibular hold on the round carapace leg, the greatest excitement prevailed among the spectators, wagers being bantered between high and low; all distinctions of caste, for the time being, were banished by the talismanic greed for gold and silver. It soon became evident the auto-dafe rites would be deferred until the result of the antfight was determined, for the order of procession was broken, the familiars with their human burdens crowding to gain a favorable position to witness the closing struggle. While the ants gnawed and swayed in hold upon the tarantula's leg, he struggled to bring them within reach of his mandibles. The partial success of the assailants and assailed was greeted with vociferous shouts by the wager partisans, unabashed by the presence of the inquisitorial brotherhood and archiepiscopal dignitaries of the church. The tribune sadly reminded us, that the brutality of the scene was the tested ægis of democratic equality based upon individual or associate physical strength adroitly used, — chance advantages, or in massed subjection to the rule of selfish partisanship, as with ants in attacking the spider whose habits and interests in life interfered in no way with their own. But as you perceive, the hinds are so stupid in the dullness of their perceptions, that with the clear analogy of the cause and result of their own condition, in demonstration from the arbitrary power of physical strength that captured and forced the ants and spider into antagonistic collision, they fail to detect in the caste orders of priest-craft and knighthood the same motor influence in mental combination.

With our attention called so directly to the resemblance, — from the pitiable condition of the poor tit victims, who had been tortured into the vestibule of death, and were awaiting the ordeal of fire to recall from their bodies the vital spark, — we could not fail to admire the comprehensive quickness of our Coliseo cousins' detective acumen for analyzing the relations and gradations of instinctive responsibility. Indeed, it would be impossible for you to imagine a more discordant or disgusting scene, or a baser use of language than these bloodthirsty humans used at every disjuncture effected by the combatants. That you may obtain an inkling of this pandemonic scene of instinctive madness, I will adventure a sketch.

Prominent among the representatives of foreign nations was a Scotch abbot of the abbey of Saint Maythedielscratch-me on the Tweed. In person and language he was tall and lank. Although engaged in "amusement" with an English prior, the incumbent of Goddamnmee on the Tyne, just over the border, they "staked" fearful invectives with their metallic coin. But the English canon was evidently no favorite, and required all the bluffness he had attained by a long course of dominant rule over a submissive brotherhood, as well as a full purse of tithes, for he was bantered with wagers on every side. But with bellowing disdain in the short and curt Durham dialect, he was ready with horned alternatives to oppose all comers. Chief among his secondary challengers, with coined words, was the abbe of Mortdieu, who mingled frothy gibes with his wager temptations. The ants were at length vanquished, to the great chagrin of the English abbot's debtors, who tried all sorts of evasions to escape payment. Oncleslydenbet, the German prior of the priory of Shufflehausen, repudiated the payment of his forfeited wager in the concise terms, "Not a bit of it!" The Irish abbot of Fivewounds said, "To be sure I have bet and lost, but it was the understanding that I should never pay unless I won!" The Dutch incumbent of Dunderandblixen, the Welsh of Sweatmyleeks, and French paid theirs; the first with fearful imprecations and sputterings, the second with an abundant flow of perspiration, and the last with strings of sacres.

But for your injunction, I should have withheld the incidents of a scene so horribly repulsive in its ferocious combinations of speaking inhumanity. Especially as it implies distracted attention from the wretched condition of the dislocated tits; but in direction we submitted to the ulterior intention of the Coliseos. The tribune, whose aid you will recognize, said, that you required an exact index of the real condition of the Giga and Animalculan races under the instinctive rule of the stomach and senses, that you might compare the evidences with those of the first falcon era, to judge of the cycle changes. The task, with his assistance, is by no means pleasant, as with increasing familiarity the scenes of brutality become more repulsive, so that you need have no fears that they will affect us otherwise than as a ferment to clarify the instincts of our bodies from corrupt humors and indwelling passions. Much time was exhausted in wrangling before the ant circus was removed to admit the papal and inquisitorial trains into the lists. The arrival of the former, while the battle of the ants and spider was pending, was exceedingly fortunate for Cardinal Buenaventura, as by wagering with the pope an equivalent, with the spider for his champion, he was exorcised of his perjury, and by remitting to his holiness certain pecuniary obligations, received absolution to date. Our sympathy for the tarantula's loss of his carapace legs, with the extreme division of the fourth, was not strongly excited, as in excess of his pope contemporary's alleged stigmatic attributes of infliction and redemption, from instinctive woes, he in reality possesses the "miraculous" power of material reproduction of lost limbs.

The arena of the lists had been rearranged for the final enactment, and furnished with stakes and faggots for the sacrificial ordeal of fire, as a purgatorial agent for the purification of the body from heretical sins. These altars were raised between the two pavilions; and the ladies', I am obliged with sadness to relate, was filled with representatives of the sex, yet they were of a class whose gratifications were solely dependent upon the organs of sense. During the converting labors of the confessors with the San Benitos the inquisitoral monks, dignitaries of the church, and knights, chanted a grand high mass to ritualistic ceremonial accompaniments. This was pronounced to be more affecting, by the occupants of the ladies pavilion, from the sobbing wails and lamentations of the doomed Irish knight, Sir O'Ham Ill Tong, and his squire, Kan Avan, who unceasingly proclaimed their innocence from perjury after they were tied to the stakes, thus adding contumacy to the ordeal and decisions of the judges. As their whinings disgraced the order of knighthood they were gagged with clouts.

When these preliminary preparations were nearly completed, the tribune gave the longed for signal of rescue. In a twinkling our Manatitlans were in the midst of the inquisitors and ecclesiastics, upsetting them in a confused pile with the knights. We liberated the tits by the time the Coliseo aids arrived, who placed the rescued in the arms of their relatives. The panic we created could not have been greater if we had been gigas instead of giantescoes. The women, who had looked calmly on the preparations for human sacrifice by the tortures of a slow fire, screamed and fainted at our appearance, their attendant knights surrendering without opposition. The pope and cardinals were in as much fear as the unsanctified until the tribune made known the source from whence we came. Then relieved from his apprehensions of bodily danger, he commenced crossing himself furiously, muttering the while anathemas against the Coliseos and their followers for the sacrilegious invasion, and interference with the sacred offices of the church. Counting upon our intimidation through the enactment of prestigeous mummeries, he would have left with his satellites; but the tribune detained him by laying his hand upon his shoulder with an inflection that made the burly little head of the church understand that it had forced a crisis, from which mumbling ritual cant could not extricate it without the assurance of a strict regard for the personal independence of all Animalculans from religious restrictions. We will give you the terms of the tribune's injunction: —

"Now Canonicus, that there may be no more acts of treachery on your part, or through the instigations of those subject to your direction, I shall dictate to you terms in behalf of our Manatitlan colonies that you must keep inviolate, for if they are infringed upon in the least degree we shall hold you and your associates personally responsible. We have heretofore offered you good will, but you have taken advantage of it to impose upon our communities and inflict personal injuries upon our adherents; showing that you are alike destitute of gratitude, and the disposition that inclines to honorable and just reciprocation. In the first place, you are to respect the privileges of all claiming a desire for our protection, whether an Animalculan of Mouthpat birth, or of other nationality. In the second place we shall hold you personally responsible for any acts of cruelty perpetrated by Animalculans under your control upon animal or insect; and for the honor of instinctive humanity we especially interdict all barbarous amusements between man and man. Fortunately, communication has again been opened with our motherland, and by the timely arrival of our cousins, we have been able to avail ourselves of their counsel when most needed. From the enactments of to-day, which have borne witness to the steadfast goodness of your daughter, Idolisima, in her adherence to the inculcations of her Coliseo education, we shall hereafter require that all the Animalculan children of Rome shall from henceforth be submitted to our censorial and educational charge. We are aware that in assuming the charge we undertake an immense and arduous responsibility from the fecundity of your animal propensities. But in time our inauguration will effect the substitution of improved quality for quantity with an affectionate compensation that with cultivation and reciprocation you have the means of realizing in fatherly association with your daughter."

Here he was interrupted by Sir O'Ham Ill Tong, — from whose mouth the clout had been removed, — with the ejaculation : " An sure, by our holy mother, what your honor is plased to say is all gospel truth, and by the same token, if there is any more, I believe it on my shoul! But if your honor will be so obbleeging as to order one of your giants to remove this gorget that binds my throat to the stake, and set me free, if yees have closed what ye are going to say, I can hear the rest more comfortably, and ye'll save from fire as good a Christian as ever was burnt for pargary, — an for the life of me, I can't just see how that can be, for I niver spoke a blissed word in the matter at all ! "

This characteristic wordy appeal of the Irishman was complied with on the moment, and when released he fell on his knees at the feet of the tribune, invoking upon his head the blessings of all the saints in the Irish calendar, offering to serve him to the end of his days, on foot, or on roachback, for nothing at all save fair wages, hursts, parquisites, pickings, and the run of the kitchen, and sich likes! The ludicrous fall, in the use of words, from knightly inflation caused a smile notwithstanding the embarrassments imposed by our interruption. But the flow of his tongued gratitude to the ascendant party, having reference to future benefits, could with difficulty be stayed. When accomplished he was advised to return with all speed to the place of his birth before a worse evil befell his body than purification by fire.

The pope and his cardinals, while subject to the tribune's reprehensions, became as servile in their bearing as they had before been arrogant. The pope's offer of entertainment for the night we could not accept, after witnessing the barbarous proceedings of the day, as we preferred the protection of our silicoth tapas under the cornice to the lair hospitality of human wolves. After the crowd dispersed a messenger was despatched to apprise the Coliseans of the successful result of our enterprise, with the desire that a falcon might be sent for us at early dawn. In accordance with our request a falcon appeared with the earliest light on the following morning, and on alighting, to our great surprise, Idolisima Canonica, in company with the tribune's parents and other representatives of Coliseo to the howdah's full capacity for entertainment, descended to receive our salutations. It appeared, from the relation of the tribune's sister, that Idolisima after leaving the lists had sought protection under the escort of Penny Song and Long Bow, at the coliseum, well knowing that the ties of daughter and sister would not save her from the dungeons of the Inquisition, recently built by the pope, her uncle. From the greeting she gave the tribune, it required no prompting of speech to apprise us that a stronger motive than fear of the dungeons caused her to seek an asylum with her foster guardians of childhood.

Without teasing our curious interest in her welfare she appealed to the tribune: " Your Manatitlan cousins, Novuotus, would scarcely feel satisfied with your interest in my behalf, if I should allow you to hold me silently exculpated from your abiding confidence in the wisdom that dictated our betrothal. As we are united, yet single, in the fulfillment of our marital alliance, I am anxiously desirous] of having the affectionate sanction of their approbation, which could not be conceded if they thought me capable of participating otherwise than by form, in the cruel enactments of yesterday. In speaking, it is my wish to vindicate myself from voluntary conformity with the chivalric usages in which I engaged, as they were then, as they ever have been, sadly repugnant to my conceptions of love and affection, which vividly impress me with the reality of an existence independent of the body's mortality. That your cousins may realize in what measure they are entitled to bestow upon me their lenity and sympathy, I will relate to them the cause that influenced my introduction to the guardianship of your people. Of my ancestral origin, I am certain that the events of yesterday will better inform you than I am able from my own knowledge. In my fourth year, when my perceptions were only advanced in the bud sufficient for impression, without enlisting the maturing aid of thoughtful judgment derived from comparison, my father, mother, and two brothers, became very sick with the plague, alike destructive to the Giga and Animalculan races. The harrowing scenes of bereavement daily recounted in my presence aroused impressions of fear for myself, and when my parents became sick, I added anxiety to my aunt's responsibility, to whose sole charge we had been left by the servants and neighbors. One day, despairing and hopeless, my aunt had enfolded me in her arms, and while wofully lamenting threatened desolation from the death of my parents and brother, I was trying to solace her with endearments, my arms were around her neck, with my lips to her ear and face turned backward, when the door softly opened and my voice was stilled by the surprise of my eyes at the entrance of a very large and beautiful woman. [Here Idolisima fondly embraced, with carresses, the neck of the Doschessa of Romelia.] My silence and fixed attention attracted my aunt's eyes to the door, and when she discovered the stranger, who appeared to be well known to her, she was much abashed that she had been surprised in useless lamentation. This made me gladly hope that the visitor was an angel sent to help us; so, growing bold with the feeling of hope, I gently withdrew myself from my aunt's embrace and approached the stranger imploring her aid for my parents and brothers! In a moment I was in her arms; then I became so quickly changed, I no longer feared, for her voice and touch made me feel so secure that I was sure she came from heaven to help us. But I have since experienced that she came from a far better place beyond the reach of gold and its selfish attractions. Urging my aunt to take new courage, she went out, still holding me in her arms, and meeting a censor, with others of her people, she returned, and soon our house became quite changed, for they brought with them your clothes that remain clean, so that from being dismal and sad everything looked nice and cheerful. Then mother reviving, she also knew the angel, who asked her if she could take me to her Coliseo home? Oh, how gladly my mother smiled with her whispered yes, so faintly and pleadingly grateful in earnest expression it has ever been present since. Striving to speak, my mother fell asleep without breathing, but still she smiled, until it paled to a shadow in the moonlight, then we were taken to the Coliseo, for my aunt required rest, where we were bathed and dressed with the care of so many sweet faced persons, that we became quite bewildered with love and slept. In the morning we were wakened with sunlight and song, so harmonious with gladness, my aunt touched me to be sure, for we both thought alike. I was placed in school where there were so many that in love we were counted as one, for we were so united in affection that tears of vexation and sadness never flowed. Then as I grew in years, I learned to grow emulous in the personal requirements of purity, and with the care of myself, was able to assist others, for the grateful meed of love. Much more I could say, but I feel that you do not require it. When my father became my uncle, Pope Innocent I., I had reached my eighteenth year, and had been betrothed to Novuotus, and we were enjoying the hospitality of my foster parents, the Dosch and Doschessa of Romelia, during our term of probation, when my father, after his election, demanded that my aunt and myself should be restored to his care. The Dosch, after a long consultation with his advisors, for he was very sorry to part with us, concluded that it was best for us to return. This was a very sad decision for us, but we were assured that if we found ourselves unhappy from being unable to do good, we could come back. We were then taken to my 'uncle's' temporal court, but none were truly glad to see us, for we were strangers, even to my father, for he would not accept the monthly invitation to visit me at the Coliseo school. Yet we were constantly surrounded with a throng of servants and people of condition, so that with them, we were mere puppets, and felt very unclean, for we could do nothing for ourselves. But all were very profuse in the utterance of complimentary formalities, which made us feel like exiles from real affection. The first process to which we were subjected was purification from heresy; in preparation our silicoth garments were removed, and our persons were redressed with woven stuffs, that in their newness smelt animally earthy, and soon became irksome in weight, and rankly unendurable, notwithstanding they had been sprinkled with holy water at the time we were baptized. Then, my aunt was made, or created, queen of Rome, and I was christened Princess Idolisima Canonica by the same process. A few days since my aunt escaped and went back to unite with her Coliseo affections; since than I have been closely guarded, as I had attempted to leave with her, but was foiled by accident. So with my father turned uncle, I became the puppet, or bone and flesh of contention for the strongest and best trained brute in the lists, and was seemingly obliged to conform to the dictated usages;

but with the determination to escape if an opportunity offered. In this relation I have traced my childish impressions in the language I have been accustomed to impart them to my loved school associates; to remodel them to suit my present perceptions would detract from memorial pledges of affection. That you saw me in the lists yesterday, came not through my own volition, as you will concede; but perhaps, from a wise wish on the part of the Dosch and his advisors to test in what I lacked for the power of resisting the wordy flatteries and material vanities to which I would be exposed in the papal court of my uncle. For I had exceeded in years your period of matriculation, as my instinctive impressions are still retained in memory, with the infantile desires perpetuated in assumed motherhood of artificial productions in baby likeness. But if you can only see me as I feel, and have felt, in pity for the barren love of the mothers of Rome, who are content with the pride of hope in the advancement of their children to material possessions, and the empty honors conferred by my uncle, the pope, you must know that with all my imperfections I am, in the current love of purity and goodness, free from my body's instinctive taint. Knowing my earnest yearning for purified worth, in my desire for the welfare of others, you will, I am sure, lend me your aid for the higher attainments achieved in realization by your primitive people for loving perception."

This appeal of Idolisima received the sanction of our affections with a warmth of genial outflow confirmed in baptismal attestation from our eyes. The Doschessa in reciprocating the fond caresses of Idolisima said that she and her aunt had nobly sustained the venture, showing that with woman love's perception, when free from selfishness, is in a measure independent of age, when subject to the constant lead of example. In closing it will be well to state that the union of Novuotus with Idolisima was consummated with her father's consent. The event was celebrated by Penny Song, with an improvization styled the Epoch Ardens, Long Bow assisting with a transposition of High Water to Fair Water. But from their lack of precedental knowledge pertaining to Manatitlan history, and of love independent of the body's instinctive materialism for the bombastic usury of imagination in word portrayal, their poetical ovations would have been deemed tame unaided by their skillful rendition in the well timed measure of song with the instrumental accompaniment of zithern and gourd. This transcript, of an event so varied in its bearings, will afford you a more correct insight into the condition of our Roman colonists, than tracings from historical records, and statistical accounts, of which you will be advised at our earliest leisure.

TITVIEW.

Nota Bene. — Our wives have not been idle, but under the direction of Oluisandria have auramented many of the Giga women of Rome, who with silent example emulate the family of Indegatus.

At the conclusion of Titview's letter, the Dosch stated that the initial voyagers of the second falcon era found the colonists of Constantinople and Jerusalem in a still more prosperous condition than those of Rome, and equally rejoiced in the fulfillment of expectant waiting. But as they presented the same characteristic features of dogmatic rule, with caste variations, having their origin in the degree of control exercised by individuals over their own habits, for the control of the masses, the opposition to colonistic example was in the main the same. "Now, in company of Mr. Welson, I will leave you to meditate upon the examples I have adduced of Giga and Animalculan humanity under the rule of instinctive habits and customs engendered from the unreason of the stomach in its control of the brain, until your letters arrive from the other side of the precedental gulph to strengthen the contrast between your past and present thoughts. In the meantime you can study, with our auramental aid, the habits of professional instinct cultivated by the adjunct members of the corps, in adverse defiance of Heraclean example and our thought substitutions. Although formulistically influenced to habits of outward conformity, neither Dr. Baāhar or the curators of sound, and artist, have changed in thought from precedental routine ; and in their present mood, would relate the events which have transpired as surface matters of fact, for publication, or scientific gossip, and the excitement of wonder and surprise in the gaping multitude who throng with open ears lyceums and public lecture rooms, for drum impressions. Your college and scientific professors can be truthfully likened to your railroad locomotives, which swiftly progress forwards and backwards on their rails, but once off their tracks are helpless, from their own resources, to move themselves or their trains. But with the stage gradations of instinct now open to your view, you can readily discover and test the animus source of present happiness with the realities of its impression as an assurance of immortality."

CHAPTER XX.

AFTER matin song on the morning of the 9th of December, Correliana announced the near approach of couriers with letters. At noon the train arrived having in charge Padre Simon as a special envoy from Captain Greenwood. His convoy brought letters from Europe and Montevideo in answer to those despatched by M. Hollydorf and Mr. Welson, also to the other members of the corps. The padre, although looking jaded with fatigue, was in his usual happy state of mental confusion, to which the affectionate greetings of the Heracleans and Kyronese greatly added, causing him to exclaim: " Well, I truly declare upon the soul of my conscience, I believe you are truly glad to see me!" His mental perplexity was increased by the nature of his mission, which will be gradually developed in progressive relation. Mr. Welson received a large budget of letters, many of which were from his correspondents in Panama, who were anxious to learn the cause of his long delayed return. Those relating to his business engagements were quickly despatched, his interest being more directly enlisted in the perusal of letters from his Montevidean and Buenos Ayrean correspondents. We will first offer the transcript of M. Baudois's letter.

BUENOS AYRES, *November* 19, 182 .

TO SENOR DON GUILLERMO WELSON:
    Heraclea of the Falls, Andean La Plata.

My Dear Sir, — Your letter was duly received, and I can truly say that its contents taxed my credulity to its utmost stretch. The discovery of remnant Latin and Asiatic races was quite sufficient for the rational digestion of wondering admiration! But the revelation of Animalculan representatives of humanity quite staggered and mazed my powers of marvelous conception, until reason had with cool reflection weighed your written evidences of sanity. Then, quite convinced that you did not design to test the extent of my gullibility with the conjurations of your imagination, I found within myself evidences of sustaining approval that confirmed my believing reliance in the sincerity of your imparted happiness. If the extinct Giga animal species are represented by living Animalculan, it appears quite reasonable that the order of continuation should embrace the human. With these deductive considerations my reason became reconciled for the recognition of your Manatitlans as real negotiable representatives of humanity. Again my wondering admiration expanded from your description of their habits and educated power of self control for the exampled reciprocation of good will. It is but natural to conceive that affectionate confidence must result from self government, for purity and goodness are the parents of unselfishness. In comparative degree we have seen that the children of our race, who from necessity have been trained for the exercise of self denial, are reliable and affectionate in contributing to the associate happiness of others. It is impossible for me to find adequate words suited for the expression of the relief afforded from the light that dawned upon my perception through the vista opened for the realization of Creative intention! Dull, indeed, must be the faculties of a person unable to realize, upon the moment, the certain effect that would result from cultivating the germ of goodness for the control of instinctive animality. In deference to the Manatitlans' loving perspicacity, which has enabled them almost innately to discover the impression of design in cultivating from infancy the germ of goodness, my body seemed to shrink with shame from its distended growth of fungus, emblematical of stupidity, under the influence of their imparted intelligence. But with a knowledge of cause my efforts for relief were attended with an instinctive oppression from animality, so apparent in its selfishness, that with my utmost effort I could only obtain partial relief, with the resolve of affording, at most, an imperfect example for the benefit of future generations. Under the presaged reality of the Heracleans' affectionate and enduring sympathy, my hereditary infidelity, begot from the adverse fatuity of sectarian delusions, faded like the Pampa mists before the brightness and genial warmth of the morning sun. Now that the Manatitlan system of education has revived my hopes, I look forward with trustful desire that my life may be prolonged to witness the full inauguration of self-legislation, for time and eternity, in freedom from the proxied impositions of instinctive priestcraft, and politic statesmen.

With regard to the search you wished to have me make for relic information: I have succeeded in unearthing collateral evidence which proves beyond a doubt that many galleys from the interior waters of the eastern continent, connecting through the Mediterranean Sea with the Atlantic Ocean, reached the shores of the La Plata estuary before the advent of our Christian era. At different dates during the last two centuries rumors have been rife attesting to the existence of a walled city inhabited by a white race, situated in an Andean valley that gives rise to the river Vermejo, or its tributary source. The reports, at different periods, caused the Jesuits to make several attempts to negotiate a favorable disposition on the part of the Indians to impart knowledge in verification; but all their overtures were treated with an evasive skill that involved the question of its real existence in greater mystery. Their missionary attempts to penetrate the country of the wild hordes, beyond the river chacas, were opposed with determined and successful hostility. The community of Pompolio has been an acknowledged fact for centuries. The ancestors of your Kyronese were the reputed founders of Mendoza, from which they were driven by the mongrel progenitors of the present inhabitants.

The remains of vessels of undoubted Phœnician, Egyptian, and Roman construction have been found imbedded in the preserving guanic and alluvial deposits that have filled with their accumulations the inlets where they were moored. Of these Don Pedro Garcia will give you a succint written account. The following notice of a relict discovery I have copied in translation from an old number of the "Gazette de Bogota":—

"Voice le passage tel qui le donnent les Nouvelles Annals des Voyages, 1st Tome, page 393, anne 1832. Au village de Dolores duex lieus de Montevideo un plantuer decouvrier une pierre tumuliare des caracteres inconuus. Relevant cette pierre il trouve un caceau de briques renfermant, deux sabres antique bronze, un casque, et un boucler, tres amphoræ par le huile, et une in terre de grand dimensions. Tout ces debris emporter au savant pere Martinez Garcia, il est parvenir a lire la pierre ces mots in caracteres grec. Voir ton Phillipi * * * * *. Alexand fils to Macedon * * *. Vasi epi tes execui * * * *. k * * ty * * * en * to * * top * * Pelatin. Ca est dire completement les mots. Alexander fils de Phillipe etait de Macedon, vers la 63 Olimpiade in ces lieux Petolemie les reste manque. Sur le poignee des epies est grave un portrait que commun etre celui de Alexandre, sur le casque on remarque un circulure representant Achille trainant le cadavre Hector ante de murs Troie. Fait il conclure de cette decouverte de contemporaire de Aristotle a fouile de du sol Bresil. Est il probable que Ptolemie ce clef bein conu de la plotte de Alexandre entraine par un tempte an milieu dece las enciens appelaient les grand mur ait ete jette les cotes du La Plata et y ait marque son passage pax ce memorial monument fait dans trois les ces fort curieux les archeoloques." — *Gazette Universelle de Bogota.*

At the period this relic discovery indicates, the proximate swiftness of the current stream setting to and from the Strait of Gibraltar to the La Plata estuary can be estimated by present calculation, which has been rated at twenty miles an hour during the height of a monsoon gale. This would give a surface speed to a floating object of four or five miles an hour, under direct impetus; and with a due allowance for counter slips from eddies, lack of direction, and other causes, a chance voyage might be accomplished in from twenty to thirty days. With the square sails used in Ptolomic vessels, with shallow prows made to imitate the breasts and necks of water fowls, a swift passage could have been made with intention and a favoring wind; but it was the superstitious custom of the ancient mariners to rely upon the direction of fate. M. Hollydorf or Dr. Baāhar could arrive at a proximate conjecture of the derivative source of the Kyronese from the Syriac root and terminals, and by like tests, if of Beberi or Morisco extraction. The fourth wonder of your Arabian Knights' discoveries is the fact that a savage of the wild hordes can be favorably influenced by exampled goodness, or in any way hold himself amenable to either reason or kindness. Yet I would urge the necessity of impressing them with a knowledge of good and evil, with a detective perception sufficient for your protection against their revengeful instincts, that you may not become the victims of misplaced confidence for the malicious injuries inflicted by our race.

In closing my epistle I will truthfully declare that

I envy the meanest capacity of your party his privilege of contributing to your common fund of enjoyment. How I have longed, waited, and despaired, for the irresistible charm of affectionate sincerity, that betokens in its reach immortality. A note from Don Pedro advised me of his family's desire that I would visit them as a consulting aid in your behalf. You are but too well aware of his past source of disquietude. Greatly to my relief I found him with his two daughters, your little favorites, Lovieta and Lavoca, filled to overflowing with an affectionate appreciation of the manifold resources of your discovery. As our meeting, under the circumstances, was characteristic, I will endeavor to render the mutual impressions of the scene in enactment. In the place of the Teutonic custodian, with the forbidding Cyclopean visage, Don Pedro received me at the open puerta, evidently in waiting expectation of my arrival. Confronting each other at the entrance we stood for a second regarding each the other with the mutual reflection of wistful eyes, until the rising flush of our united emotions bespoke like impressions; then with an impulsive disregard for formalistic dignity, and greatly to the surprise of watchful neighbors, we embraced with the allied warmth of our new role of sympathy. Shrinking from the curious gaze of strangers we entered, Don Pedro surrendering his post to the portress who received from me, for the first time, a courteous greeting, which was returned in kind. In the patio we were met by Lovieta and Lavoca who, with the quick interpretation of infantile affection, discovered the prompted source of our unusual cordiality. Receiving permission from their father they at once commenced to unravel the tale of Don Guillermo's wonderful discoveries, with frequent halting questions for elucidation, and commentaries upon the startling information imparted from your letter to them, which was premised with the confidential acknowledgement:

"Why, M. Baudois, it's queer we love you so much this morning, and were so afraid of you before, which kept us from liking you!" Then for an hour or more they prattled of affection with such a clear perception of the Manatitlan rendering, that I felt acutely the poverty of my own resources. In testimony of affection's inexhaustible attractions, I remained with them three days before the prime object of the visit engaged our attention, and then, with reluctance, we turned our vision back to trace with our material guide-posts the progenitorial evidences of Heraclean advent.

You urge me to visit Heraclea. I can assure you that there is not in the wide range of thought a prospect that would afford me a tithe of the pleasure; but I am ashamed to acknowledge that I feel within me the old leaven of instinctive dislike for the members of the corps, from nationality. For personally, as you are aware, I have no acquaintance with them. From this cause I am so doubtful of self control within that I cannot with sincerity venture the experiment of voluntary association with them, even under Heraclean auspices. My feelings, as a Frenchman, are still instinctively patriotic, notwithstanding the reproof administered in every sentence of your letter. I am well aware of all you would urge in favor of the trial, and my reason sanctions with desire all you could say; and it is with humiliation that I am candidly obliged to avow myself unfit for an association with the German members of the corps, although they have received the prestige of Heraclean adoption. If my disposition would allow me to enact the part of affectionate sympathy, the treacherous disguise of hypocrisy would not shield me from Manatitlan auramentation, with whom I would fain hold myself, in sincerity, worthy, to the extent of my freedom from instinctive disability. However much you may regret my lack of the noble qualities, which allow your unbiased passions the privilege of repose, I am certain that your native Gallic infusion of clannish prejudice, will, even with your new light, appreciate the honesty of my motives. To merit the esteem of the Manatitlans, I will use every possible means for the subjugation of my sectional prejudices; having already been obliged to acknowledge to myself, from the revolting impression still retained of the commune massacres of Paris, that with the murderous spirit in train, we should scarcely have shown as merciful a record as the Germans, if we had been victorious. With better generals as players, in the deadly game of war, the weak movements of our imbeciles were forestalled and checkmated. This causes a furious undercurrent for reprisal, especially as the war was precipitated by a tyrant without the provisionary tact to foresee and provide for the tottering destiny of his throne. Of course, the disastrous results of war reflect from the ruler to the subjects, alike in imbecility and determined intelligence. Our soldiers were driven like sheep to the slaughter. In fact, there was no real cause for the war on either side, and great reason why it should have been firmly opposed on the part of the French people, who have in the game enacted the part of pawns and are paying the penalty. The victorious can afford the dole of generosity to the defeated. This retrospective glance I offer as a specimen of my instinctive forbearance, and until I feel the sincerity of my self-control, I must avoid temptation likely to arouse hatred and revenge, the chief constituents of patriotism. With sincere gratitude for the happiness I have enjoyed from your intuition, you can rest assured of its strong impression in reciprocation. H. BAUDOIS.

N. B. Since the above conclusion of my letter, I have visited and opened a tumulus raised over the remains of a Roman woman. The inscription engraved upon a stone covered by the mound, had been rendered nearly illegible from detrition, but we were enabled to decipher, with an approach to certainty, the following detached words: COR. * * * *, * * * AUGUSTA * *. IB UXOR ⚲ * * * * * * *. ALLISSUIS. * * *. The distaff was an emblem used for the commemoration of industrious habits when graved in an upright position, but when trailed from a spun thread, it indicated a gad-about reputation. Don Pedro will write you a particular description of our joint labors. I will now stay my still overburdened pen, to give my thoughts maturity for better expression. H. B.

As Mr. Welson read M. Baudois's letter aloud with a slow, clear, and distinct accentuation, it was intelligible to all, and elicited warm encomiums, with a strong desire to listen to the reading of Don Pedro Garcia's, which we transcribe.

BUENOS AYRES.

QUERIDO DON GUILLERMO, — Since we received your letters describing your marvelous adventures (which we believed, because we could not doubt the truthful sincerity of your affection), we have been in a constant flutter of joyous exhilaration, which has served to clear the murky atmosphere of our household from its time-honored odors and rites of instinctive religion. While bestirring ourselves for the relief of our bodies with the labors of purification for domestic entertainment, we (M. Baudois is now a member of our family) have employed our thoughts in trying to anticipate the effect of your revelation in stemming the tide of ritual selfishness. Also in measuring the extent of opposition and consternation it will cause among the ceremonial adherents of sects congregating for the worship of self-preservation while preying upon each other. We have already felt a foreboding of its practical effect, in demonstration, from our family incubus, Padre Molinero. By depriving sects of their material heaven and hell, — upon which in positive and negative entrance fee priestcraft has issued policies of insurance for the soul of instinct from time out of date, — you will lay an eternal embargo upon their selfish schemes of praying premiums. Indeed, in sanguine forecast we can now see the gasping flutterings of saving grace in its last ritualistic struggles for salvation from inevitable oblivion, giving place to the glorious effulgence of an affectionate immortality. As you can well imagine, I have but little to say in extenuation for my past infatuation, other than that my reason and reverence halted with the dullness of indifference, causing me to accept forms, from the fact that my veneration could find no hopeful resting place. We can now scarcely endure the reflection, that through life we have remained so dull of apprehension, as not to have discoved from self-intuition, that purity and goodness could alone fulfill the indications of Creative intention for the assurance, in life, of an affectionate immortality. From the moment I read your letter I became subject to an awakening translation from self, and in relief from the dread incubus, become overjoyed with the prospect of affording aid to others. That you may more fully realize the effect of the transition, it was quickly discovered by the watchful expectancy of your chiquita favorites Lovieta and Lavoca, who were waiting for the confections of love they knew you would not fail to send for their affectionate regalement. When the rays of gladness began to dawn in my face, they interrupted my reading by nestling their arms about my neck, while they whispered, as if fearful of disturbing the joyful emotions, "O father, how happy you look, there must be something good for us, — do let us kiss you and then read it for we long to hear what Querido Don Guillermo has written to make you look so alegre. With this appeal, seconded by an affectionate assault, I commenced from the beginning of your letter, — but half unfolded to my own view, — and as I read explained to their wondering comprehension the marvelous transitions of your experience. With the introduction of Correliana my reading became interjectional from the staccato inflection of kisses telegraphed to her goodness, with the exclamations, "How beautiful," "Marvilloso!" "Oh, if"— but the wish remained unfinished in voiceful expression, yet the conscious flush of momentary sadness plainly interpreted the burden of their thoughts in hopeful appeal. These emotions, advocated in truthful sincerity the sway of goodness, during the infantile period, if unprejudiced by deleterious example, — when my unfortunates have retained its impression so perfectly, notwithstanding their exposure to its adverse influence. With the ready perception of such youthful neophytes the Manatitlans' demonstration of exampled direction in purity and goodness must succeed with our race. For older appreciation, how could its truthful impression be more clearly defined for the comprehension of common sense, or more agreeable to loving veneration directed to the Supreme Source of all good, than through the attractive avenue of infantile perception? Practical, or exampled purity and goodness in attainment, are in substance, to my understanding, the length and breadth of Manatitlan "theology," if I may be permitted to use a word so devoid of intrinsic meaning, for the expression of the highest possible conceptions of realized achievement. I feel certain that I have not misconstrued the Manatitlan "Code" from the enthusiastic approval of the above named theologists known to your loving sympathy.

How the sectaries will dispose of the Manatitlan method of perfecting their children in loving affection cannot be solved by anticipation. But my household has been "blessed" with a partial solution of the Catholic method that will be adopted, by a demonstrative denunciation of you and your Manatitlan exemplars as infernal innovators, by your old "friend" Padre Molinero. Forgetting in his wrathful displeasure that anathemas were vicarious oaths in fiendish transposition for priestly cursing, he dispensed them freely for the final disposal of all innovators, and in personal designation included those that I hold most dear in my affectionate esteem. With a self-control, that made me feel for a moment exultant, in view of my former frailties of temper, I coolly reminded him of the formulistic rites established by society for association; assuring him that I should sustain the sanctuary privileges, and stable rights of my roof, recognized by civilized humanity, against the intrusion of any and all persons refusing to hold themselves amenable to the unobtrusive rules of instinctive propriety. This admonition so enraged him, that his malignant intolerance burst forth in demoniacal ebullition, heedless of my direct hints that he was overstepping the limits of patient endurance. But as he continued to inveigh, I with authority withheld his further speech, with a determination that overawed him, and then, while directing his way to the outer gates, stated in plain terms my desire to hear him express a determination to absent himself from my house henceforth and forever. This final ultimatum, after years of undisputed sway, caused his former expression of vengeful hate to appear, in comparison, like the mild gleams of summer electricity, in contrast with the fierce flashes uttered with the deep mutterings of the full charged thunder cloud. Indeed, when the portera discharged him with her absolution, his visage became visibly expanded with a toadish expression of ire, and his throat with a sack constriction resembling the cobra's when about to strike for venomed injection. Failing in speech to intimidate, he had recourse to the fierce ritualary crossings of excommunication, which formerly caused kings and emperors to tremble with instinctive fear. This impotent effort must have summoned to my face a contemptuous expression, for Teudschen, the portera, made a significant gesture of questioning inquiry with her foot, as he passed over the threshold, which I negatived with a decisive shake of the head, else, I should now have to bear the stigma of sanctioning an act of celerity she was desirous of communicating to Padre Molinero to expedite his exit. Instead of denying her impulsive intention of rendering pedal acceleration to the padre's outward movement, when I reproved her for the meditated unfeminine act, which in consummation would have given rise to great scandal, she innocently asked, "what else could a woman do when there was no broom handy?" Then she continued, in extenuation, "If men come into the houses of good peoples, dressed like women in petticoats, and don't behave properly, as they ought, it's right that they should be served by a woman as a man would treat them if he dared!" This Hibernic style of pleading, with its touch upon the mild nature and lack of decision in my exercise of authority, at once dispelled my ill humor; and I questioned her, whether as a good Lutheran she was familiar with the text that taught the returns that were to be made for despiteful treatment? After a little thoughtful hesitation, she said, "I don't exactly remember, but I believe they were kicked out of the house." This answer closed my catechismal interview, but however remote its orthodoxy was from the inculcation of the text, its validity was loyal in the sound doctrinal expression of instinct. For the humorous method of Teudschen's style of speaking, with the broken wabbling tones of her voice, in variation from Low Dutch guttural to the harsh grating rasp of High German, I will refer you to your memory of the gratification you derived from conversing with her. Leaving the patio with my feelings of anger unruffled by Teudschen's patriotic simplicity, which had long been aggravated by Padre Molinero's French sympathies, I returned falteringly to the salon, fearing that the mother of my children might view the expulsion of her confessor as an unpardonable sin; still I could not help congratulating myself upon the manner in which I had rid my household of the traitorous fomenter of misrule and family discord. Entering her apartment fully prepared to meet her scornful and defiant glances, ill masked under the disguise of indifference, I was startled out of my assumed composure by the unexpected greeting of tearful eyes, and soft pleading glances bespeaking self-deprecation. These premonitions of repentant affection brought back, with the loving glow of gladness, the happy impressions of our early wedded life when I was all sufficient as a confessor without the aid of priest or mother-in-law. Determined to make good the step I had taken for freeing my family from the prying curiosity and dictation of droning priests, I met her advances with affectionate warmth; but after listening to her expressions of revived sympathy, I with conscious power, never before realized, asked her if she was willing to seek another priestly adviser; if she still considered it necessary to bar her husband from his affectionate privilege? Feeling a sympathetic tremor in full assent with my wishes, as she silently embraced me, I expressed, with endearments, the hope that I had proved myself worthy of her love and confidence, except from occasional displays of temper provoked by the influence that had caused her estrangement. An answering sob, with its regretful pressure, confirming the favorable advantage of the moment, I questioned whether her unbiased perception had ever discovered in me a willful deficiency, or one of indifference, that I could correct for enhancing her present happiness, or advancing her preparation for a future state. If I had been so unfortunate or neglectful in my intention, with her affectionate reciprocation, I would make it my constant text for amendment. In vain she tried to give expression in words to the welling revival of loving emotions, but, although in voiceful effort she failed, her lips with truthful impression absolved me. Blinded with repentant tears she indicated her desire to be left alone; obedient to the unexpressed wish I left her.

Lovieta and Lavoca, who had witnessed the scene of happy reconciliation, sat with arms entwined about each other's necks, mingling their tears in grateful sympathy, otherwise holding themselves aloof with wonderful discretion, as if with the understanding that their participation would divert the full measure of love's revived reciprocation. But as the door closed upon their mother, I felt their hands caressingly raise mine to their lips, while in relieved vent of consolation and childish dislike, they exclaimed in sobbing accents, first Lovieta, " The — the — ugly — o — old zapelote!" [carrion vulture] to which Lavoca appended, " Who — who — comes — to — to — our — house,—when — he — he — knows —we —we — ain't dead."

Upon you, my dear Don Guillermo, rests this irreverent title, and knowledge of vocation in application, and as you see the retentive rendering my children have made of it, I hope that you will not sin away the day of grace offered for repentance! If you, a stranger, found yourself unable to resist the pleadings of their affectionate natures, you can judge of a fond father's partiality, and will excuse his frequent introduction of their quaint comparisons, especially as they are largely indebted to you for their capital ideas? They have proved the choicest of our blessings, and in love's arbitrations have ever been the pure mediums of affectionate reunion. After a sufficient season had been allowed from my anxious desire, the children were sent to add their weight to the favoring balance of their mother's affection. Quickly returning, with guarded steps prompted for affectionate surprise, they approached silently, — as I sat with bowed head, hopefully musing, with the desire that Consolata might be changed into unwavering semblance of your Heraclean matrons, — and the velvety wreath of their arms again encircled my neck, while Lovieta and Lavoca's Manatitlan voices whispered, in joyous emulation, "Go to mamma!" Dear Don Guillermo, you will rejoice to know that she met me at the door with a fond embrace, and the sobbing supplication, "Pedro, can you forgive me? If you can, and will let me love you again, your desires shall ever be mine!" The coveted appeal required no repetition, for there was in her words an expression of anguished regret, that surprised conviction with the assurance of our mutual amendment. You will, I am sure, forgive me for obtruding my uxorious prolixity, in giving vent to the expression of our united happiness, a boon long coveted without the hopeful expectation of realization. Consolata (I trust that I shall never again have occasion to revive the old name of Malaspina, as a household term of endearment addressed to my wife, once so painfully familiar to your ears), rejoices that your forced adventure has terminated so happily, for she insists that her willful vagaries caused you to accept Captain Greenwood's invitation. In delegating our representative pen to ask your forgiveness for the discomfort she caused, she promises an amendment that shall be addressed in requital to the comfort of others. Should you return, you will be surprised with the change wrought in the appearance of our household, even with the advantage this will afford you for anticipation. Could you but note the placid enjoyment of Lovieta and Lavoca, imparted from the blending and calm repose of parental example, your present joys would be greatly enhanced. Even Teudschen, in the wondering admiration of her phlegmatic stolidity, clasps her hands with surprised inertia, while subject to the active direction of Consolata in the busy avocations that ever delight the tidy housekeeper. In truth, there is a strange mystery, which puzzles us in accounting for her inventive resources, and their apt adaptability for the conservation of purity and comfort. In the style of her own and childrens' dresses, which in apparent devisement originated within herself, we discover prompted aid, as well as in the selection and preparation of material. But for comfort, cleanliness, and beauty of adaptation, they are a constant source of congratulatory admiration, although *outré* in regard to the prevailing fashions; but as they bear a strong resemblance to the Heraclean costumes, description would prove gratuitous. If this great change in Consolata has been wrought through the reproving self conviction of the unlovely contrast she presented to the description you gave of Correliana Adinope, I can almost feel grateful that her ill temper, under the sway of her confessor, was carried to the extremity of forcing upon itself a remedy that has proved so salutary in effect; and the fear of a relapse, which was at first entertained, is gradually passing away. M. Baudois, without hesitation, suggests Manatitlan influence. If it is possible that they have vouchsafed us direction, as guides and instructors in our extremity, they are certainly aware of our grateful emotions. Impressed with this belief, without fully understanding the process by which auramental thought-substitution is effected, Lovieta and Lavoca, when subject to transient scintillations of temper, will firmly close their mouths to guard against the utterance of words prompted by anger. The change in M. Baudois presents many features equally remarkable with those of Consolata. You will recollect that he excelled as a pianist, but of that class whose talent resides in the mechanical use of the fingers and eyes in execution; now he holds you enrapt with the pathos of harmonized sympathy. He often exclaims that his seemingly impromptu compositions are a marvel to himself. Has it not amazed you greatly, from the ready realization of immortal impressions through the avenue of unselfish goodness, that humanity has continued heedless through the lapse of so many ages, blinded with will o'wisp infatuations? Although still full to overflowing with grateful reciprocations of happy experience, we will forego their written expression for the present, to give place for the description of our discoveries of relict mementoes brought by the ocean currents and wind wafts from the Eastern continent, decades of centuries ago. They will certainly afford Mr. Dow material aid upon which to found his conjectural history.

While dredging the Laguna Fecal in the year 1852, for ammoniacal guano in its crystalized and mixed combinations, the sieve grapnel brought to the surface several pieces of fashioned wood of remarkable appearance. From their shape and peculiar method of union, the curiosity of the laborers became excited; and as my devotion to antiquarian research was known to the Padrone, a message was sent to me expressing the desire that I would attend personally and direct the labors of the workmen in accordance with my judgment. Before my arrival, an anchor of hard copper alloy was raised. After carefully removing the ammoniacal incrustations, a clear impress of its form was found stamped on the shank beneath the ring. This indenture was the maritime seal of the early Phœnician cities. Knowing the high scientific value that I placed upon the relict vestiges of past ages, the Padrone and workmen voluntarily surrendered all that had been recovered.

After carefully sounding to obtain a knowledge of the imbedded extent of the detached portions, and to learn the dimensions of the vessel, if its planks and timbers were yet retained in position by their fastenings, it was inclosed in a coffer-dam, and the retained guanic admixture with water was pumped out for evaporation. When cleared, the trove, with the aid of dredge and shovels, was in a few days fully exposed to view. As we had anticipated, the prize proved to be the remains of an ancient galley. Calculating from the keel, which remained nearly entire, the extreme length from its heel to the stem of the prow must have exceeded, free from overworks, seventy feet. The keel was stepped for prow, main, and stern masts. The former and latter were respectively placed within a few feet of the extreme ends of the keelson, or its semblance, which strengthened the true step in the keel. The septum support of the main-mast united the after and forward decks, separating the banks of oarsmen, with ample space between for passage fore and aft. Beneath the lower bank-pits a portion of the deck remained, showing the foot-wear of the rowers in their forward and backward steps of reach. The hold beneath was of sufficient capacity for the storage of provisions for a long voyage. The run-planking on the starboard side was sufficiently well preserved to show the columbares for a single bank of eight oars, as well as those in the stern designed for the rudder blades. The fact of there being only a single bank of oars in the true planking, afforded presumptive evidence of commercial intention, as in a vessel so large their propelling power would have proved insufficient for ordinary progression, but as adjuvantic aids to the sails, in light winds, they would prove valuable. The seams between the planks still retained the papyrus with which they had been caulked; this had been introduced with pitch or melted resin; the combined effects of age and ammonia had changed the paying substance, so that in appearance it resembled amber. The remains of the galley presented for the study of the antiquarian a double interest, historical and mechanical, the latter, with its material indications of skill in art adaptation, affording a clue to the periodic stages of progression as the head and hand mark of coeval intelligence. The wood used in its construction was the red, aromatic cedar of Lebanon, which gave indications of large growth, many of the planks reaching the entire length of the hull, perhaps determining at the time the size of the vessel. During the early ascendancy of Roman rule, the Appenine pine, or fir, was brought into requisition for ship building; but the cedar of Lebanon still retained its reputation founded upon intrinsic value, as it exceeded all other woods in elastic toughness, lightness, durability, and unattractive freedom from parasitic accumulations peculiar to the Mediterranean and ocean. The rostrum or beak had been detached from the prow and set upon an altar aft, among other memorial lares. In form it blended the graceful curve of the swan's neck with the repulsive rugæ of the serpent's expansive skin, characteristically sustained with the mythical "figure-head" of a dragon. In addition to the holes mortised in the true keel for the reception of the ribs, they were secured in place and rendered steadfast by a clamp attachment to an overlaying substitute for the modern keelson, to which they were firmly bolted, with a workmanlike precision that had defied the lapse of time, and decomposing agency of salts and exposure, a chance portion only of the metal being exposed. But to the lost art of hardening copper with a non-erosive alloy, its preservation was chiefly to be attributed. The larboard portion of the prow's planking still retained the eye, consecrated as the watchful guardian of course and detector of danger.

When the hull was raised from its long repose in

the bed of the lagoon, which had formerly been an inlet of the La Plata, and throughly dried, it was restored, in a measure, to its former lightness, from which it was easy to conceive its swift progress over the waters when impelled by oars or a favoring gale, as it but slightly taxed the strength of twenty men to bear it upon their shoulders to its present resting place, in the outer patio of my museum built for its reception.

Your description of the Kyronese lineaments favors a descent from the primal union of the Phœnician with the North African races. I have relics exhumed remote from the shores of the La Plata, of undoubted Morisco fabrication, some of which bear a date corresponding with the second century of our era. From these material evidences, we have conclusive proof that the currents evoked from the disgorgements of the Mediterranean's tidal surplus, through the Strait of Gibraltar, and the large river-drains of South America, by the La Plata estuary, in monsoon reciprocation, have proved the accidental highway of tempest-sped vessels, from the period when they were first built with a carrying capacity sufficient for the transportation of merchandise, and the free navigation of the inland seas with sails. The diversity of color, facial contour, and structural art of fabrication, plainly bespeak an intermixture of European, Asiatic, and African races with the aborigines of America. That the supply was accidental, and limited to the recurrence of causes happening after the lapse of long intervals, is apparent from the numerous depopulated cities, whose inhabitants, like those of old Heraclea, presumed upon their ritual intelligence to enslave the natives and barbarously treat them, until the arrogance of folly and overindulgence opened a way for retributive judgment in total annihilation. Hence the mottled appearance of the natives in the neighborhood of the ruined cities of Mexico and Yucatan, derived from a relapse after a sparse inoculation of the lighter shades of color. The same effects are apparent in all the coast eddies confluent with the inter-oceanic currents from the disemboguement of large rivers, as in the fruitful valleys of their interior tributaries, ruins indicating exotic races are generally found.

As all these conjectural evolutions of deductive thought, relative to the transition events of the past, are unprofitable, in comparison with the cultivation of the conscious elements of goodness, for the prospective happiness of future generations, through the avenue of educated self-legislation, you must excuse the little attention that I have devoted to research for the elucidation of Heraclean derivation.

Our desire to see you, under the impressions imparted from Heraclean example, has increased with our happiness, until it has become almost irresistible. Strange as it may appear to you, Consolata consulted me this morning upon the feasibility of making a trip to Heraclea. The idea was an infinite source of pleasure to us all, M. Baudois alone expressing regretful sorrow that there was a prejudiced obstacle in the way over which he held doubtful control, and until he had reduced it to kindly subjugation he would not adventure himself as the possible cause of a lack of freedom to the genial flow of sincerity. The nature of this "lion," still at large, barring his way to Heraclean enjoyment, we can surmise, but we think from the present cordial relations existing between him and Teudschen that it would prove a whimsical prejudice rather than a reality. Lovieta and Lavoca have set their little heads together in council, with our consent and approval, to solicit your aid and intercession for their admission into the Heraclean school. Notwithstanding our children are to us the solacing light and warmth of affection in personification, we wish to advance their happiness, for transmission, upon a sure basis, even if we shall be obliged to defer our own visit for a season; for we have full faith in their increasing love, and shall find ample consolation for our temporary bereavement in the prospect of a joyful reunion. In the meantime, — if our petition should prove agreeable to the prætor and tribunes, — while waiting for Captain Greenwood, the children specially desire that you will extend to the Manatitlan volantaphs an invitation for them to make our house their abiding place whenever they visit Buenos Ayres. M. Baudois is now engaged in devising means for their accommodation on our roof, having already completed the architectural designs for the falcon mews, and colemena for the phæton bees. As yet his invention halts in calculating the furnishing requirements suited in capacity for the size of our anticipated guests. Lovieta and Lavoca jointly solicit one or more of the Manatitlan donecellita giantescoes to act as auramental governantes in preparation for their entrance into the Heraclean school, if the Dosch approves and will favor them with a suspension of their school regulations. "If the request is granted, they promise to attend personally to their comfort, and will try to prove obedient to their direction in everything." Consolata hopes that your present powers of self-control will enable you to banish from your memory the remembrance of her unworthiness, promising, with the opportunity, to give full expression to the sincerity of her supplication with practical evidences of her amendment. Our endeavors to impress "our little folks" with the relative size of the Manatitlan giantescoes, mediums, and tits, has been but partially successful, from our own deficiency of subject comprehension. At present they seize upon representative minutiæ, in suggestive similitude, for mental comparison. For the material illustration of head capacity, pins of different sizes have been brought into requisition, and for the eyes those of needles; still there is a lingering vagueness in all of our conceptions of Manatitlan proportions in bodily endowment.

If consistent with Manatitlan propriety, in the economy of time, it would afford us great satisfaction if they would permit us to entertain a sufficient number of their people to keep us well directed, for we have become vividly conscious of our instinctive frailties. In anticipation of a favorable answer to our joint requests I have had a flag-staff raised upon our roof surmounted with the letters P. G. In closing we wish to inquire if, as with us, in our household association, you, in your intercourse with the Heracleans are disinclined to speak unless you have something useful, solacing, or mirthful to say? We have certainly grown chary in speech, but with a flowing increase in the current of loving communication, with the prospect of reducing language to a nearer approximation to the truthful intention of its manifest devisement. With gratitude for the inexpressible happiness you have been the means of conferring, we shall ever esteem it a favor if you will permit us to supply your material wants, foreign to the resources of Heraclea. PEDRO GARCIA,
for household adherents.

P. S. Will it surprise you to learn that Pedro Garcia, whose vanity delighted in being esteemed learned in past usages under the patronizing titles conferred by the garnered wisdom of colleges and societies, and M. Baudois, the corresponding savan of the French Academy, have consigned to the elementary combustion of fire all their theoretical works? On the 27th prox. the uniformly bound works of my library, expatiating upon the theory and practice of theology, medicine, and law, in company with those of M. Baudois treating on glacial and other theories of the earth's transitions and destiny, were carted to my quinta and consigned to the flames.   P. G.

CHAPTER XXI.

AFTER the letters of M. Baudois and Don Pedro had been read aloud to the Prætor's family, the Dosch remarked that the Animalculan Mouthpats furnished confirmatory evidence of the provincialism of the Asiatic Heraclean emigrants, as they were undoubtedly parasitic companions of the involuntary voyage across the ocean. Their Scythio-Celtic jargonic idiom corresponds with the Latins' incursive invasions into the Klappish and Celtic territories, while in habits and customs they show the marked impression of instinctive traits peculiar to the old Heracleans in their irruptive stages of progression. With the classical tendency of history to transmit evil, the Mouthpats have retained the traditional impressions of their ancestors' instinctive association with the Gigas, in addition to those derived from direct inheritance in kind.

In compliance with Don Pedro's request for Manatitlan aid and protection, the Dosch announced his intention of sending two giantescoes to act as auramental governantes for Lovieta and Lavoca, preparatory to their entrance into the Heraclean school, which were to be selected from the reserve maiden fund for the relief of widowers. In answer to Mr. Welson's inquiry with regard to their age, the Dosch assured him that there was no need of anxiety on that score, for the funded maidens were not of the soured acrimonious kind represented as " old maids " by Giga bruits, whose passionate instincts of affection had been denied marriageable reciprocation, but of matured kindly dispositions to whom the controlling influence of animal passion was unknown. "Yet," he continued, "they possess the common inheritance of womankind, which delights in making the tongue vocalize thoughtful inspirations of affection; but with our Manatitlan matrons and maidens its use for detraction is also unknown."

Correliana's usually quiet composure had shown evidences of happy transition to fluttering excitement, after an interview with the padre, which was heightened when the Dosch, after the preliminaries for Manatitlan correspondence with the family of Don Pedro had been arranged, asked her to show the material cause of her gratefully glad perturbation. Upon this hint she produced from her bosom the photographic likeness of Captain Greenwood. This he had presented to her on board of the *Tortuga;* but while engaged in packing, the day before their departure for Heraclea, it had mysteriously disappeared, and after thorough search it was supposed to be irretrievably lost. The wind was charged with its abstraction, and the waters of the river as its receiver, but Correliana was confident in the belief of their innocence from the absence of the first party, as the day was perfectly calm, and she recollected of placing it beneath a book when the Captain required the aid of her hands. The book, unfortunately, was the "Art of Confession Made Easy," by Fray Manuel de Jaen; and belonged to Padre Simon, who in one of his "fits" of abstraction recovered it, and used the photograph as a mark, unobserved by Correliana; and as it contained his polemical stock of knowledge for quotation, he was guarded in withholding it from others, and immediately placed it in the transom locker of his stateroom, where it remained lost to his own memory until found by Antonio on the boat's passage down the river. As the loss of the photograph had been the cause of continued anxiety to all, from the inconsolable regrets of the loser, the Captain determined to summons the padre to appear with the couriers at Amelcoy, to bear the treasure back to its owner, in penance for the sorrow he had caused by his heedlessness. Correliana's conscious blush of happiness, as the semblance of her chosen was passed in review, imparted its impression to the invisible as well as the visible, for the Doschavita with her coterie of companions were anxious to judge of the selection made by their favorite.

The Dosch remarked that the odd fancy of the premeditated surroundings, was in kind characteristic of the jumble of gold, charity, and redeeming grace, as the barter conditions of salvation in the Giga mythology. Professional craftsmen, and mechanical members of societies and orders, parade their badges and insignia as the vain-glorious emblems of exclusive selfishness, while they preach a universal heaven free to all without distinction of persons. Your democratic orators, styled the "heaven born," assume attitudes for portraiture suited to their special assumptions of vanity for self-inflation; but with the evident fear that the beholder's perception will fail to engross the reflection of their eloquent ability, they hold in their hands appropriately labeled books or manuscripts. The doctor's idiosyncratic pose is defiant, as if he recognized death in a successful rival. With his nose scornfully upturned he consults vacancy with his eyes, in search of prognostic symptoms for diagnostic antagonism; his left hand advanced, is raised aloft inclosing a vial wand labeled Nostratic Viaticum, while his right with feeling expression triumphantly grasps the skull of a patient who had tested the value of his prescriptive vise. The lawyer, brigand, and priest, assume attitudes as characteristically expressive of professional vanity. The vanity of the fashionable "lady," with the mythological signification of intention, adverse to her impressions, rests with her hand, in studied negligence, placed upon a volume of popular sermons, allowing the gilt label, "Christian Virtues," to appear with an array of ringed jewels upon her fingers.

Captain Greenwood in keeping with the advertising disposition of selfish vanity, in emblematic signification of vocation, but with a humorous variation, has perched himself upon a pile of bagged paddy (the Siamese apply the term paddy to everything unclean), while a companion with the evident design of expressing the gambling tendency of speculation, is engaged in dealing "hands" from a pack of cards. In the background of the photograph the religious view exposes a pagoda and mill for cleaning paddy. The face of the captain expresses an impression of saturnine and cynical appreciation of the rare combinations entering into the tout ensemble of the picture. His dress, of Siamese fashion, is also in alliance with the counterparts of the scene, and his cognomic designation of Truly Rural Greenwood. The portrait reveals, with all its incongruous constituents, the "sterling" qualities concealed beneath the acrid asperities of the outer husk. These were discovered by our waiftly Heraclean cousin; and her amused study of his germ, divested of the rough externals imposed by customary habits, attracted his attention to the cause, which led him to intrust the Kyronese to the care and direction of Mr. Dow, and the vessel to his subordinates, while he devoted himself to the removal of his civilized artificials, for the weft of his thoughts with the proffer of her own. His quick appreciation of her artless worth and purity directed his thoughts to self investigation, with a perceptible improvement in all that pertained to the ruling power of her influence. Notwithstanding the growing strength of the attraction was open for the observance of all, not a word of surprise was uttered, or a quizzical manifestation of instinct to insinuate motives other than those of the purest nature expressed in an alliance of the sexes. Even the "hands" and sailors' tongues found no prompting encouragement for gossip, each rendering homage to her power with imposed reverence suited to their capabilities of perception, while in evidence of the controlling influence of her example, there was a marked change in their habits in all the essentials of purity.

Correliana submitted to the retrospective review of the Dosch, as a matter of fact relation, in freedom from other emotions than those of joyful gladness for the recovery of her treasure, and the appreciation bestowed by the Manatitlan matrons upon the wisdom of her choice. Then Mr. Dow smilingly offered the Dosch and Doschessa the privilege of reading the letter of his wife and children; but as it was one of reproof we will simply state, that in writing home he had been so much absorbed in the prospective grandeur of his elevation as the precentor of the discovery, that the remarkable traits of the Heracleans and Manatitlans had scarcely been noticed, but in a sufficient degree for the excitement of intense curiosity. This unpardonable oversight had caused, in the place of congratulations, a letter filled to overflowing with catechistic questions relating to the habits and customs of the Animalculans, to the entire exclusion of a remark touching his agency in the discovery, inasmuch as it would contribute to his personal fame. The reflex action of Mr. Dow's omission greatly amused the Dosch and his wife. After the mirth had subsided, occasioned by the reading of Mr. Dow's family letter, which left him as void of home information as he had left them in his letter communication with regard to the habits and customs of the Manatitlans, M. Hollydorf proposed to read aloud the reply of the R. H. B. Society to his letter.

CHAPTER XXII.

BERLIN, *September 3d, 187–.*

To M. HOLLYDORF, *Director of the Animalculan Corps of the R. H. B. Society, at present conducting their explorations in the newly discovered city of Heraclea Doweri, in the country of the wild hordes in the Andean region of La Plata.* — Greeting, in behalf of our patrons and members of the Society, with personal congratulations ! Wonder and surprised amazement are terms of too weak invention to express the emotional excitement caused by your letter of discovery, which reached us with unaccountable despatch. The introduction, — although conducted with consummate skill, that our credulity might not be overstrained by the relation of concurring events in a manner too abrupt, — was of a character so startling that our mouths unconsciously retained the smoke of our pipes, to be emitted in the full volume of accumulation with an involumed sigh of relief, when impressed with the conviction of your sanity. It appears from your relation, that your progress was attended by concurrent events of the most surprising description, which by a conjunction of circumstances, when subject to opposition, were readily overcome, with an effect that tended to the final development of the Animalculan race of humanity. The train with the slightest descriptive discrepancy would have consigned your letter to the basket as a maniacal production. As it was, with the first announcement, all eyes became fixed upon the reader in mute astonishment, too deeply impressed for skeptical thought, or incredulous comparison; the valves of respiratory emotion continued closed to voluntary exercise, with singular endurance, until the last sentence of your letter opened their vent with a prolonged whe-w accompanied with a volume of condensed smoke commensurate with the capacity of each member for marvelous inflation. Before proceeding to the theoretical discussion, and precedental comparison of your revelations with those of past ages, the members indulged in a variety of extra-scientific ejaculations decidedly foreign to your Manatitlan code of educational ethics.

In discursive consultation after the members had become sufficiently restored for the exercise of their wonted mental equilibrium, by potations obnoxious to the cultivated habits of your exemplars, the discovery was subjected to an analytico-cosmogenerical evolution of ideas. During the discussion a variety of theories were advanced by the prominent members, embracing involution and evolution, to account for the infinitesimal size of the Animalculan race of humanity, and causes of reduction. Some advocated psychological condensation of the souls of our race after death, the degrees of perfection being indicated by the gradations of size distinguishing the castes of giantesco, medium, and tit. Others that they were the concentrated essences of our human vitality in happy translation exonerated from the corruptions of organic support. But a majority were inclined to subject you and your associates to a thorough test upon the score of optical illusion; a few, however, contended that you were subject to necromantic agency. In fact, your mental and physical condition was analyzed with the nicest tests of scrutiny that could be brought to bear from the manifestations developed in the composition of your letter. From the extremes of experience, and extensive resources of the members for tracing the influences of climate, with the moral effect of the eventful transitions through which you passed, they soon arrived at the definite conclusion that your discoveries were legitimately compatible with a sound mind, notwithstanding the precedental lack of parallel examples for comparison. But before final adoption, the effects likely to be wrought in mental, moral, and theological philosophy were elaborately discussed. But from a lack of data to establish the fact of their complete and separate existence as a race, or whether they are representative soul iotas of our own race undergoing the process of refinement, the question was held in suspension for the bias of well attested information. The discovery, however, was unanimously indorsed as an unprecedented matter of fact, tangible to your own and the initiated senses of your associates, and in no way improbable, although reduced to the extreme limits of believing divisibility.

From the insatiable desire of your personal friends, and the few court magnates admitted to the secret, to see and read your letter, in proof of the reality of the discovery, it has become extremely dilapidated; besides, the encroachments of memento clippings have in some parts reached the text. From the almost realized probability of its becoming illegible it is the general wish of the members, expressed in a series of resolutions, that you should retain the duplicate to be filed in the recorded transactions of the Society on your return. Of this you may be sure, from the reverence bestowed upon the one received, in after years it will be esteemed as a relic of attraction sufficient for the liquidation of any emergency to which the Society may become subject from revolution or invasion. By the adoption of this advisable precaution the discovery will be preserved for the perpetual honor and benefit of the Society, as an index of its preëminent claims above those of rival societies and associations in foreign countries. We shall not for the future presume to offer you advice for the direction of your investigations, but in your next communication the Society would be pleased to learn from the Manatitlan naturalists the number and species of the Giga animals that are at present represented in Animalculan life. If your researches have developed other racial peculiarities in the interim they will serve as a digestive stimulant to the doubtful.

The society has elected, by a unanimous vote, Mr. Welson, Mr. Dow, and Captain Greenwood, corresponding members, and the Padre Simon, Jack, and Bill (you neglected to write the names of the last mentioned in full), honorary fellows. This is a step toward liberalism quite unprecedented in the annals of the Society. Their election, however, was held as a politic necessity to prevent them from advancing claims of priority under the patronage of foreign societies, as Prussia now intends to maintain her position as a — if not the — leading nation of Europe, which the reputation of her Krupp's cannon have gained. I am recommended to hint the necessity of precaution in keeping watch over their movements, lest by surreptitious publication of the discovery, the honors of the Society dependent upon priority might be imperiled. Your personal friends send three household tympano-microscopes, suited for dining-room entertainment and post-prandial speeches. Herr Dollynitzen, the eminent toy architect, to whom the discovery was communicated as a state secret, has exercised his utmost skill in the erection and adornment of the palaces intended for the Manatitlan Dosch and his advisors; also in the arrangement of the Court suites for the production of magnificent effects. The appointments of the lesser buildings and accommodations for attache's attendants, are in admirable keeping with the grandeur of the design. The buildings are supplied with all the modern improvements and appliances for the distribution of gas and water, for illumination and lavatory purposes. The brewery, stables, and distillery, are without the chief inclosure in the rear. The punch bowls, bier glasses, and state table service of plate, if found too cumbersome for use, will serve as monuments for the memorial attestation of our artistic skill. In fact, you will be surprised with the skill exhibited in the accomplishment of the undertaking, in all that appertains to durability and finished taste, when you consider the short space of time allowed for its completion. Imperial majesty, and the prince of diplomats, have promised to find time to offer you their autographical greetings for the honors you have conferred upon the crown by your important discoveries, which will be sanctioned in acceptance with the double approval of our august bird of prey, as token of recognized merit.

Yours truly,
Per order of the R. H. B. Society,
BUGWITZS, *Sec.*

N. B. (Entre nous.) If you can send us (liberals) a codified formula of the Manititlan system of education (under cover to me), it might be used as a reformatory basis for a revolutionary movement to effect the rescue of governmental power from the arbitrary sway of the legitimate few. Anything new, with the reputation of an experienced trial of six or seven thousand years, would serve as a subject for public speaking and talk; for as the Manatitlans say, the liberal and radical democracy of our race are attracted by the sonorous bellowings of the physically strong lunged leaders of herd, and amused in dalliance with the softly toned melody of the lowing kine. B.

M. Hollydorf, after reading the secretary's letter, would have suppressed the autographic missive; but the Dosch called his attention to it,—laughingly adding, that he had been advised of the contents by a third party, who was present during the process of dictation. Observing the flush that mounted to M. Hollydorf's face, he said, "I perceive that our system of espionage is not fully sanctioned by your thoughts. But as our object is devoid of instinctive curiosity and malice, and solely devoted to the emancipation of your race from the impositions of selfishness, you will upon mature consideration approve of our course. We are fully aware of the difficulty you experience in divesting yourselves of the reverential awe inspired by the sounding appellations of king, emperor, prince, and other titles bestowed for self-gratulation in the flights of vanity. But if you will analyze the charter privileges conferred with these vapory titles, you will find that patents of nobility are the real talons of your standard emblem of nationality, which allows the grantee to become a participator with the imperial or kingly beak, in rending the spoils of oppression. In truth, the whole structure of your mythological and classical literature, upon which the anointed supremacy of kingly and noble power rests, is as vague and shadowy in its reflection, as a source of awe, as the sun photograph of an ass's ears upon the ground for the intimidation of their owner.

"The privileged follies of the upper ranks, rather than their wisdom, is, from the contrasted meanness of self, the instinctive cause of reverential fear with the poverty stricken. The man who will accept the direction of others, when obliged to dissemble his own follies, not only contravenes the manifest indications of Creative intention, but demeans the natural honesty of his instinct below brute capacity. It is also equally evident that a man who will not deal honestly with himself, is not only unworthy, but will betray the trust reposed in him by others, and as an apostate to his human privilege demeans his instinct as far below the reach of the lower orders as his capabilities are above. With this class, who ape the privilege of ruling others when lacking the will for self-control, our espionage is no treason, but the study of instinct, devoted to selfish gratification, in search of means for emancipation. The craft of the diplomat, whose foxy instinct endeavors to fix the incentive stigma of a causeless war upon neighboring nations, as the precursor of slaughtered millions, for the absorption of coveted territory, should prove a source of reprobation, rather than praise, to the peaceful perceptions of instinct.

In illustration of the covetous nature of the letter, which from the patriotic sympathy of shame you would withhold, we will state from the basis of auramental experience, that the victory of Germany over her Gallic neighbor, who lacked the leading energies of a man capable of controlling with inspired confidence her armies, will prove far worse than a defeat for the continued prosperity of the country. This is especially evident to our perceptions, as it has stimulated the policy of preparing the means while lying in wait for a pretext to absorb the coveted northern seaboard, under the present national control of Belgium, Holland, Denmark, and Russia. Flattered by the prestige she has gained from consolidation, she forgets that her Gallic achievement was solely dependent upon fortuitous circumstances. With a mind capable of commanding unity of action, France would prove more than her equal in the battle field, of which the elder Bonaparte gave evidence in controlling the powers of Europe. You will perceive by the tendency of this prelude that we are fully prepared for the propositions contained in your autographic letter, which will of itself attest to our protective right of espionage, and will render it : —

KAISERLANT, *Aug.*, 187–.

MY DEAR HOLLYDORF, you will be pleased to hear that we hold the French disposition still handsomely in check, by fomentations skillfully applied by our prince of diplomats. But his mind is too expansive for the frivolity of cultivating the natural mushroom tendency of the Celt for intestine irruption. More anon! we have a conception in the womb of the future! Your discovery is truly astonishing! Many of the scientific scarcely credit it! Why not? It's easy to believe if you only think so. A glance at Alsace and Lorraine for instance. Will the Manatitlans acknowledge fealty to Prussia in recognition of our rights of priority? Push the Dosch gently on the subject. It will be of advantage to the Dosch to subscribe for our protection, as we shall soon assume the leading role of Europe, which England can't gainsay. You are authorized to act as our vice-imperator for the execution of a protectorate annexation. In case of obduracy how large a force of our veterans would you require for their subjugation? Answer in your next. The prince advises expectant treatment, with such placebos as your better knowledge of the characteristics of their blind side may suggest. As the giantescoes could be made especially useful in diplomacy and warlike operations, which with our progressive enterprise may soon occur, the enlistment of two or more corps would enable us to anticipate the moves of the enemy. The prince thinks it advisable to enlist a corps that understands the French language, as we are obliged to keep a sharp eye in that direction. We have also thought it advisable to keep the discovery a secret among ourselves for the present. Ask the Dosch if he can approximate in calculation the number of our animalculan subjects in Prussia, and learn where their chief cities are located! Would it not be well to have an animalculan survey of the empire, under

Manatitlan engineers, for its topographical division into departments? You have, by the advice of the prince, been enrolled as a candidate for the honors of knighthood, and will hereafter be designated as the Count Palatine of Heraclea, and Viceroy of Manatitla. The king herald of the royal commandery only awaits the transmission of their national escutcheons for incorporation with the Prussian, and quarterings of your family, before the announcement of your full investment and title will be proclaimed. The prince of diplomats advises gentle dissimulation in the inceptive stages of your negotiations with the Manatitlans; and in the second insinuations with non-commital or evasive attachments; and boldness when fully prepared to offer your ultimatum. This plan worked admirably with the French, who are probably far more accomplished in the diplomatic art than the Manatitlans. If you had experienced the advantages of a married life, you could better appreciate the benefits of the preparatory stages proposed, from their successful adaptability for the quiet managment of domestic affairs. We shall anticipate with increased interest the arrival of your next letters. Please present the Kaiser pipe to Dr. Baahar with its accompanying sack of Latakia and Shiraz ammunition, as its smoke will aid him in resurrecting many interesting items from the ruins of old Heraclea. The gift you can render more acceptable by an appropriate presentation speech, at the close of a public banquet given at our expense. If you think it will enhance the acceptability of the gift, you can allude to its dedication by my own and the lips of the prince, which will be sure to impart to him a politic shrewdness that will outwit the Jesuits.

YOUR KAISER.

M. Hollydorf acknowledged with a flush of shame the correctness of the Dosch's verbatim rendering of the Kaiser letter. But the Dosch rallied him with the assurance that the effort was above the general average of the kind, and really acceptable as an aid in demonstrating the indifference of potentates for the real welfare of their subjects. "If," he urged, "Newton, Humboldt, Arago, and other scientific celebrities, had occupied but half of the time in study for the practical relief of their race from the potential rule of selfishness, that they devoted to theories strained from débris gleanings of earth's attritions, from counter elementary action, they would have secured in grateful reciprocation an ever enduring immortality, that would have lived with the endowment of their living impression, in the current of affection, to the garnered end of the allotted term of mortal representation. Instead of using their gigantic endowments for the development of man's knowledge of himself, in privileged relation to creative design, for the fulfillment of indications vouchsafed for his affectionate direction, they endeavored to illuminate the precedental path of irrational delusions, and left their immense labors, under the seal of acknowledged greatness, as barren of sympathy for immortal direction, as the sands of Sahara are for the support of the Arab wanderer. We do not disclaim the collateral benefits to be derived from the cultivation of practical "science," but to devote one's energies to exhuming relics, tracing glacial tracks, chasing butterflies and other insects, for capture, without other motive than for classification and the gratification of curiosity, is as void of beneficial result as the youthful Gigas' antiquarian search for postage stamps. The obstinate perseverance of the Scotch Animalculans in imitating the fanatical absurdities of their Giga exemplars, has become proverbial with our Manatitlan colonists, who render it "scratching for miracles to cure evils of easy prevention." A reputation founded solely upon entomological pursuits will deservedly prove as short lived, and lackworthy of sympathy, as the ancient family hunts of the Gaels, who left a breeding cause for wasted time spent in the pursuit of their bodies' parasitic foes, when with ease they could have rid themselves of the pests, and added to their comfort by cleanliness. In our aerial study of botanical adaptation, we have observed the date palm, indigenous to the eastern shores of the Mediterranean, growing at the mouths of all the large American rivers of tropical latitudes, corresponding with the current streams of the ocean setting from the Strait of Gibraltar. In Mexico and Yucatan, the date palm grows as luxuriantly, and yields as abundantly in exotic transplantation as under the favoring influences of native soil and climate. Even in the latifundium with its exalted altitude, you will observe that with cultivation it yields fruit of a quality far superior to the Egyptian, and has in reality proved the bread of life to the Heracleans. Now if the scientific of your race would but study these vicarious indications of nature for transplanting increase, and cultivate them in extension with intelligent zeal for affectionate bestowal, war, and its indigent charity sequence of doles, would forever pass away from the surplant of confiding love. But your race are now so wedded to the fruits of precedentalism, that if you, on your return to the haunts of civilization, should attempt to promulgate your present happy thoughts, without forestalling them with the substantial relation of your animalculan discovery of the Manatitlan race, your own relatives would denounce you as lunatic Utopiasts.

THE MANATITLANS. 293

CHAPTER XXIII.

As the four were returning from the auriculum to the quarters of the corps, a week or more after the padre's return, he overtook them and listened to their conversation unperceived. As each entertained one or two Manatitlan auralists in his ear, the conversation was strangely diversified in irrelevancy, which would have caused a stranger to the events transpiring, possessed with the least taint of superstition, to have supposed them insane or bewitched. The padre listened with wondering attraction to catch the drift of their mirthful sallies, hoping to learn the cause, or obtain a clew to their mysterious convocations. Their incoherent address of questions, which although unanswered, appeared to provoke outbursts of merriment, in one, without attracting the least notice from the others, caused him at first to think it was the prelude to one of Mr. Welson's practical essays of humor designed to entrap him. But the earnest manner in which the conversation was conducted, and the unmistakable evidences of genuine mirth, put this conjecture to flight. His next suggestive impression was ushered in with a shudder; could it be possible that they were subject to a spell of enchantment, and that the seeming city of Heraclea was the abode of enchanters, the spirits of darkness against which the fathers had especially warned the heedless? This frightful ghost of a suspicion received such evident confirmation that he immediately had recourse to his rosary that had been bestowed by Fraile Gallagato, of Amelcoy, for numbering exorcising prayers. The rattling of the beads, with the muttering sound of his Ave Marias, attracted the attention of Mr. Welson, as he turned aside to allow the others to enter the puerta in advance, and he for the first time became aware of the padre's presence. The anxious dismay of the padre's countenance revealed the source of his emotions, prompting Mr. Welson to play upon his fears, but the Dosch auramented the inquiry: —

"Well padre, what is it that causes you to look so frightened?"

Padre. "Wha-wha-what does it all mean, I should like to know? Are you bewitched or leagued with the devil [crossing himself], or what are you doing at any rate? I wish to goodness I had gone back to Montevideo!"

Mr. Welson. "Are you not comfortable, padre? Just tell us what you lack, and we will endeavor to supply your wants. The prætor has within an hour made particular inquires for your welfare."

Padre. "It's not that, I have everything my body requires, — but my conscience, — my mind, — I declare upon the welfare of my soul, I can't endure the thoughts of subjecting myself any longer to the temptations of the evil one."

Mr. Welson. "Why padre, there is nothing to my knowledge that should alarm your conscience, or soul, for we are only holding intercourse with human beings, and as you must feel from your own thoughts, we are farther from evil than ever in our lives before. Ease your mind from alarm, and suspicious fears, for in good and seasonable time everything that now appears mysterious will be explained for your privileged understanding. For your assurance and relief from imaginary fears, you have only to turn your thoughts to your own improvement, both mental and physical; which should convince you that from whatever source derived, the influence is good. Do we not appear far more happy here, than on board of the *Tortuga?* If you would but think, and give heed to the promptings of your thoughts, you could not fail to realize that the source of your happiness is derived from an example of purity and goodness, and of necessity, in direct opposition to evil."

Padre. " But I have had warnings clear and distinct, as from the voice of a spirit, in a still small voice, as if coming from afar. Then at another time, I felt like one possessed with thoughts that were not his own, and could not do as I had been taught, without self-reproof, and was lead away from parental instruction, and my Christian education. In fact, as it were, I have been prevented from keeping company with my own conscience, and could not pray and do as I liked."

Mr. Welson (losing his prompted direction). " But you did attempt to do as you liked, when the viper offered visible objection to your taking the dried tobacco leaves in the garden of the old mission of Amyntas, in passing on your way to Amelcoy."

The padre's consternation when exposed to the reared head of the viper, — which had in fact darted from its coil upon a leaf beneath the one the padre's hand was approaching, and struck its fangs into the loose sleeve of his coat, — was not greater than from this display of " second sight," on the part of Mr. Welson, which revealed a scene that he felt confident was only known to himself and Fraile Gallagato, to whom he confessed in Amelcoy. Staring upon Mr. Welson with eyes aghast, he staggered backward with hands upraised, in repellant attitude, as if deprecatingly warding off some dangerous influence that had possessed itself of his personal embodiment.

Mr. Welson (laughing). " There now, you have tempted me to play with your superstitions, or rather I have been tempted. Be content for the present, and in time all will be revealed to you in freedom from supernatural agency."

With this parting admonition Mr. Welson entered the house. The padre, after he had sufficiently recovered the use of his faculties, uttered in self-defense an abjuring protest of two Marias, kissing in addenda the beads and cross with transubstantial desire for their seal of effectual grace, then soliloquized: "They can't convince me that they are not leagued with the spirit of darkness; and if I live to see the morrow's sun I'll shake the dust, — well, if they had any, — off the soles of my feet, if I am obliged to traverse the paths of the wilderness that separates me from civilization alone."

As if to put his intention into immediate execution he walked rapidly down the avenue of the latifundium and out of the gate; but when skirting the copse of the temple grove he met the Heraclean herdsmen and their wives. Their jocund mirth, sportive with songs and gladness, withdrew his thoughts from self by their grateful tokens of affection bestowed in the full outflow of joyful greeting, which caused him to forget his impressions of their enchantment from supernatural agency, and he was soon engaged, with Manatitlan aid, in the laughing exchange of Latin and English terms of idiomatic phrase. On his return to the quarters of the corps suspicion had been banished from his memory; but his doubts and fears were again revived, when on entering the dining-room he encountered the same mysterious impression of a communion with the presence of unseen spirits. The entrance of Dr. Baāhar, with the buzz and genealogical curators of sound, dispelled the influence, but they, as well as the padre, had questioned the source of its power. After the evening meal the padre sought the opportunity of renewing his petition for permission to depart in the morning; anointing it with grateful acknowledgments for their kindness to him personally, while in the style of exorcism he urged the necessity with the quotation, " What shall it profit a man if he gain the whole world and lose his own soul ? " The effect of this appeal, for instinctive self-preservation, was so comical in its misplaced application that the four were obliged to join the auriculars in giving gleeful vent to their mirth ; this, however, was as suddenly checked, when their supposed derisive lack of sympathy affected the padre to the extent of producing tears. Unable to restrain his contempt for the self-imposed shallowness of the padre's perception, Mr. Welson, —under the auricular direction of Corycæus, the familiar who attended him on his journey, — gave a final touch to his victim's superstitious fears, by asking : " Did your conscience or soul find themselves in a purer atmosphere, or in less suspicious companionship, when subjected by Fraile Gallagato to the sacramental spirits of a Rosario punch, and the fumes of tobacco, than with us who have abjured their use ? You need not answer upon the impulse of the moment ; but if, after a night's reflection, your fears for your soul's safety still prompt you to leave us, and the affectionate interest enlisted in your behalf on the part of the Heracleans and Kyronese, the means for your conveyance to Amelcoy and deliverance into the keeping of your noble compadre shall not be wanting. But in bidding you a personal farewell, from your self-will in adhering to delusions that require proxied aid granted from confession and absolution administered by a being so manifestly corrupt as the Fraile Gallagato, we shall be obliged to forego the hopeful retaining interest that we feel in your welfare, unless by the contrast, your thoughtful eyes are opened to see and feel the great loss you will have sustained in the sacrifice of truthful and affectionate sincerity."

Padre. " But why, Mr. Welson, have you kept

from me anything that it was proper and useful for the rest of you to know."

Mr. Welson. "In the first place, you was not particularly interested in scientific investigations, or book lore, else you would have participated in the discovery that has puzzled and alarmed you. In the second place, as you represented the instinctive class of ritualistic habit and creed followers, who believe in what they have been taught without questioning palpable absurdities, you have been exhibited to us as an illustration of the unthinking characteristics of our race. As you have subserved the purpose of showing the irrational subterfuges of sectarianism for shirking the responsibilities of honest example, for the delusive indulgence of instinctive desire in excess of reasonable gratification, I will now inform you that shortly after our arrival in Heraclea we were introduced, through the reflecting aid of the tympano-microscope, to a race of human Animulculans, by Mistress Correliano. These had been known to her Heraclean ancestors for many centuries, and were the originators of their system of education. The largest, or giantescoes as they are called, are perceptible in form to our unaided eyes; but, with a few exceptions, they belong solely to the Manatitlans, the race to which we were introduced. But the lower grades, which are styled mediums and tits, are in human resemblance indistinctly visible to the naked eye. The falcons, that visited Mistress Correliana on board of the *Tortuga*, were guided by individuals of this race, which accounts for their wonderful sagacity, and the mysterious intelligence which she had gained of the transactions of our race while immured within the walls of Heraclea. When M. Hollydorf commenced his investigations for tracing the relation of the dry, animalculan species with the representatives of our gigantic orders, the thought never occurred to his imagination that by any possibility the initial type of humanity would be discovered. You will now be able to judge, from your impressions, our emotions when we were obliged to recognize under the powerful reflection of the tympano-microscope our own representative embodiment in minute miniature; but with a perfection in beauty that put the pretensions of our race to shame. The object of the R. H. B. Society, was, as you have heard explained, — to obtain a knowledge of aboriginal animalculan dryad life for comparison with kindred species within the influence of civilization to judge of its progressive effects. As Dr. Baāhar was too much occupied with his naturalistic pursuits to attend on the day of discovery, he with the other members were held in probationary ignorance of the new race for experimental effects, allied to those to which you have been subject. You can now ponder for the night upon the revelation I have given you, and if to-morrow you wish for tangible evidence of its truth you can accompany us to the auriculum.''

The padre had directed his attention chiefly to Mr. Welson during his admonitory explanation, with eyes amazedly questioning the faces of the others for confirmation, and mouth agape, which at certain passages of the rehearsal contracted with grimaced efforts to swallow; but with the closing invitation he relapsed into a ruminating mood of fitful cogitation. In this condition he remained, scarcely noticing the return of Dr. Baāhar and curators of sound, notwithstanding they were subjected to auramental impression for his especial detection, and proof edification, from the incongruous lack of method shown in their conversation, which the doctor, with aptness, styled the languaged lowing salutations of the herd at nightfall. Corycæus, the padre's auramental familiar, reported his ruminations at the hour of retiring to Mr. Welson, who sought the opportunity to give them direction by reading to him a postscript of Captain Greenwood's letter, containing information that he was too much vexed to give him by word of mouth at Amelcoy. Calling the padre into the colonnade, after the other members of the corps had retired, he read the captain's announcement to him by moonlight, which we will render verbatim: —

"P. S. The padre's appearance, with the knowledge that he had passed the major portion of the night in drinking, smoking, and chewing, with Fraile Gallagato, who conducted him in a state of inebriation to the house of a woman of unmentionable fame, so annoyed me that I held no communication with him, and write that which it would have afforded me unalloyed pleasure to have imparted to him, by word of mouth, if he had been in a worthy condition. Please inform him that our success in collecting gold on the spits of the Pilcomayo, during our return trip down the river, so far exceeded our personal requirements, after equally sharing with the absent in Heraclea, that we have, at the suggestion of Jack and Bill, — who rightfully aver that he was the discoverer, — set aside a sufficient amount for insuring his family an ample token of his fatherly remembrance and desire for their welfare. The amount will be forwarded to the address of any reliable person he may name, subject to his childrens' order. T. R. G."

The padre at the close of the message gave one audible gulp of choking shame, and sank down upon the pavement in groveling attitude, exclaiming in broken accents of woful misery, "unworthy brute that I am!" In this condition Mr. Welson left him, with a simple parting salutation. Corycæus, in the morning, reported that he continued prone and imbecile in thought, until lunar impression caused his scattered faculties to become wild in intention, causing him to utter vehemently the talismanic Giga word, Reform! But as its suggestions encouraged moderation, as the source of saving grace, he wandered forth into the herald darkness of the morning's dawn. Directing his steps in the gray light to the summit fora, his thoughts were led to view, — in the emergence of day, from the chill sombre darkness of night, as the first radiant rays of sunlight appeared above the horizon, — the bright perspective of Heraclean affection which began to dawn with its warmth and purity, beckoning him from the gloom of the past, with the determination that his example should contrast with it, as a day of light for the guidance of future generations in the path of happiness. As he stood in the rolling mist wafted by the air current of the falls, on its nourishing mission to the latifundium, Mr. Welson joined the prætor's family to unite with them in their morning salutations. In turning their eyes upward to catch the first rays of the sun on the brink of the falls, the head of the padre appeared enveloped in a cloud of mist. When first discovered, his attention seemed to be attracted to objects beyond the walls; but with the first strains of the morning anthem he removed his hat and united his voice in the song of praise. At its close, he beckoned them to join him on the summit terrace. The prætor understanding the invitation, challenged Mr. Welson and Dr. Baãhar to a trial of speed, with his wife and daughter, up the ascent. Accepting the gage they started, the civilized competitors taking the shorter and direct avenue from the city gate. At the word, up the crescent avenue Correliana and her mother sped with equal steps, gliding upward in the pathway with graceful motions, and swiftness rivaling the fabled Camilla's, the prætor following with a steady movement of practiced ease, content to hold his starting distance good. The padre's past and future, as umpire of this novel race, quickly merged into the present. With jubilant mirth he urged the doctor, with hand and hat, to greater speed for the honor of the corps. But from the weissich of the falling water his words of encouragement failed to reach the object of his admonition. In the abandon of momentary excitement, his gestures were of that comical cast that we should expect from a man who had been aroused from a slumber that had continued from youth to age, and installed, when suddenly awakened, to preside as umpire over a scene like the one in review, without comprehending his growth in stature. On all, except the toiling object of his exhortations, his pantomimic gesticulations served as brakes to stay their speed. The diminution in speed of the prætorial family became quite apparent from their mirthful checks; and Mr. Welson, who had for a time maintained an increasing distance in advance of the doctor, came to a full stop at the foot of the summit incline, where the latter passed him, attributing his disability to shortness of breath from overhaste in the beginning. Congratulating himself upon his own prudence in reserving his strength for the last stretch, the doctor reached the summit, but was chagrined to find the prætor and family awaiting his arrival at the goal. The padre, forgetful of his night's vigil, and the cause, bantered Dr. Baāhar upon his signal defeat; but an inquiring look from Mr. Welson reminded him of his petition, and he became silent until asked, when descending, why he had beckoned them to the summit fora?

"Well, I declare," he replied, "your race put it out of my mind altògether; but I wanted you to see what a beautiful effect the morning sun had upon the scenery."

Correliana referred him to the visit he had paid to the summit in company with Cleorita and Oviata on the morning after his return from Amelcoy; a reference that caused him to become blushingly silent. Mr. Welson then informed the prætor and family that the padre and Dr. Baāhar had already been initiated into the object of their secret convocations, and that from henceforth there would be no reservations in conversation.

CHAPTER XXIV.

The padre on his return to the quarters of the corps, found Mr. Dow alone, and questioned him upon the sincerity of Mr. Welson's revelation. "You know," he said, "that I am a sort of orphan waif among you, in the matter of science, which Mr. Welson, Dr. Baāhar, and others, with the exception of yourself, have taken advantage for their amusement; not that they have treated me unkindly or disrespectfully; but when they saw me really anxious from fear, which they could have relieved, it was hard that they should tease and add to my perplexity."

Mr. Dow assured him that the revelations of Mr. Welson were strictly true, and that whatever was at variance with their former selves had been effected by Manatitlan wisdom. He then asked Corycæus, if present, to give a joint pull upon the most sensitive vibrilla in the padre's ear. His sudden start, with the tearful winking of his eyes, gave evidence that the Manatitlans were still there. Seeing that the old mythical idea of spirit possession still lingered, he asked them to sing in chorus Old Lang Syne, as that was the only tune the padre could recognize. This was so well rendered in sympathetic harmony that the padre beat time with both hands, and at its close exclaimed, "My goodness gracious, I never heard such music; why it thrilled me through and through, yet the voices seemed small, and far off, as if they came from the heavenly realms of bliss!"

After the morning meal the padre was escorted to

the auriculum by all the initiated, including the mayorong and his family. The padre having been placed in a favorable position for hearing and seeing, at an appointed signal a large number of giantescoes, mediums, and tits, suddenly appeared on the reflecting platform of the tympano-microscope, with a movement so quick that neither action or source of emergence could be detected. The suddenness of their appearance caused the quick adjournment of the padre's hand to his hair, its usual place of resort when his faculties were surprised with doubts requiring the aid of counter-irritation for elucidation. The Dosch and Doschessa advanced to the front of the platform, and after a reciprocal introduction to the padre, the former plainly stated the reason why a portion had been excluded from a knowledge of their existence. It was not, however, until a full hour had passed in the discussion of various topics relating to Manatitlan influence, that the padre ventured to speak. He then timidly inquired of the Dosch, "Do you and your wife and the rest of the Manatitlans feel quite like men and comfortable?"

Dosch. "You can rest assured that we all feel like men,— except our women,— and really comfortable!"

Padre. "Are your women in being smaller less comfortable than the men?"

Dosch. "As you perceive, they hold the same relative proportion with regard to size as the females of your own race. But if they were larger, perhaps the men would feel less comfortable. You know from experience that women wield a strong influence upon whichever side they lend their weight."

Padre. "Do you cook your food or eat it raw?"

Dosch. "We are, like your own race, omnivorous, but select and adapt our food to the healthy requirements of our bodies, using fire for its preparation."

Padre. "But how can you make such small fires?"

Dosch. "We hold ourselves in advance of your race in that respect, as we are not dependent upon material combustion for our fires. You are, of course, laboring under the impression that our small size must embarrass the organic functions. But in the intensity of the spark you will find an apt illustration of the vital energy that we have been enabled to preserve in the purity of its brightness. Although you may esteem it a spark of egoism, we can, with truth, assert that we feel free from the vagaries of appetite and lust, and an infinitesimal concentration of vitality that imparts purity to our impressions. This exemption from the ills to which you are subject in the flesh, we have obtained by the consistent cultivation of our perceptive endowments bestowed by the Creator as an heirloom independent of the body's material tenement. For your enlightenment with reference to the physical coöperation of our educated perceptions with the body, I will say that we possess, from transmitted cultivation, a nervous and muscular energy, with a sensitive perception, that enables us to detect and guard against dangers while yet distant. In sudden emergency, by our agile presence of mind, we can, without extraordinary effort, avoid impending danger, that would inevitably prove fatal to your slower faculties of apprehension. That you may appreciate, in a measure, the quickness of our movements, Corycæus, the padre's familiar, will pass from the platform to his ear."

Corycæus. "Yes, and he may catch me if he can."

Quick as our eyes were turned on the padre, the change in direction was anticipated by Corycæus, for he was back on the platform in time to see with laughing glee the padre's hand reach his ear with a clap that jarred his head. But certain of his capture, the padre, without noticing the return of Corycæus to the stage, cautiously introduced his forefinger into the cavity of his ear, with his thumb on the alert to secure his prisoner when raised to the surface. So certainly intent was he of the capture, that he was deaf to the suppressed laughter provoked by his movements, until after the removal and cautious separation of his thumb and finger; then his surprise was greeted by a genial outburst. Discovering Corycæus on the platform, one of his old furtive glances of superstition crossed the bridge of his nose, the laughing jeers causing him to exclaim, "Well you can think what you like, but the devil's in it!"

Corycæus. "If you mean in your ear, the compliment is not intended for me!"

The rejoinder, and mirth, caused the padre to propose a second trial; this proving as unsuccessful as the first, he exclaimed, "You are altogether too spry for my catching; I'd sooner try to catch a flea on the watch! But the fact is, I can't quite make you out to my mind. You seem to be what you say you are; still there is neither sacred or profane authority for your existence, unless we take the Fathers' assurance that it's possible for evil spirits to assume any form, or shape, or preach any doctrine they choose, for the purpose of temptation, of which numerous instances have been recorded by Frey Manuel Jaen, and other sacred authors."

Mr. Welson (impatiently). "The fact is apparent, padre, that you are either stupidly incorrigible or there is a prompted method in your mythological vagaries, to show us how loath your kind are to give up animal indulgences that can be absolved by confession. Once for all! Why is it that you have been endowed with the power of discernment, which you style conscience, to judge between right and wrong, except to assert your probationary privilege to a higher destiny than sensually begot animal life that is subject to compostic defilement and corruption? Or why should you be preferred to a material heaven above the beasts of the field, who have lived in accordance with their special capacities, while you have defiled yours with beastly indulgences? Like yourself we have been subject to auramentation, but have thankfully accepted the promptings bestowed for the enlightenment and correction of our perceptions. We feel that although bodily present, you are not with us."

Padre. " It was yourself, Mr. Welson, that made me cautious ! If I have doubted the evidences of my senses, your deceptions have placed the stumbling block in my way. Since we have been in Heraclea your mysteries whenever I was present kept me on the lookout that I might not be caught napping. Besides all the Kyronese children had disappeared, except the infants, and those over ten years of age. While among the Heracleans there was not a miss or master, except Correliana; or a married man or woman less than twenty-four years of age. When I inquired the cause, you said they were at school, which did not seem consistent, for I could find no signs of one within the walls. Then as there were no churches or signs of religious worship, my fears were excited; for you rose with the sun and welcomed its rising with songs, something you had never attempted to do before; for even after a jovial night, spent in drinking, playing poker, and lansquenet, you appeared more sorry than glad to see it. After breakfast you avoided me and betook yourselves to some strange place, so that through the day I scarcely saw you. These and many other strange freaks made me feel as though I had really strayed into an enchanted city; which impression was strengthened by my own contradictory thoughts. (Looking curiously at Corycæus.) Often a distinct small voice, but as plain as my own, would dispute the number of my Ave Marias, although I had numbered them on my rosary — there, now, is the same voice asking me from what the beads were made. Well as there can be no secrets kept here, I might as well own, that plug tobacco was the easiest thing that I could make them of. Indeed I scarcely knew the half that I was thinking, I became so confused and bewildered. Sometimes I thought my impressions were caused by your scientific tricks, played upon me when asleep; but then there was no authority for that; so you will see from all that I have said, I am hardly in a condition to give credit to my senses, upon the question of these apparitions, that you call Manatitlan Animalculans, as to whether they are real humans, or spirits of evil conjured by the devil to betray souls to damnation."

Corycœus. "But you know, padre, you tried all sorts of exorcisms to get rid of me without effect, until you traced my exhortations to your ears, and supposing they might arise from defect, you smeared your ears with tobacco spittle, which proved an effectual remedy."

Padre (laughing). "You must allow then that tobacco is good for ridding one of an annoyance?"

Corycœus. "If you call good advice an annoyance!"

Dr. Badhar. "He is wedded to his delusions, and with eyes and ears, refuses to see and hear."

Padre. "You may call my religion a delusion, and a budget of traditional superstitions, or whatever you like, but I shall never become an apostate until I can find a better, under proof to supply its place."

Dosch. "It is better that he should be left to draw his own inferences from our example and teachings, as words of reason will prove futile to disabuse him of his bead ritualism. Possibly, Corycæus may have yielded overmuch to his humor from the obstinacy of the padre's infatuations."

Corcyra (wife of Corycæus). "With permission, it would please me to suggest in behalf of our auramental labors with the padre, that we had not the power of controlling him with the privilege of exampled pleadings, so we were obliged to have recourse to stratagem to rescue him from the toils of Fraile Gallagato and his own weakness."

Padre (gaspingly). " Wa-wa-was — "

Corcyra. " Yes, we always attend our husbands, as our bonds of affection are inseparable, and independent of bodily duality. But you need not be so much alarmed, although we are free to acknowledge that we were greatly shocked to see your kindly nature self-betrayed for its own degradation, in a manner so revolting to our impressions of purity."

The padre bowed his head to conceal his face, flushed with shameful self-reproaches. The Dosch diverted attention from the padre by introducing Codecio, who proposed to give a synoptical description of the advantages imparted from their system of education, which would be exemplified by a visit to the Heraclean schools.

" In rendering our homage of grateful affection to Inovatus Desiderata for the inestimable boon of an educated power of self-control over the body's instincts, founded upon unselfish reciprocation, we also with equal fervency correspond with Analogius, his successor, who perfected the founder's system with the censor's safeguard.

" *The Censor's* duty commences immediately after birth, at the completion of the nurse's midwife assistance rendered to the mother, as upon her this aid naturally devolves. In no instance has there ever occurred the necessity of man's intervention with this function, which innate delicacy declares repugnant to modest purity. The censor then in conjunction with the nurse, who remains as a constant guard, directs the parents for the adaptation of affectionate solicitation for the welfare of the child. With constant study the natural inclinations of the child are led and trained for the healthy reciprocation of purity and goodness; also for the recognition of cause and effect in progressive degree sufficient for the enlistment of truthful confidence. But a few generations passed after the censorship was matured by Analogius before the querulous whimperings of infancy had ceased altogether. When at the close of the second year they entered the nursery department of the national schools, the children were as self-capable and independent of aid for the adjustment of dress to their persons as though they had been to the manner born. In like respect their practical appreciation of cleanliness was as actively demonstrated in purity of intention, as with their more experienced elders. Your people have been taught to believe from precedental prejudice, founded upon the selfish arrogance of ancient exemplars, that the word censor signifies an arbitrary agent for the restriction of liberty under the rulings of tyrannical power. But as with the tribunes of Heraclea, who act as censors under the direction of the prætor, our privileges extend in an advisorial capacity through all the gradations of life, from the child to parents, and in their collective capacity, styled the people. As the censor's vocation is to study and cultivate, for good, the mental and physical capacity of the child from birth, you will readily understand the advantage we obtain for direction in all that pertains to health and the unselfish display of goodness and purity. From the same source in reciprocation our cultivated knowledge obtains a clue to the predilections of instinct for vocations and variations of employment necessary for the supply of food, and the sustaining comforts in currency for the reciprocation of affection. But, above all, we are enabled to perfect the union of the marriage ties by the selection of compatible respondents. You will, however, better comprehend the method of attainment by the rehearsal of our process of education.

" *Our Manatitlan System of Education* commences

at birth, in giving direction to the dawning perceptions of the nursling, that its desires may be toned to its healthy requirements. For the achievement of this important object the exaggerated and inconsiderate fondness of the parents requires the close attention of the nurse and censor, that the material attractions may be strictly adapted for the development of the child's real necessities, in direction for its future mental and physical welfare. At the age of two years the child is placed in the infantile department of the national school, but still continues under the special care of its nurse and parental censor. When the child reaches the age of five years full matriculation takes place, as with the expiration of the infantile term, self-care as well as self-control have become sufficiently impressed for emulous improvement under the exampled lead of their elders. With the full accomplishment of ten years, the youthful term commences, with an easy initiation into the life sustaining responsibilities of community association. But from the earliest stage the children are familiarized with the pastime labors of vocation. These never assume the repugnant features of tasks, but are adopted as useful amusements, from choice, as compulsion and disciplined correction have no part in our exampled system of education. This electic plan of imitation enables them in after life to render needed assistance in association, without novitiation, which would embarrass the continued uniformity of household regulations, rather than aid in their easy dispensation. Within the inclosures, of both the male and female schools, all the appliances required in the pursuits of vocation, and the conduction of domestic affairs, are self-supplied after the first installment of foundation. Indeed, from youthful invention in the school departments we are often indebted for the enhanced comforts of affection. As the mercenary selfishness of morbid craving is unknown, there is an affectionate solicitude with each for the others' welfare in joyous reciprocation. Our grade distinctions of giantesco, medium, and tit, which are usually determined in the seventeenth year, although in premonition from the age of ten, are those of confiding reliance and mutual aid, in freedom from instinctive envy and arrogance.

"At the age of twenty-three the male graduates from the school inclosure into the active degrees of life's associate coöperation. In premonition, the connubial censors have studied and kept a record of instinctive traits, and characteristic blendings of affection, of the male and female matriculants, for comparison and the selection of coaptives in the unity of predisposition for the fulfillment of marriage intention. On the morning of the day that accomplishes the full term of school graduation, the man is introduced to his future wife, who has been returned to her parent's charge for the three month's probationary test of full compatibility, during which her intended enjoys their hospitality. The adjudged unity of these marriage selections has been so perfect in conception, that there has not been a single instance of misapprehension, or one that failed in fulfilling the complete assimilation of affection for sole representation, independent of attaint from the lustful vagrancy of desire. A day of visitation for each school is set apart for the monthly reunion of parents and children. The happiness imparted in anticipation and realization from these visits exceeds by far the utmost capacity of word description; but once enjoyed they give maturity to conception for the full assurance of an affectionate immortality.

"After your visit to the Heraclean schools, we feel assured that 'argument' will not be required to establish the all powerful efficiency of the system in securing affectionate coöperation for the perfection of self-legislation. Your governments for the compulsion of untutored instinct, by arbitrary enactments surprised from the impulsive vagaries begot from excessive indulgence, will then appear self condemned as lunatic monstrosities conjured from and transmitted by hereditary indigestion."

At the close of Codecio's exposition the padre, who had listened attentively, could not withhold his approval, which he characteristically expressed. "I declare to conscience, upon my soul, I believe you are right! But how are we to get on with our national mixtures, when the stronger prey upon the weaker, without laws and government? If we are not able to govern ourselves just now, individually and collectively, I think you must allow that it would be hard to find a better constitution than that of the United States for liberty?"

Codecio. "In answer to your inquiry, a clear demonstration of facts, derived from auramentation, will prove all sufficient for your comprehension of the real governmental status of the republic in question. We have traced the progress of the United States from their earliest date; at first with the hope of influencing the adverse experience of the Spanish, French, Dutch, and English colonists for lenient consolidation, despite their religious tenets, which had been the cause of counter oppressions and expatriation. This was attempted with the ulterior intention of effecting a cohesive tendency from a united education, adapted to their practical requirements for real progression in freedom from precedental imitation. But the repulsive elements of instinct continued, under the aggravations of exile, to grow more rampantly rabid in inveteracy, offering but little hope for the encouragement of our efforts for kindly reconciliation. With the one remarkable exception of the colony founded by Penn, — who tested and proved the trustworthy natures of the savages, when subject to honorable treatment, — the labors of our auramentors were void of effect. Of all the colonists those of New England waged the most ruthlessly relentless war against their aboriginal benefactors, seemingly intent upon offering them as a grateful sacrifice for their selfish 'freedom to worship God.' The next inevitable stage of instinctive fanaticism, was the inauguration of sectarian persecution, in direct rebuttal of their own claims for sympathy, and freedom of privilege. But a few generations passed before the old leaven of hereditary intolerance flourished its sceptre of arbitrary compulsion with renewed vengeful despite. Then a new bone of contention for future generations was introduced by the mother country, who bestowed the African troglodytes as bond slaves to work out the heavenly salvation of the colonists.

"The northern section of the country was alternated by seasons of extreme heat and cold, that rendered the negro an incumbrance rather than an aid, so that gradually the colonists with frugal policy emancipated their slaves, as the second grateful sacrifice to God But the southern department with a semi-tropical climate, and a vegetation spontaneously fruitful, requiring for the production of cotton, rice, and sugar but little cultivation in comparison with the labor bestowed upon the detrite soil of the north, was well adapted for the propagation of the physical inertia of the negro. Also for the indulgence of their masters' *otium cum dignitate* derivation from the English cavalier, the buccaniers of the Spanish main, and subordinate admixture of Huguenotic blood tainted with the religious fanatic absolutism demonstrated by Calvin in friendly bestowal upon Servetus. Under these favoring auspices, the institution of slavery flourished with the southern department, until the increasing herds of mongrelized humanity, and a ready market for their staple productions, brought into full play the old leaven of arrogance. In demonstration of our maxim, that indolence is the hatchment of vice and hot-bed for the enforcement of evil, the southrons began to plot for ruling supremacy, stigmatizing the northern laboring classes with comparisons that discovered in forecast the inveteracy of premeditated hatred. In practical demonstration of intention, after the republican era was well advanced in its first century, they inaugurated the trial of brute force as an argument in the national halls of legislation, not however, with the chivalric challenge of the lion's roar, but with the sneaking approaches of the tigress who dares not brook manly opposition. This overt act, which plainly indicated the design of taking piratical advantage of the supposed pusillanimity of the northerners, for the purpose of subjecting them to dictation as plebeians, set the doors of ' Janus ajar,' until with opposing provocations, concessive on one side, and in degree aggressive on the other, they were finally opened wide for the inception of civil war. The comparatively healthy stamina of the laborer gained the victory, and the slaves were liberated. Then came the problem, ' What shall we do with them ? They are natives of the soil, and if we act consistently we must extend to them the privileges of citizenship, and a votive voice in governmental affairs.' This was accorded, without any initiatory proviso for raising them from animal disability, to an instinctive perception of the responsibilities incurred.

" With this new element added to the antagonistic contributions of Europe and Asia, the attraction of cohesion became more widely separated. Yet with blind infatuation, the progressive stability of a republic of incompatibles was still proclaimed, in defiance of your 'sacred' proverb, which says, that ' a house divided against itself cannot stand.' In fact, we have looked upon your country as a cosmopolitan insane asylum, which naturalizes foreign lunatics for the election of the most desperate bedlamites to office, that in the confusion of governmental discord the kleptomaniac democracy may obtain its votive share of the spoils. Whereas, if the foreign elements had been treated as guest-patients, until their monomanias had been reduced to a condition for the legitimate appreciation of sane example, their children's children, of the third generation, would have realized the benefits of votive unity. Or if the 'pilgrim fathers' had not been blinded by the fanatical infatuation which inculcates the doctrine 'that it is not of ourselves, but through the intangible labyrinth of redeeming grace that a clue to salvation is to be obtained for heavenly citizenship,' they would have extended to the children of their benefactors the privilege of uniting with their own in the advantages of a school education. This course which we have adopted in our colonistic settlements in foreign countries bespeaks for itself an abiding harmony. However honest the infatuation, there should be few sympathizers with the exterminating prayer of the veteran Miles Standish, sighted over the barrel of his musket, in voiced inflection to the report of powder-sped bullet, in behalf of its victim, 'May God have mercy on your soul.' This petition, which he negatived in act, was raised in reverential gratitude for a home with the privilege of worshiping God according to the dictates of his own conscience. This glance, in answer to your question, will enable you to realize the impossibility of adverse elements abiding in concord together, after instinctive habits and customs, with their prejudices, have been confirmed in practice by long usage."

Padre. "If I rightly understand your system of education, it deprives the children of parental care when most needed? This seems to me like refined cruelty, approaching barbarism in its tendency. With all your hopes I think you will find a decided opposition from motherly affection against its adoption by our race. Besides, it is opposed to sound doctrine, which urges children to obey parents in all things."

Codecio. "Bethink you of your wet nurses, and the practical usage of your fashionable mothers, who intrust their infant's nursery education to hireling instinct, and you will find your objections answered conclusively. In addition to the mercenary example of servants, follow the children of your race to the formulistic teachings of your schools in which a majority of your female teachers are yet in their teens, with an experience founded upon precedental rehearsals of the most repulsive description. Then for the illustration of the sordid inconsistency and treachery of your people toward their children, visit with your knowledge the boards of education and you will find ignorance the least objectionable trait, for they make a mart of their influence with dealers in school books that in the display of instinctive selfishness utterly ignores the real advantages of your own system of education. If you will, in addition, review the incidents of your own infantile period, you will discover that you was the cause of more anxiety than comfort to your parents, from the constant rebellion of your instinctive desires against what you then supposed to be arbitrary restrictions. Remember this injunction when you visit the Heraclean school; recall all the events of your past life within the scope of memory that you may be able to place them in the balance opposed to our method for Heraclean behoof. But if you do not make an avowal in commendation, as frank in acknowledgment of the children's contented affection and the wisdom shown in their seclusion, as you made of their parent's worth to Fraile Gallagato, I shall feel greatly disappointed in my estimate of your perceptive goodness."

This allusion of Codecio, to the sacramental night scene passed with Fraile Gallagato at Amelcoy, suffused the padre's face with the scarlet mantle of shame, from which he was relieved when the Dosch petitioned for his aid in adjusting the tympano-microscopes that had been presented by the members of the R. H. B. Society to the Dosch and prætor. Upon trial, when adjusted to the dining tables of the corps and prætor, they were found to exceed in reflective power the larger field instrument brought out by M. Hollydorf. The avenues of the instruments, surrounding the field platform of reflection, were margined on the outer side with a façade of palatial residences, appearing to the unassisted eye like a decorated moulding, with cornice beads elaborately carved, while with microscopic aid each building stood out in bold relief, exhibiting, as a whole, the grandeur of the architect's conceptions. The minutiæ was also clearly exposed, showing an adaptation of intention for convenience and comfort that plainly declared the superintendence of interested parties. The convenient adjustment of the interiors, described by the Doschessa, was a source of unthinking surprise to the padre, until Fabricatus, a Manatitlan architect, announced that he had been commissioned to superintend the buildings while in the process of erection, not only with thought substitution, but actual labor, that in result surprised the superstitious awe of the German workmen. "The palace of the prince Dosch," as it had been labeled in emblazoned advertisement, occupied the obtuse curve of the centre, midway between the tympano-auricular and microscopic reflectors. Its resemblance in external configuration to the ancient kingly palace of San Souci, was immediately recognized by the members of the corps. The Dosch enjoyed a hearty laugh when he read the inscription emblazoned on the central shield surmounting the architrave cornice of the portico. The arms of Prussia were united with an empty shield, evidently designed for the Manatitlan herald's record, over which, and anchoring both, was the banded scroll of hope, with the legend, "In God we trust," while underneath was inscribed, Palace of Prince Dosch of Manatitla.

To M. Hollydorf and Dr. Baāhar it was a source of special annoyance and chagrin, each urging that it should be erased, notwithstanding the genealogical curator of sound proved by quotation that it was legitimately proper and well designed for the expression of ancient usages. But the Manatitlans begged that it might remain intact, as it would prove an ungracious act to receive a present so apt and valuable in aiding personal intercourse between the races, and then mar it because the donor's thoughts were prompted by custom to give it a sounding dedication, in ignorance of the recipient's peculiarities.

"It matters little," the Dosch urged, "what formulistic words are used in the bestowal of a gift, however selfish the insinuation may prove, when void in the possibility of attainment. But the implied arrogation bodes ill for the future peace of Europe, in despite of our auramental warnings, as it indicates a disposition to seek a pretext for the absorption of northern seabord, adequate for the commercial representation of the coveted reputation of the leading continental power. As with Russia, who feels and understands that her power will be ranked second or third-rate, with its almost illimitable stretch of inland empire, without a seaboard with harbors approachable at all seasons, she covets the possibility of attainment. But as fealty to a government enjoins the adoption of laws and usages, the nominal bribe of an empty title will prove hardly sufficient for the encouragement of an instinctive adoption of a diet of kraut, sausage, lager bier, and tobacco smoke, as a viaticum of heavenly translation. However, we hope in reciprocation to bestow upon diplomatic master and pupil an impression that will induce them to aid in an effort for the kindly consolidation of a universal government of self legislation, under the sway of international schools, for the inception of a common language, and reciprocal interests in preliminary course for the introduction of our code of education."

The padre, after the conclusion of his labors, occupied his eyes and newly aroused inception of thought in watchful meditation upon the scene in progress, endeavoring to believe the conjoined evidence of his senses; but, as a test, was trying an exorcising Ave Maria, when Manito and his choir added to his perplexity by singing the following stanzas: —

> "From the maze of superstitions wild,
> Behold the padre a new born child!
> His thoughts from the body's bondage free,
> The viper's fangs will no longer see.
>
> "When he can without the 'fathers' think,
> From the body's grossness he will shrink.
> Then his thoughts set free with joys supreme,
> Gracious love will be his daily theme.
> With goodness gracious we give him hail!
> To immortal joys that never fail."

The padre could not help joining in the merriment the song provoked, while his glowing face attested to the aptness of the conception; still he offered the smiling remonstrance, "According to your own creed, I should be lacking in honesty if I pretended to believe what I can't understand. If the Dosch can explain to me how a man's soul can be saved that was lost by transgression, without the efficacy of pardoning grace, it would please me to learn how it is to be done."

Dosch. "Well, padre, I am afraid Manito's inspiration was a little premature. But let us now dispense with mythological themes for those from which we can truly realize tangible impressions of happiness, as we are now able to add to conversation facial expression, and, in a limited degree, exampled effect."

Padre. "Pardon my interruption. I hope that you do not think me willfully obstinate, for I had much rather you would think me stupid in perception and weak in determination. I know that habits contracted in youth rule in age, but aside from my lack of self control, I can appreciate the fact of the Heracleans' real happiness, and that it is derived in source from purity and goodness. Indeed, I have wondered how they could tolerate me; and can also see the vast improvement made in the habits of Mr. Welson, Mr. Dow, and the members of the corps. If I have been beset with doubts, it cannot be strange to your experience. If I waver again, I hope that you will pity, rather than chide me with vexation; for I shall try to act honestly, according to my impressions. It would be presumptuous folly for me to uphold my frailties and inconstancy against the clear evidence of my perceptions, which cannot fail to realize the truthfulness of all you advocate in the example of the Heracleans. It was the transparent purity of your bodies that made me think that you might represent in translation "the souls of the just made perfect."

Dosch. "We possibly expected more from your perception than our experience warranted; but we thought that your natural goodness could be revived for an appreciation of the success we have obtained with your companions, that would lead you to think, instead of talking from the impression of your senses. But the Heraclean parents have concluded to anticipate the day appointed to visit the schools, and for your advantage have set apart the morrow for the monthly reunion with their " boys."

CHAPTER XXV.

THE prætor and his family called at the quarters of the corps shortly after breakfast on the following morning, to escort the members to the Adolescentium. Instead of proceeding up the foræ avenue to the temple gate, the prætor conducted them to the edificos sacerdotium, and from the court of the centre building led them to an intermural stairway, that commanded the only means of ascent to the temple walls, which were higher and distinct from those of the cinctus enclosure. The prætor in ascending explained that the houses abutting the wall upon the inner and outer faces, now occupied by the teachers, were designed in building to facilitate the mysteries of the temple ceremonies. Reaching the parapet we passed in its walk to a septum wall of an elliptic form, uniting at its distal extremity from the falls, with, but was in altitude higher than the cinctus. The outer and inner walls of the lower enclosure contained an oblong piece of ground of considerable extent. The interspaces between the walks were planted with fruit trees and vines, the mist of the falls veiling it from the brink of the precipice. The priests had undoubtedly availed themselves of these natural aids for the furtherance of their mysterious impositions; its counterpart of the northern temple being subject to the same interposed screen, which closed in the view to all beyond the walls. The undulating upward lift of the misty veil disclosed the familiar blossoms of the apple, pear, peach, and apricot, with other exotics of the temperate zones that the misty atmosphere favored; discovering to us the flavoring source of preserves which we had attributed to artificial production. The prætor informed us that the germs of these were of Manatitlan transportation. The fruitful view on either hand — for the temple garden was also under kindred cultivation — called forth expressions of admiration.

The prætor, addressing Dr. Baāhar, directed his attention to a pyre in the centre of the orchard enclosure. "That," he said, "will answer your question with reference to the disposal of our dead during the siege; although it has been long disused for incineration, we still continue the practice in a less objectionable way. Opposite, at the extreme outer curve of the wall, you observe turrets rising above the parapet; these are the vents to ovens or chambers of incineration, and the urns bordering the garden walks are the family receptacles for the united ashes of the deceased. Our present method is of Manatitlan devisement, and it enables us to reduce the bodies to their material ultimatum. The northern garden is used for the same purpose, the alternation being dictated by the direction of the wind draught in its waft from the cinctus enclosure. We were advised by the Manatitlans that your people practiced inhumation, and supposing that you were prejudiced in favor of the burial rites of your ancestors, with the padre's tenacity, we withheld our method of disposal until your objections had been anticipated by Manatitlan influence. As you have been impressed with the body's corruptibility in diseased materialism, and adjunct manifestations of instinctive vitality, of voluntary and involuntary source, you will now regard with horror, akin to our own, the putrefactive process of decomposition which of necessity imperils the well-being of the living from the entombment of the dead. How have you been able to escape the conviction that your practice of inhumation is cannibalism in a double sense, as you virtually live on the products of recomposition derived from the decomposition of a dead ancestry, and are subject to corrupt inoculation from the putrefactive emanations of decay. The very fact of the festering incorporation of a dead ancestry with the earth from which you derive sustenance, has conveyed a shock to our sympathies, in your behalf, that exceeds our powers of expression, as it is so directly opposed to the current realization of purity. Have you never thought of the material analogy sustained by the bodies of the present generation's reincorporation with the future, in resemblance to the ancient Egyptian theory of transmigration, which led them to associate their embalmed relatives with the bodies of reptiles similarly prepared? The bright array of vessels you see arranged in the colonnades on either side of the ovens are the body receptacles for incineration, but they were designed for bath basins, and used by the luxurious old Heracleans, when they visited the City of the Falls, for recuperation from the effects of excessive indulgence. Their massive thickness and primitive design, with the resistant qualities of the metal, has rendered them proof to wearing attrition through the ages they have been in use. The Dosch, on your first arrival, cautioned us not to be over hasty in making known to you the extent of our utensil resources in this metal, as he said you worshiped it as the god of your salvation, the largest possessors being esteemed the most godly, without regard to the means used in obtaining it. But what could we think of the sanity of your race, when they averred that this god of their worship was the inciter of envy, hate, and revenge, the ministering demons of murder, and its tributary types of woe? Still, with your ready appreciation of our affection, we can scarcely imagine that you were ever ready to sacrifice honor, honesty, and all the endearing ties of instinct to possess, as a devotee, its favors for aggregation, in excess of the requirements enforced by custom, which has made gold an equivalent for an endurable life with your race."

Padre (excitedly). "He does n't mean to say that they are made of gold? Why there is enough to make the Jews believe that Heraclea is the New Jerusalem, and the prætor the promised Messiah!"

Prætor. "One would suppose from the padre's excitement that he had been a worshiper?"

Dr. Baāhar. "A far off worshiper. His sympathy was excited for the failings of a race who were known in their prime as Hebrews. And it is recorded in legendary lore, that one of their number, named Judas, betrayed a person who declared himself to be a son of their god; but they scoffed, derided, and crucified him. He was the originator of the sect to which we belonged. But with regard to your process! are you able to reduce the bones as well as the flesh, without trituration or chemical aids?"

Prætor. "We first eliminate with a slow desiccating heat every evaporable compound of the body, restoring to the air its contingent elements in comparative purity. When desiccation is fully accomplished, the heat is increased for reductive calcination. This stage achieved, calcareous earth is placed in the niches of the oven for residuum absorption of its vapor, then the ovens are hermetically closed, until with the gradual increase of heat complete degradation leaves the organization of the body in ashen representation; through which can be traced, in opaque outline, the silvery white of the nerves, and all the corporate elements, from variation in form and color; but when gathered for the urn, the whole will scarcely exceed a deunx in weight. The urns, as you perceive, occupy allotted spaces beneath the trees of the avenue, without tablets, or chiseled inscription in memorial epitaph."

Dr. Baáhar. " So, so, — certainly your method as a sanitary precaution recommends itself for universal adoption ; while to the doctor of a sensitive disposition, it would prove a great source of relief, as it will abolish the useless investigations of the coroner, founded upon the reslaughter and ghastly exposure of human remains to the gloating vision of the horribly curious. Also the undertaker's advertising exhibitions, and processional pageantries, alike abhorrent with the shambles of the coroner from the reek of contagious odor. And last, but not least, the lying addendas of eulogistic instinct, bestowed in sermons, prayers, and epitaphs charged with heavenly recommendations for the unworthy."

Mr. Welson. " Aside from the negatively politic advantages suggested by the doctor, there is to me something touchingly reverent in mingling the ashes of the good in a family receptacle, common to all in its memorial expression ; and in safety from the desecration of glacial selfishness in track of gold, that, ' for improvement,' substitutes living tenements for those of the dead."

Padre. " But not in safety, Mr. Welson, if the urns are of the same material as the furnace doors and ovens ? "

Mr. Welson. " You are fearfully right, padre, in your suggestive amendment, and a substitution must be adopted before your thoughtless confessional exposure to Fraile Gallagato elicits the prying espionage of his order. Nay, but you need not color so deeply, for we well know that in intention you were guiltless of wrong. Nevertheless, you should learn from your heedless derilection, that the vagrant tongue of confession is lost to judgment and discernment of the rights of self, for you exposed the really good to danger ! "

The silence of the padre showed that he sorrowfully acknowledged the justice of Mr. Welson's strictures.

Having made the circuit of the oblong enclosure devoted to incineration, and the orchard cultivation of vine and tree, our party descended into the school enclosure, the garden of which was planted upon the more abrupt incline of the temple hill. From thence by an ascending avenue, we gained an esplanade overlooking the " court of the foræ," within the temple gates, where the children were congregated with their parents who had already arrived. The prætor and Correliana, each holding in restraint an arm of the impatient mother, whispered their desire that we should remain silent, that unobserved we might witness the unalloyed happiness of parents and children.

The eager impatience of the prætor and mother of Correliana, in joyful manifestation, proclaimed that they, in the protective solace of the second union, had been blest with sons. Looking through the fissures in the rudely constructed doors, two youths, one past, and the other verging upon puberty, were seen standing upon the pedestal plinth of one of the pillars of the court colonnade, nearest to the gates, with eyes fixed in expectant gaze upon the closing portals through which had been admitted the groups of happy parents around whose necks were clasped the arms of loving children. In their appearance, as they stood motionless in the trustful support of each other's arms, watching for the entrance of their primal source of affection with eager eyes, we discovered their relationship from the remarkable resemblance they bore in likeness to Correliana. Although strikingly preëminent in the distinctive halo that becomes inbred from the hereditary impression of matured judgment in parental bequeathment, they did not greatly excel their companions in personal beauty. Tall and graceful, they possessed in common with their companions complexions of clear transparency, which disclosed the movements of expression under emotional control, in freedom from speck or taint. As the portals closed their eyes questioned each other with a shadow of curious inquiry, not in doubt or anxiety, for the welfare of their parents, but for the cause of their unwonted delay. Without being heedless or lacking in sympathy for the happiness of their associates, or unmindful of the cheering salutations of parents and children, it was easy to trace in their faces emotional changes akin to sorrowful disappointment. To restrain the mother's yearning longer was impossible; pushing wide apart the inner gates she stood revealed, uttering the call, "Plautus — Adestus!" But affection in premonition had beckoned their eyes to the source before the words reached them, and the eager parents had hardly overstepped the threshold ere they were clasped in their arms. The consummation of this greeting gave a freer flow to the general expression of joy; the scholars, old and young, soon clustered around us, eager to become known and recognized in the current reciprocation of affection by name, bestowing in love such endearments, that for the moment, with sadness, our own youthful impressions, barren of their cheer, reappeared in contrasted desolation. But translated back to the reality, by the warmth of glowing sympathy, with its unspeakable thrill of tender emotions, the void of our past lives was relieved of its selfish regrets. The teachers we had frequently met, and had found in them such worth garnered with experience in the practical dispensation of exampled goodness, that our nearest of kin stood afar off in comparison with the reverent warmth of affection that these guardian exemplars of youth attracted with the genial current of their sympathy. Well did I interpret from my own impressions the retrospective thoughts that brought frequent flushes to the faces of my companions when the mirrored past was contrasted with the present.

After an hour spent in sweet communion with their

parents, the children were summoned by their teachers to guide us through the school departments. The culinary dependencies were first visited; in these the morning's quota of children were engaged in the preparation of food for our entertainment, with such cleanly decorum that our appetites were revived in expectation. In the "workshops" and garden detachments exhibited the useful combinations of labor, exercise, and amusement, which practiced in communion, gave a sportive air of cleanliness to their employments. During the infantile period, educational impression was intrusted to the nurses, who while inculcating lessons of self-control over the appetites and passions, attracted the affections above the cravings of instinctive animality. Their assurance that goodness was intuitive with the Heraclean children was fully sustained, for in their intercourse they were altogether free from the petulant exactions of selfishness. The teachers informed us that the Kyronese children, on their first introduction, felt the loss of parental association, but were soon weaned by the loving attention of censors and nurses, whose experience enables them to attract, while increasing in strength the ties of parental affection. After the first monthly visit of their parents they became not only reconciled to their association, but emulous of gaining the loving influence that relieved the Heraclean children from petulance and selfishness. This appeared to us strange, as they resembled the children of our own race, whose instinctive selfishness is ever on the reach for more, from its first dawn to the dim vision and palsied mumblings of extreme age. But in explanation, the teachers said, that during the first days, their cravings could only be satisfied by advancing a peremptory claim to everything they saw in the possession of the Heraclean children; who were amused in supplying their insatiable wants, and wonderingly curious in observing the effect produced by their accumulations. When all the material resources of the Heraclean children had been exhausted, the Kyronese were scarcely able to move in their dormitories, which were nearly filled with the miscellaneous collections that had been contributed for the gratification of their miserly dispositions.

"Our own, as well as the donor's curiosity was on the tiptoe of expectation, to learn the next phase in this unexampled manifestation of greediness. For a time, after they found that every portable article of their entertainers had been transferred to their possession, they employed their senses in handling, arranging, and nibbling, until tired, satiated, and nauseated with the changes and selfish gratification of taste. Then they began to look about for some new source of instinctive pleasure ; a view of each other's treasures soon begot a covetous desire for counter possession; this led to exchanges, and haggling endeavors to overreach each other with infantile chicanery ; this practice soon led to squabbles that required our interference, which in turn rendered the trading art unpopular. Next, in course, they commenced purloining, and when the loss was discovered they used disparaging invectives which led to a trial of strength for the recovery of lost articles. They next proceeded to fortification, and constant guard, with occasional sallies for reprisal, the skirmishing calling for our arbitration, and restoration of the articles in dispute to the original owner, caused this method of appropriation to be discontinued, at least in non-edible articles, that could not be disposed of by the mouth. But at night their accumulations of eatables were subject to each other's encroachments, and from over eating, to prevent robbery and discovery, they made themselves sick, which called for the censor to enact the part of doctor, with such success that food in excess of their wants became decidedly distasteful. This diversion produced a thoughtful stay of their selfish propensities, which in train caused them to look upon their accumulations as incumbrances, and at first a somewhat reluctant restoration of the least coveted articles to Heraclean proprietorship. But as the kindly impression of goodness in bestowal began to expand, the petals of affection opened for the full clearance of vagrant covetousness. The grateful impressions of reciprocation soon brought into play, with the elder, their hereditary mechanical resources, which have since proved to them a revenue source of gladness. Of course we aided in the advancement of the selfish fermentation for the removal of the lees in the remedial process of clarification, and reaction of covetousness for the exemplification of its effects to the Heraclean children, to whom its impressions were new."

The padre's smiling face, already known to the Kyronese children, soon ingratiated him as a particular favorite with the Heracleans, and in their charge he soon disappeared, and was afterwards found in the workshop demonstrating the advantage of paneling for strengthening and rendering doors less cumbersome, the parents of the children regarding his handywork with curious admiration. In the neighborhood of twenty acres of land on the southern slope of the hill enclosure were cultivated by the children as a garden and orchard, as well as for the field growth of cereals, with an emulous desire for parental commendation. The distinctions in size being mainly dependent upon age, the Manatitlan gradations were of course impracticable, but the smaller children were constantly under the supervision of their nurses and censors, although not from necessity, as there was an affectionate disposition on the part of the elder and larger boys to offer their backs as steps, and hands as aids to assist the young and weak whenever an opportunity offered. Indeed, the effect of their example, after a few weeks of arbitrary sway, effectually cured the Kyronese children of their fagging dispositions.

Having witnessed the children's proficiency in a variety of useful pastimes, we were invited to visit the culinary department a second time, to see the food in its prepared state ready to be served. The prætor observing our admiring surprise at the ease with which the various manipulations had been accomplished, without bewraying with dust and adhesive mixtures the persons and clothing of the youthful principals and aids, said that each, with intuitive perception, felt that purity within themselves was necessary for the sanction of confidence in associate reciprocation. To be not only cleanly, but pure, without a questioning thought of subterfuge, was clearly the motor influence of every enactment, with the special desire that their personalities should reflect the refinements of reality. "In all the departments the children are taught by example, that their personal individuality may become responsible to itself for acceptable purity to others in current association; so that in health all their wants of instinct are self supplied, although rendered facile by household cooperation, without unnatural exactions that would beget impressions of meniality. From these exercises of self dependence, the spirit of emulation has proved an incentive to invention.

"So you will perceive, that instead of the classical renderings of murder and its congeneric inhumanities, which the Dosch informs me obtains the highest grade of your collegiate honors, our accomplishments and refinements all aim to an increase in affectionate purity, and confidence in association, for real perfection in living assurance of immortality. He also informs me, that this evidence of maturity in judgment would be looked upon with superstitious awe, as of supernal agency, indicating a moribund state of precocity, while with the Manatitlans and Heracleans it is esteemed as a necessary manifestation for the fulfillment of Creative indications. But withal, it has been hard for us to conceive how you have been able to avoid the impression of the absolute cause and tendency of your misery; with the extremes of want and superfluity in your midst, it should have warned your people that they were receding from happiness. In like manner we are puzzled to conceive upon what they found their present and future hopes of happiness, when they are constantly at variance with their own kind."

We were spared the full sum of his wondering inquiries, by Plauto and Adestus, who came to announce the hour of refection. In mustering, the padre and Dr. Baāhar were missing. The padre was found surrounded by the children and their parents in the workshop, having just completed a drawing shave, from a copper alloyed pruning knife, he was in ecstacies from the keenness and permanency of its edge. Looking up, in questioning appeal, to learn the nature of its alloy, his eyes met the prætor's, who answered that all their cutting instruments and tools were made from old Heraclean swords, spears, and other warlike arms. "But of the metals entering into their composition I cannot inform you, as all the armorer's records were destroyed in the sack of the old city; but I am pleased if you have found them serviceable."

"Serviceable!" exclaimed the padre, with astonished admiration, " why, man alive, if it will hold the edge and work like this, you can make your city the richest in the world, according to its size, by patenting the combination, and live like princes upon the royalty!"

" If it will prove serviceable in advancing the peaceful prosperity of the world, I will endeavor to learn the character of the metals and method of composition," answered the prætor; " but in the mean time lay aside your implements, and join with us in partaking of the refection prepared by the children."

Joining in the search for Dr. Baāhar, he was discovered in a natural grotto, engaged in sketching in outline a statue garlanded with fresh vines. When aroused from his penciling meditations, by Correliana, he accosted her archly in the apostrophic style. "Ah, ha! so, so, Mistress Correliana, I have caught you at last? I see that your young gentlemen still pay their garlanded respects to Sieba the Vendic goddess of love! Moreover, in the future I shall claim a sort of cousinship with you, for your Roman ancestors in borrowing the Arconan goddess of Rugen isle to associate with their Venus, accepted a German as well as a Slavonian deity. But where are the associate representatives of your borrowed Nemisa — Flyntz, Zernbog, Iphabog, and others of the fraternal godhead — which should be in company? I hope, for relation's sake, your people have not enacted the part of iconoclastics? for they were wont to hold near association in Vendic mythology."

The doctor's illusive antiquarian nest was here robbed of its cuckold eggs by a laughing exclamation of the mayorong, who in apologizing explained, that the supposed garlands were vine disguised Kyronese mousetraps, which were woven with leaves and flowers to prevent detection from the instinctive caution of the little rodentian marauders. This revelation collapsed the doctor's enthusiasm for his discovery, which he supposed to be a sure indication of the Heraclean's surreptitious worship of Pagan deities. Upon questioning the lad who had fabricated them, he stated that they were made to capture the destructive pirates of the banana patch, and that he had selected the head of the grotto image to keep the leaves and flowers fresh until night.

His denouement was a bonne bouche for the padre, who was in feudal arrears with his Irish bulls begot from hybrid mythology. His mirthful thrusts caused in the doctor's mood a show of testiness, until Correliana reminded the exultant padre that it was hardly generous to pursue his advantage before strangers. With all his reverence and submissive obedience to her will, he sottoriously muttered in thought, " Does he think that a turban will make a turk, or a wreath upon an image declare it to be an object of worship?" The mirthful flashes of the padre's eyes from beneath the wreath of Kyronese and Heraclean children surmounting his shoulders, with the frequent checks he placed upon his tongue, enhanced the humorous infection, to the evident discomfiture of his snuborian foe. Naturally endowed with the elements of strong affection, his habits had stimulated misplaced confidence, which had placed him at the beck of imposition and negotiable friendships. Of genial warmth, when the object was present, but with absence, his remembrance would relapse into hibernating torpidity. These superficial traits had subjected him to impositions without lessening his susceptibility to repetition. The Dosch had recommended Correliana and the prætor to observe his peculiarities closely, as from his superficial range of impressions they would obtain an idea of the leading traits of representative democracy, peculiar to the civilized races. Although in the manifestations of innate goodness he was not only an exception to the majority, but a rarity with the minority, still the evanescent durability of his affectionate impressions, depending upon the superficial current of precedental routine, that delights in the sensational excitement of the senses, was a typical reflection of the masses. " You will find a majority of those who patronize the legendary motto, ' What shall I do to be saved,' like the padre's original self, when first encountered by Correliana. With a quid of tobacco in their mouths, and a pipe projecting therefrom, and a glass of demonizing spirits in their right hands, while from the effect produced ' they cry out in the anguish of spirit, What shall I do to be saved from the wrath to come?'"

The refection was dispensed by the children in the garden colonnade, who waited upon the requirements of their parents and guests with such joyful alacrity that affectionate reciprocation reduced the limits of food to an availing necessity, which caused the padre to exclaim with impulsive fervor, " I wish to goodness gracious Jimmy and all the rest were here!"

The day was far advanced, when the chief censor, in behalf of the children, expressed their gratitude to the members of the corps for their deliverance from the inveteracy of savage hatred. Then as a closing memento, Correliana read the nuptial record of the few that were about to graduate, that the members of the corps might hold the traits in memory for personal comparison and selection of candidates in their next day's visit to the female school. At our departure, after evening song, in which it was the children's special delight to join with their parents, we were made sensible of a grateful share in their affectionate memories; but the padre's kindlier, yet vagrant disposition, had been discovered beneath its artificial mask of entailed habit, so at parting he attracted the warmer flow of their sympathies which suffused his eyes with kindly moisture. When he was finally permitted to overstep the foræ threshold of the temple portals, he exclaimed with glistening eyes, " My conscience' sake alive, I feel as if every soul of those boys had passed through me with gladness ; and I can truly and thankfully say, that I feel in the purity of their loving goodness as if they had offered me the only object worth living for. What joy there would be, if our Sundays could be spent in communion with parents and children free from the alloy of selfishness?" The earnestness of the padre's implied petition met with a hearty response from all.

CHAPTER XXVI.

ON their way from the temple school, Correliana invited the padre and members of the corps to pay a twilight visit to her garden. Passing through her father's into her own garden, while yet the upward slant of the sun's rays reached above the Andean peaks, the party were surprised and startled by the winged hoverings of a cloud of birds of every feather, accompanied with the vocal salutation, — " Well, my goodness gracious, if here ain't the padre ! well, I declare, aha, aha ! " — pronounced in variations of tone peculiar to the raven, starling, and parrot. With a confused fluttering, twittering, and tonguester terms of speech, they with encouraged familiarity, alighted wherever a perch was offered. Correliana tried to still their clamors by calling upon the leaders, but only effected a change, all uniting in the word " Musick ! " To stay their noisy importunities she beckoned her visitors to be seated, and then under the escort of her feathered choristers she brought forth an instrument of music which Captain Greenwood had presented to her as a parting gift. On opening the case she presented the instrument to the curators of sound, who were known to be excellent pianists. From its resemblance to an accordion they started back with horror, without touching it, which caused the beaming face of Correliana to become overshadowed with disappointment. But the humorous smiles of the others relieved her from sudden apprehension, by suggesting, as the cause of their

15

dismay, some foregone amusing event. In explanation, M. Hollydorf described the mechanical affinity of the accordion with the primitive bagpipes, which to the modern musician were a nightmare revelation of the past ages of discord. "Except in its improved capacity of breathing sweet harmony in the hands of an experienced musician, the accordion has the same monotonous drone of its ancestral relative. The source of Signor Pettynose and Herr Lindenhoff's chagrin had its origin on our voyage hitherward from the annoyance caused by one in the hands of the *Tortuga's* cook, which they purchased and threw overboard, and its ghostly resuscitation in your hands has given rise to their expressions of horror. I perceive that the instrument in your hands only bears an outward resemblance to the accordion, and the moment its tones are revealed, I am sure my impressions will be sustained, and the artists will be more enthusiastically retentive in its praise, than they have been in pantomimic rejection."

While M. Hollydorf was soothing the wounded enthusiasm of Correliana's affection, the instinctively sensitive curators passed the case, with its instrument, from one to the other, with an expression, kindred in acting translation, to the effect likely to be produced upon two civilized or savage bachelors in the armed disposal of an infant which had been subjected to their inspection, for commendation, by a fond mother. Finding that their former criticisms of Heraclean music had placed them in a dilemma that required vindication, they questioned each other's ability for extrication. Pettynose having used an accordion in boyhood as a dernier Alma Mater for the nursing of his musical faculties, offered in acknowledgment of his debt of gratitude, with manifest reluctance, the tribute of his experience in expiation of his long neglect and indifference to the rudimentary ties of affection. With the first outbreathings

of the foundling, as his fingers deftly caressed with
familiar touch the well known features, he became
conscious that the ties of relationship had been rendered harmonious by a foreign marriage. Reassured
by this discovery the petulant asperities of his face
vanished; then after a short wandering prelude for
thoughtful familiarization, he lapsed into a musical
reverie of the past, that gradually caused his disem-
bodiment from the petty assumptions of instinct,
leaving his natural spirit of goodness to soar in flight
upon the wings of sympathy. In a few moments he
became lost to material impressions, other than from
the imposed invocation of his fingers, causing the
colonnades and courts to become tremulous with the
lulling concord of sweet sounds. Correliana with
hands reverentially folded over her breast leaned
against a vine-wreathed pillar, regarding his face and
fingers with her large luminous eyes overshadowed
with a misty veil of thoughtful inspiration, as if in
search for the mazy source of the mysterious influence
that held her entranced within the spell of inwrought
concord. But the motor spirit of memory in reviving vision bore upon its talismanic wings the artist
far away from self to roam among scenes bright with
the revels of childhood in the land of his birth, on
the banks of the swift flowing Amaril, whose cascades
embowered by the tropical hill groves of Brazil had
inspired with the rippling flow of their echoes his love
for music. The reveried air of "Home, sweet home"
surprised his listeners with a responsive echo, that
held them immovable in hushed silence, with a controlling power that banished self. Even the harsh
discordant screams of the parrots, calling for the ves-
per notice of their mistress, were hushed, as if suddenly made aware of their voiced defects; while birds
with voices attuned to song in cadenced time swayed
silently listening upon their sprayed perches, eying
askance as if in search of the new songster from

whom the sweet notes came. As minutes unheeded winged their way into the current of the past, and the night shades of twilight deepened, stronger grew the charmed bondage that held Correliana and her mother dumb and motionless, bound by the sweet chords of melodious inspiration. But alas, as if to typify the ephemeral pleasures of sense, the spell was rudely broken by the grosser instinctive impressions of the unfortunate padre, who recalled the wandering spirit of Pettynose, by asking, " Can't you play Yankee Doodle,. Jim Crow, God Save the Queen, or something we know ? "

The reader has undoubtedly felt the chill of sudden obscurity when the mellow light of a declining summer sun had been intercepted by a thunder gust, and the startling effect produced by the lightning's dazzling gleam that makes murky darkness palpable after its transient blaze. This gleam the aroused Pettynose darted on the padre, as he thundered with quavering voice : " You soulless son of a paddy ! are you so dead to the divine influence of harmony that you could not feel that I was moved by an inspiration beyond the reach of the time-serving twaddle of national humbuggery and the idol worship of sectarian selfishness ? "

As the rumbling growl of the enraged musician ceased, the soft expression of Correliana's face was for a moment lighted with an expression of reproach directed to the reproved and reprover. The padre, whose lack of thoughtful impression had invoked this outburst, turned with flushing winces from face to face, questioning the source of his error, but only met frowns of pitying, or disdainful reproof, which prudently inclined him to silence. Pettynose, restored to his instinctive self, examined the instrument to discover the attachment that had contributed by its aid for the production of sounds of such pure accord, in freedom from the drone of its prototype. Sliding back the key-board his vision was introduced to a novel mechanism, bearing but a slight resemblance to that of the accordion, except in formulistic fabrication. In the place of a reed-board of wood it had one of glass. Covering the openings were reeds of bamboo answering to the stops of the keys. Raising the plate he discovered on its under side longitudinal fossæ corresponding in length and form with the string attachments of the harp, which it represented in miniature; over this the peculiar strings were strung. The wind in passing from the bellow's font through the open slots caused an æolian vibration, which was increased in volume and sweetness of sound by the vibrating accord of the reeds. The spirits of the two curators of sound were highly elated by the discovery of this rare musical waif; at the same time were surprised to learn that it had been the companion of their river voyage; but readily accounted for its concealment with the supposition that Captain Greenwood withheld it from the idea that it was to them an object of aversion. Pettynose, when leaving, would, in his heedless selfishness, have taken the instrument with him, but M. Hollydorf, anticipating Correliana's anxiety, bade him recollect that it was a gage of affection.

When the music ceased, a raven and parrot, who had perched upon the padre's broad-brimmed hat, commenced a gossiping promenade, backwards and forwards upon the diametric extremes of its circumference, alternating the rise and descent of its rim from his nose to the back of his coat-collar, to his great annoyance, which added to the comical effect. But the padre continued silent, notwithstanding the birds' quotations of his familiar phrases, " Well, I declare! my conscience' sake ! " and the like, which seemed to be prompted by the changes wrought in the position of his hat. At length Correliana became mindful of their annoyance, and despatched them to their roosts;

then she apologized to the padre for the liberty she had taken, by saying, that they had imitated her when she was repeating his phrases to familiarize her ear with the intonations of the English language. But I am not alone accountable," she said, "for that demure personage," pointing to the prætor, "has largely contributed by his patience and perseverance to their proficiency. The Doschessa wishes to have me remind you that their imitations are an apt example of ritual observances, classical educations, and fashionable accomplishments, which are styled the progressive features of Giga civilization."

"It pleases me to hear you try to make the padre understand the difference between a practical education and one of words," patronizingly added Dr. Baāhar.

This assumption of the doctor's dispersed the depressing cloud that weighed upon the padre's spirits, who replied, " Ah doctor, you forget that to-day you were unable to make the material distinction between an ancient goddess of your fathers, and a Heraclean statue crowned with Kyronese mousetraps, even with the advantage of your wordy education?"

Mr. Welson laughingly commended Correliana and the prætor for their successful essay in the professorial art, offering to recommend their talents to the Dominican College of Gautemala, or its Jesuitical propagandic rival institution of Chinandagua of Nicaragua, which were devoted to the education of parrots for the dissemination of their tenets among the people, if disposed to enlarge their sphere of usefulness. Declining, in the same vein, his intercession in their behalf, the members of the corps were invited to join the family of the prætor at the table, where they could have the advantage of seeing and hearing the Manatitlans, as the Dosch was desirous of joining in the conversation. The voice of Correliana aroused chirping murmurs from the leafy coverts of tree and bush, but with such drowsy pipes of recognition it was easy to discover that the notes were muffled in the head's feathery couch beneath the wings. When seated, the Dosch addressed the padre as follows, " Your race claim that the chief object of their lives is to obtain present and future happiness; now I would like to ask you whether your ' system of education,' founded upon the parrotic rehearsal of progenitorial self-inflicted woes, has any tendency for the fulfillment of their hopes? Then answer, with thoughtful consideration drawn from your day's experience, whether you have ever approached so near the shrine of an enjoyment, so pure and unadulterated, in joyful impression, as that of to-day?"

Padre. " I declare to gracious, never!"

Dosch. " But would you not relapse at the sight of a priest, as the converted Jew did at the chink of the shekels?"

Padre. (Scratching his head.) " Well, you know it's a hard thing for one to abjure his religion, altogether?"

Dr. Baáhar. (Testily.) " He's doomed to blind martyrdom in defense of his idols!"

Padre. " But I know those that belong to my own creed, and am not willing to be caught with mousetraps! Besides, when we were together on board of the *Tortuga*, after our good luck, I asked you to advise me about the education of my children; and you replied, ' Let them glean all the knowledge they can from the schools of their country, until they arrive at the age of twenty; then send them to Germany, and they will then be able to appreciate the rudimentary principles of our philosophical course of study, if they are well posted in physical gymnastics and sword exercise.' Of course the Manatitlans are well acquainted with the gigantic self-esteem of the French ideas of education, which is expressed in the proverb, ' Live in Paris a year, and then die, content.' "

Dosch. " The philosophical self-complacency of the Germans with regard to the benefits of what they are pleased to style their system of education, is dependant upon habit rather than merit, even in the scale of civilized estimation. But when reduced from its superflage to reality, the course of study pursued in a German university, of sufficient celebrity to attract foreign students, commences with the majority, — from the boasting authority of their own statements, — by a test of the body's capacity to hold beer, saur kraut, and sausage, seasoned with tobacco smoke, for the encouragement of philosophical ruminations. When the freshman devotee's body has become adequately distended, ethics and the art of disputatious war are inaugurated, with the premeditated intention of testing antagonistic skill in the gentle art of tattooing. Although lacking in the graceful designs of the more primitive races, the facial carvings in these friendly encounters indicate a nosological taste for depiction, characterized with boldness of touch on the part of the successful aspirant for honors. After having had his passions slaked by sword indenture indemnification, for the aggravations of guttural opprobriums, swilled at the collateral troughs of learning, the candidate for high collegiate honors, with his initiatory degree tattooed in autographic commendation of his skillful attainments, enters upon his second term. This completed, his body has become seasoned to saturation with the constituent elements of lager bier, schnapps, saur kraut, pickled herring, sausage, and like condimental retainers, which are incorporated with tobacco-smoke in preparation for the study of the Oriental languages. At the expiration of three or more years employed in hard smoking, drinking, — or lageration, as it is classically termed, — with concomitant study, and wrangling, he reaches the acme of Teutonic elaboration; becoming to all intents and purposes as useless a casket of automatical movements and articulate sounds, as the feathered theologians educated in the Jesuit and Dominican Colleges of Chinandagua and Guatemala. His sympathetic impressibility can be truly likened to the saw of the surgical sawyer, in feeling for the suffering it imparts to the integuments and bone in separation. The French proverb, which you quoted, is certainly apt. For those who have engaged in the follies of Paris for a year, of self accord, we have found so utterly absorbed in the vanities of selfish gratification, that even the legendary memory of an instinctive soul has been lost, as well as all thoughts of purity and goodness in devisement for reciprocation. The real fact must now be apparent to your impressions, that the generations of your race have turned their backs to the realities of the future, — which could be secured for those that are to succeed them, — in pursuit of the will o' wisp phantoms of the past. To-morrow, you will be able to realize how we preserve and improve the germs of purity and goodness for transmission, in freedom from the frivolous vanities of sense. In parting for the night all united in expressing their appreciation of the Dosch's truthful portrayal."

CHAPTER XXVII.

WHEN the family of the prætor called in the morning to escort us to the scholia puellulitas, Correliana received the attentions of M. Hollydorf with marked pleasure; indeed, the happiness of the prætor and her mother was so joyfully exultant, that it attracted the attention of the Kyronese as well as our own. The temple on the north, occupying the esplanade of the second foræ, was the counterpart of the southern in architectural design. But its site was more commanding, embracing in the view obtained from the parapet walk, the latifundium, the grove of the temple beyond the cinctus gate, and the river expanse in the Bœotian vale below. On the south the terraced road could be traced in its upward windings to the brink of the basaltic cliff. To the north the view was in like manner circumscribed by the precipice and its outjutting flank of wooded hills. Within the enclosure of the temple walls the hill-slope to the north was more abrupt and shaded, and from its cooler temperature it was better adapted for the culture of fruitful shrubs and trees. The weissich of the falling water, and ring of the basaltic cubes, was much more distinctly impressed in their ever varying intonations, rendering hearing upon the parapet difficult, while in the colonnades of the eastern courts of the enclosure conversation became irksome and wearying. These effects were produced by the larger concavity in the southern face, by an inclination given to the main body of the water, from a northern curve in the direction of the river current above; a jutting columnar screen, without the falling water, formed a reverberating chamber that reflected the sound northward. Mr. Welson, remarking the effort required for speaking and hearing, asked the prætor the reason of the founder's preferring the shady courts of the northern temple, for the tender female plants, with its greater disadvantages from the louder sound of the falls, when the warm mellow rays of the southern were so much better adapted for the development of the motherly germ of affection?

Before answering, the prætor turned his eyes upon the questioner with a quizzical glance, then replied, — " What you have observed is, from present appearances, true, and we learned that the prætor Indegatus made the selection in accordance with the judgment your discernment has expressed. But, in referring the reasons of his choice to the Dosch Giganteo, he reversed his decision, sustaining his judgment by urging the special adaptation of the supposed objections for counteracting the then prevalent disposition of the Heraclean women for invidious gossip, and their initiation into a staid, thoughtful mood, necessary for overcoming their hereditary inclinations for tongue talk. As I perceive that the question of Mr. Welson echoes your common curiosity to learn the influence of the choice, I will notice some of the effects in their course of development. Yesterday you remarked, while upon the temple walls of the boys' enclosure, that the whirr of the falling water scarcely interfered with conversation, after you became accustomed to its counter resonance upon your voices, while here we are obliged to seek the screen of a turret, and then speak and hear with difficulty; not so much from the overwhelming loudness, as the confused blending of sound, that renders articulate modulation tiresome. Although partially overcome when the ear becomes accustomed to the impression, still the monotonous replication of variations in kind, without order in sequence, is too close, as the Dosch informs me, in its resemblance to the unmeaning plash of words, to distinguish in utterance those void of affectionate sympathy. This toning influence imposes a thoughtful silence upon those inclined to speak in freedom from the sympathetic direction of thought devisement, encouraging a mood for the study of individuality. The Dosch advises me that a prattling novice, from your race, would soon discover that expressionless words became involved in the wish-a-washy plash and whirr of the falling waters; and with repetition would feel the reflection, in burlesque effect, for the enlightenment of her understanding, when fully sensible of the vague implication, and of necessity would be obliged to limit her speech to the honest expression of affectionate emotions. This, he says, has rendered the cataracts of your country unpopular with your fashionable ladies, after the sight-seeing impression has been gratified; still springs being preferred as a place of fashionable sojourn, as they neither confuse or rival in noisy revel their tongues. In proof of what I wish to convey to your understanding, you will perceive that upon useful subjects of enlightenment, the Manatitlan voice is readily heard by our accustomed ears. But when I pronounce fashionable dress, society, public opinion, theatre, and like synonyms of multitudinous expression, the sound becomes confused with the noisy repetitions of the water, requiring labored and labiate vocalization to make you comprehend their import. Indeed, I perceive that your ears are at as great a loss to recognize the familiar words, as I am to judge of their meaning from the mimicry of sound. But how quickly your perceptive attention is attracted by the sympathetic tones of my voice when attuned in approximation to an affectionate wish! Our scholastic ninietas never turn a deaf ear, or a void eye from an expression in word or emotion prompted by kindly affection, during the heaviest roar of the winter's flood. But folly invested with the blatant mechanism of Demosthenic oratory, or the rhyming jointure of poetical numbers, could not be distinguished in the faintest weish of a season of drouth. You will perceive from these hints that the elements favored the choice of Giganteo; and will now be able to test the wisdom of the preference that subjected our females to the restraining influence of this water power, so effectual for the suppression of a voiceful tongue, hollow in the resonant expression of truthful sincerity. We have been informed by Manatitlan auramentors, that your women are almost universally afflicted with a gabbling epidemic of the tongue, beyond which, and the ear, the impression of their utterances rarely reaches; and we are truly glad that you have an abundant supply of large waterfalls provided as successful aids for the inauguration of a thoughtfully silent era. The shades of the northern colonnades and courts have, by the reflection of this wise choice, been made luminous with the rays of affectionate goodness, for woman's sympathy in its purity and brightness can illumine the darkest night with enduring warmth proof to the vicissitudes of time and place."

Dr. Baāhar. " Since the days of Archimedes there has certainly never been a hydrostatic invention for the practical use of water, that can compare with the beneficial result you proclaim."

Prætor. " The Dosch desires me to give you the assurance that the hydraulic power of the cataract has been so well tested for tempering in infancy and youth a tendency to volubility, that with the least inclination to fanatical superstition, the globular form of the earth might be esteemed the result of providential intention designed for the regulation of woman's tongue, as it necessitates the waterfall in the flow of rivers."

This humorous interpretation of design excited a smile; but Correliana assured the members of the corps, that the effect produced by the sound of the waterfall had been but little exaggerated, inasmuch as it directly induced a thoughtful mood, and disinclination to speak. After a moment's thoughtful silence, she asked her father if the selection had in reality been made with the provised intention of inducting thought by interrupting speech; and if the women of Heraclea had at any anterior date given cause for the constant reproof of falling water, to chide them for the heedless use of the tongue? To which question the father replied, "You must recollect, Correliana, that many centuries have passed since the temples were dedicated to educational direction. Then, as you are aware, Indegatus had been subjected to traitorous annoyances, from which the Manatitlan Dosch of the period relieved him, enabling him to cope successfully with disaffection which had been fanned by woman's tongue. The Dosch also desires me to remind you of the lessons you have been taught of the commune degradation of civilized women in Giga countries."

Padre. "I have often heard of hydropathic treatment of scolding and gadding women, but this is certainly a great improvement, as it obviates by anticipation the ducking-stool."

Descending from the temple walls into the garden court, the necks of Correliana and her mother were suddenly enclosed in the arms of a surpassingly beautiful form, whose face was concealed by a profusion of golden hair, which floated in glancing sheen, like the floss of the silk-tree, over the heads of the united three, closing from view the caresses, which seemed to impart to the atmosphere a reciprocal flow of pervading affection, causing each member of the corps to stand transfixed with emotions transcending by far the highest attainments of admiration. M. Hollydorf stood like a statue fully enravished from self, for he alone had caught a glimpse of the sunbeam's features, as its rays darted from their concealment, animated with a glow of gladness, that had been lying in wait for a joyful surprise. Bewildered with amazement, he was seemingly lost to his personal identity, for he remained heedless and motionless, until recalled by the prætor's salutation. "Luocuratia, my evoce, you must not forget the presence of our deliverers. This is M. Hollydorf, of whom your mother and sister have so often spoken." Then leaving M. Hollydorf, with herself absorbed, he proceeded to introduce the other members of the corps, individually, the names of each Luocuratia pronounced mechanically, in repetition, without even the accompaniment of a furtive glance in diversion from the object of her first attraction. With her arms still encircling the neck of her mother and sister, she looked out from the veil of her hair, regarding M. Hollydorf with changing flushes of perplexed emotion coursing beneath her transparent skin, like borealian flashes beaming in a moonlit sky. Mr. Welson, whose quick perception had caught the source of the spell's inspiration took the prætor's arm, and then beckoning his companions, they joined the happy parents, who added to the fullness of their joy by introducing the members of the corps to their daughters. After enjoying the mutual flow of unbounded affection between parents and children, for a short time, as the centre of attraction, the prætor conducted them through the gate alcove of the garden screen, to an acacia hedge, through the interstices of which they could observe, undetected, the scenes of affectionate endearment, in animated, but silent flow, passing in the conscious enactment of thoughtful impression, between the clustering family groups.

At the conclusion of a pæan song of thanksgiving, they engaged in various pastimes, improvised from the joyous promptings of the occasion, in which both old and young participated. All their movements were so replete with the affectionate expression of gleesome mirth, song, and frolic wit, the paucity of lingual accompaniments was scarcely noticed. The impression of our own feelings, in unison, the padre recognized, who declared, upon his conscience, that he felt a brighter glow of conscious affection than words could convey, imparted from their silent expression of joyous reciprocation. He soon became so wrought with the intensity of affectionate participation, that he could not resist the attraction, but darted from ambush, exclaiming, "Upon my soul, I know that I shall be like a bull in a china shop, but I must be with them," and was soon in their midst, with face aglow from smiling excitement. The young Kyronese maidens, from toddling infancy to seven, — the first stage in the course of instinctive life, — soon took possession of the padre by the right of preëmption, holding him captive from its conferred privileges of priority in discovery; but permitted the Heraclean parents and children to participate in their joy, although holding him as a special bondsman to their arbitrary sway. Detaining Cleorita and Oviata as interpreters, they enlisted the padre as the representative champion of his race in their pastimes. But as an agile athlete his career was more successful for the enlistment of mirth, than for either grace or speed, for he fared worse than Dr. Baāhar in his trial with the family of the prætor, as he was unable to hold the shadows of an old man, of an hundred and sixty years, and his wife. Indeed, his movements and appearance indicated that he was their elder in age, for with graceful steps of equal pace, they kept their shadows from his feet, when in the eye of the sun they were lengthened in the rise of the hill. The merriment caused by his defeat cast no shadow over his happy face, but with buoyant smiles he challenged

Dr. Baāhar's badinage with the desire of testing his right to criticise. This accepted, he was again defeated, without other evidences of chagrin than the frequent use of an apologetic if, in disjunctive evidence of his ability to outrun the best, when free from its restraint. The swift action and graceful motions of the Heraclean women, maidens, and men in running exceeded by far the highest descriptive flights of poetical imagination devoted to woodnymph disportings upon the velvet sod, or those of the sea upon its margin of sand, in derivation from Grecian fable and song. While bestowing the warmest encomiums in the honest expression of admiration, the curiosity of the corps was excited to learn the means by which the graceful uniformity of the women had been preserved, in disengagement from the ungainly inheritance derived from the impression, supposed to be inherent with their first estate. For, with our civilization, a broad expansion of pelvic continuations, with the angular articulation of the lower extremities, are esteemed as a progenic provision for ease in the functional speciality of procreation. The prætor answered from the dictation of the Dosch: "Our censorial guardians have, from the earliest date of Manatitlan direction, recognized the body's unlimited capability for improvement, under the restrictive advisement of an education devoted to the kindly reciprocation of experience. Admonished by the negative effects, described as the resulting cause, that had produced with the women of your race unwieldy obesity, with a consequent lack of animus power for current communication independent of language, they studied to perfect themselves in the Manatitlan art of quality improvement, for increase in affectionate transmission, from the impress of exampled alliance, without words. The Doschessa invokes you to conceive in imagination the impression that would be made upon our women, if, without previous description they should discover a flock of your Giga belles swinging up the avenue of the latifundium, with the longitude, latitude, and circumference of their dresses in oscillating sway from the movements of their limbs in semi-revolution, at an oblique angle from their broad pelvic axis.

Mr. Welson. "Fear would certainly be the first emotion, and I doubt if upon nearer acquaintance they would be able to discover in them qualities of merit sufficient for the stay of disgust. Unless, in their kindly pity, they should look upon them as samples of a female species of humans, who had in penance for stupidity been made to assume the role of jennies, self-condemned as beasts of burden to bear the material emblems of folly. Indeed, when fully impressed with the utter dearth of their conceptive intelligence, beyond the formulistic rites of fashionable instinct, and rote rehearsals of prayers for selfish preservation, from the goading effects of self-immolation styled conscience, even pity would be likely to suffer in trembling hesitation upon the verge of abhorrence."

We will now leave the prætor and Dosch to entertain their guests in the courts and colonnades, while in reversion we complete our description of the garden tableau. After the prætor's departure with his guests, Luocuratia, unmindful of aught else, gazed through her flowing veil of hair upon the face of M. Hollydorf, with the wondering daze of the fawn when surprised in its leafy covert by the gentle presence of woman. With one arm still encircling her sister's neck, yet seemingly unconscious of her presence, she was recalled to herself, from the dreamy maze of her vision, by the voice of her mother. Then she asked with fluttering hesitation, "What is it?" Correliana caressingly removed the arm from her neck, then gathering her sister's flowing hair from her brow, bore it back from her face, and while her mother bound it with a silicoth fillet, whisperingly, with the prelude of a kiss, replied, "It is yourself, Luocuratia, be calm, and to-night you shall know." M. Hollydorf, who had attended Correliana like a doomed shadow, from the day they left the *Tortuga*, thinking and acting from her prompting, even in matters pertaining to his professional avocations, had with the first glance that he caught of her sister's face, stood like one transfixed, his eyes alternating from one to the other, until the attraction of Luocuratia's involved his own. Placing Luocuratia in her mother's charge, Correliana took M. Hollydorf's hand and directed him to a vine-covered alcove in the lower garden walk. When seated, she said, " We are so thankful, for we are now saved from the inherent misery that broods like a pall over your people. You will now be happy, but not yourself again! If I should allow you to recover from the amazement of your surprise, without an explanation, you might think me lacking in truthful sympathy, which we hold, under direction, as the privileged source of our affection. Advised, from the first, of the instability of instinctive 'love' founded upon personal attractions, which is the ruling incentive for marriage with your race, I withheld from you a knowledge of my sister's existence, and our twin resemblance, that her affections might not be invoked with peril; for as you have felt, we are endowed with the censorial essentials of perception in premonition of cause and effect. The long delayed visit to our schools was deferred, for the proof of your susceptibility to our current flow, and constancy in affection; and we are happy in being able to feel the assurance that the transfer of your allegiance to her keeping will be free from regretful reflection. Notwithstanding the long endurance to which you have been subjected, and the severity of the trial for the cure of your self-imposed humiliation, the result not only compensates for your suffering, but confirms the wisdom of the judgment that prompted the restraint, by enhancing the zest with the security of a happy fruition. The relief to me is unspeakable, for in my assisted study of your peculiarities we have learned that from your appreciation of our unselfish affection the idea of returning to your people has become repugnant beyond the endurance of thought. Your sensitiveness so well corresponds with Curatia's in nature, that we are sure her influence will aid you in transferring your sole reliance for happiness to Heraclean keeping, but not in forgetfulness of your responsibility for the welfare of your people. But it is well for you to understand her inability to cope with selfishness, which we are informed, holds supreme control with your race. Even my bolder nature that dared almost inevitable capture by our savage foes, from the physical weakness of our people, from want, shrinks with the thought of incurring the instinctive abuse they are said to heap upon the good and evil alike, who oppose their gainful lusts."

M. Hollydorf's countenance was at first moved with reflective embarrassment, from the self-impressed accusation of inconstancy, but as Correliana made no allusion to his defection, except for the expression in grateful relief, his spirits gradually revived from self-imposed oppression. Yet in attempting to express his appreciation of the remarkable resemblance of features, his tongue refused logical utterance. In anticipation of what he wished to say, Correliana bid him rest easy on the score of the past, as a full relation of all that had transpired would in no way impair the confidence of Luocuratia, but would rather tend to increase the development of her affection from the preference you have shown for her resemblance. This tacit sanction, for the transfer, restored M. Hollydorf's grateful impressions, which raised his spirits to an unwonted degree of elation. But a serious shade of thought having settled upon the brow of Correliana his apprehensions were again startled. Observing the relapse she hastened to reassure him, by asking, "How is it that you, and Captain Greenwood, have remained so long under the rule of selfishness, with natures so quick for the appreciation of our example?"

M. Hollydorf thoughtfully replied: "It was undoubtedly with us as with thousands of others, whose thoughts in association were under the control of evil example, in following the educated usages of the past with unquestioning and reverential reliance, expressed in the fatuous motto of society in all its grades, which contends that 'what has been, will be, to all eternity.' This willfully blind abrogation of creative indications for self-reduction to brutality, has been fostered by a religion that directly encourages evil by offering the means of redemption to the vilest, by rights and ceremonies which ignore the practical evidences of purity and goodness. Offering in substitution, vague terms which lure the stupid masses to present misery and a hopeless material end. A modicum of these prestigical word combinations, the padre has furnished for the education of your tonguester birds; but if you should pass through the streets of our cities, with every step, your eyes, nose, and ears would be saluted with defilements that would cause you to shrink with shame from your kinship with civilized humanity."

"Alue!" exclaimed Correliana, with sadness, "we are so puzzled in our endeavors to understand the source of the misleading infatuation; as the means of happiness is so evident and easy, and their rejection so labored, inconsistent, and unnatural, pardon my sincerity, that we are constrained, from the testimony, to believe that civilized enlightenment, with your other vague terms, are in fact the wordy hallucinations of precedental madness. In the review of our past lives, under the impression of your example, we have absolutely acknowledged the impeachment, replied M. Hollydorf. Even Dr. Baāhar's fantastical ideas of precedental 'virtue,' derived from the vicarious nursing of a maiden aunt, whose celebic worship was devoted to the curative inspiration of a pill-box, which imposed upon him the humors of medical study, has at last in so far yielded to the affectionate sincerity of Heraclean example that he secludes himself when he worships, with the smoke •offerings of the pipe dedicated by imperial and princely lips, as a reflection of worldly honors."

Correliana. "But your women, M. Hollydorf? Do they no longer feel within them the current affection bestowed for transmission with an increase from happy usage?"

M. Hollydorf. "Here, in besieged seclusion, you have had but little opportunity, even with Manatitlan teachings, to learn with a realizing impression the besetting temptations of envious vanity, which have beguiled our women from their natural inheritance of unselfish love; and if their more extended and practical experience has failed to open for understanding vision the vista of civilized woman's folly, my efforts will prove a bewildering aggravation to your already puzzled perception. But if you persevere in your colonistic intention, and are able to sustain the shock inparted from the degradation of your sex from all the hopeful endearments that should render life desirable for transmission, you will, I fear, despondingly lament the hopeless nature of your undertaking. Then, you will, I doubt not, shed tears of bitterness more acute from baffled symyathy, than those bestowed in memorial tribute for your relatives when triply besieged by savage foes, famine, and pestilence."

Correliana. "But you have ruined cities, like old Heraclea, scattered broadcast over the surface of your continents, which bespeak in as plain language the end of folly, envy, hate, and revenge?"

M. Hollydorf. "These are visited by pilgrims of curiosity, who in retailing their conjectural wares of relic origin, give no practical heed to cause and effect for the inauguration of an era of educated prevention. Yea more, on their return to the haunts of civilization jostle with indifference living memorials of a misery as abject in servile dependence upon drones, as the slaves who passed a laboring and starving existence in rearing these ruined fanes of delusion for the gratification of ambitious bigotry and despotism."

Correliana. " But you, as men, represent the different nationalities considered to be the most and least susceptible to kindly intelligence; yet each of you, in your degree, have held yourselves, from choice, with few exceptions, amenable to our example. All of your adherents have acknowledged themselves better and happier than they ever expected to be in life. Still, you doubt our ability to enlist, with the simplicity and purity of our example, the affectionate reciprocations of your women? Surely you speak in riddles of enactment and theory, as perplexing as if in discourse you should say, empty barns full of grain. Are there not many others among your learned men equally able to distinguish that purity and goodness are in reality the source of happiness; and from their own experience, that evil results in misery and woe? Then why do your anticipations forbode for our kindly sympathies a distress so dire?"

M. Hollydorf. "There are undoubtedly many thousands, if not millions, who would hold themselves as gratefully amenable to your affectionate example as the members of the corps, if they could be subjected to the same experience. For we are in no way better than the well disposed commonalty, and were as heedless before we were attracted by your example, as the generality. Speaking honestly, in my own behalf, for my own disparagement, I rarely, if ever, became disengaged in thought from the instincts of selfishness while in association with the most exalted of our kind. In truth, I never felt in the remotest degree that there was a reality in the reputed second existence advocated by our mythology, and was in no way impressed with an assurance of immortality, until we were imperceptibly led to recognize its impression from the example of yourself and people. But you must recollect that our meeting was under peculiar auspices, which enlisted and absorbed our sympathies to the exclusion of self, as if in premonition of the eventful recompense following in train. No favoring circumstances like those transpiring for our introduction, will be likely to prepossess our people in your undertaking, for their own behoof, if we except the sensational announcement which will herald your origin, in connection with our Animalculan discovery. The impression that will be imparted from your exampled exposition of the effects of Heraclean education will prove as evanescent in the substitution of purity and goodness for the material excitements of instinctive gratification, as the opening imitations of the popular humorist, or lyceum lecturer, who attract the attention of their audiences for an hour with quips and snaffles of idiomatic license, or theories as valueless as shadows. If the proscriptor's compilations should fail to awaken their thoughtful interest, in their own behalf, with a realizing desire for the inauguration of a system of education for the benefit of succeeding generations, then I fear that your treasured hopes will find in recognition a tardy requital."

Correliana. " But are not the emotions expressed by your word friendship, the talismanic offshoots of affection ; and will they not aid our example enlisted for the inauguration of a system of education that will bestow upon their children a living realization of immortal impressions?"

M. Hollydorf. " Better by far that you rely upon your own unaided example, and in no way venture your hopes upon the hazard of its trial! For there is not in the word catalogue of instinctive delusions, one so hypocritically heartless and treacherous. Friendship in demonstration with our race, is, as the Dosch has informed you, a 'marketable commodity,' as variable in expressed quality and price as the puff stocks founded upon the gambling exchange of gold. It extends its material aid upon like security in kind, and gold as the medium, is the equivalent of grateful reciprocation. In fact, gratitude and friendship in manifestation with us may be truthfully expressed as an ambuscade of expectation lying in wait for the surprise of future favors. It grieves me that I have no truthful resource from which to impart consolation and assurance, in solace for the encouragement of your proposed adventure; for, to our judgment, the sanction of the Manatitlan auramentors offers the only hopeful warrant of its feasibility. But for the better exposition of the instinctive heartlessness of our race, I will endeavor to give you a true representation of the result of our discovery, if the golden deposits of your mountains and rivers should be revealed to the students of our colleges. Abandoning their studies they would lead in the tide of adventurous emigration, and on reaching your city, heedless of your example, they would take advantage of weakness as a license, that in gratification would defy tears, pleadings, and expostulations advocating your rights of local option. The Englishman would hold it as his sovereign right to do as he pleased, with the certainty that his government would hold you responsible for any resistance to his acts, and with the pretext of an alleged affront, the ocean cormorant would plume her wings and sharpen her beak and talons for your engorgement, esteeming you and your city 'lawful prey.' Emigrants from my own, and nations of kindred habits, would claim the philosophical privilege of corrupting your fruits and grains, by brewing and distilling them into strong drinks; which Tacitus, a historian of your race, alleges was the practice of the Germans from the period of their earliest settlements. But a few days, or weeks, would pass, before your city's present cleanly freedom from the evidences of detrition, would be changed into a sty reeking with filth and saintly odors, and your temple schools into progenic beer nurseries for the instinctive propagation of liberalism, and sogdonian classics, peculiar to the transition period of the incursive pot pourri invasions of the northern, eastern, and western hordes, into Germany. In usurpation of the current flow of affection, that responds in grateful songs of praise to the Creator, the hoarse croaking of maudlin revellers would make night hideous with strepitant grunts of liberty and instinctive patriotism; while in vindication of hereditary privilege, they would exhibit their memorial 'love and friendship' by sword emblazonry tattooed upon each other's cheeks, chorused in medley with oaths from English, Irish, French, and other idiomatic mouths as accompaniments to their manuals in the art of self defense. If your people should adventure affectionate expostulation in behalf of their children, they in reply would exhibit their bloated and bleared visages as the fatherly source of a new and regenerate race of freemen, delivered by the democratic efficacy of saving grace from the pulings of puritanism. Well aware of my inherited defects and unworthiness for the privileged enjoyment of your people's purity, I shudder with the reflection that the current of your affection could be stayed, and forever turned backward, if by rumor the golden treasures in utensil use should be bruited in the civilized purlieus of our cities for the attraction of their troglodyte grovellers hitherward."

Correliana (with clasped hands and tearful eyes.) "May goodness forefend us from a calamity so dire! Better by far the consummation so long urged by our savage foes! But we must still cling to our hopes founded upon your ready perception of an affection that enables us to live away from human bodies with habits such as you have so wofully described."

As Correliana uttered in fervent appeal her invocation, the prætor called M. Hollydorf to indicate the selection he had made from the young maidens to fulfill the marriage intention with the verging graduates of the male department? In answer to this quizzical request, he acknowledged that the only maidens he had seen were Luocuratia and Correliana, but with his happy impressions would endeavor to make amends for his selfishness. All, with the exception of the padre, confirmed the censor's choice, but he with his usual uncertitude of thought made such varied and liberal selections that in consummation they would have proved sadly polygamous. The Dosch had already explained that the education of the Heraclean children had been limited to the practical requirements for the supply of family wants, in conducive aid for the perfection of happy association. So that in the educational department of letters the variety had been of the most meagre description, the quota of information relative to the affairs of the world at large having been supplied by Manatitlan auramentors. Accomplishments and ritual formulistic ceremonies were unknown.

We were more than surprised with admiration, when we visited the kitchen department, in which the manipulations were conducted with such ease and purity, that our previous ideas entertained of housekeeping were quite confounded. During our inspection of the kitchen, the busy hands of a detachment of young maidens were engaged in the preparation of food for the midday collation; their faces the while were beaming with the rays of unspeakable gladness, and their eyes in condimental purity imparted luxurious joy, as a relishing foretaste to the edible results of their culinary pastimes. In keeping the bright glow of the unique utensils, of beautifully alloyed gold, reflected in the convex and concave radiance of their disks the lustrous embodiment of maidenly proportions, with faces comically imaged in grotesque contrast with the reality. The dwarfed reduction of their graceful forms and faces to a semblance in breadth of visaged mouth, nose, cheek, and eyes, to the chattel biddy instincts who hold untidy supremacy in the kitchen departments of civilization, gave a mirthful vitality to the metallic expression that heightened the ludicrous effect, so that under our watchful gaze, it would occasionally culminate in the voiceful melody of a laugh. Purity and order reigned supreme, so that there was neither odor or speck for insect attraction. The effect of this ruling self-dependence was heightened with vivid impression, from the expression of grateful pride that beamed with the emulative exhibition of their " useful accomplishments." In their personalities they were so free from adhesive taint, that the atmosphere seemed pervaded with the clarifying transpirations of beings exalted above the grossness of mortality.

Our own unworthy selves, reflected in contrast from the clear transparency of their bodily investments, caused us to shrink abashed from the hallowed precincts that bespoke in their immaculate purity a perfection that we had supposed beyond the reach of mortal attainment. M. Hollydorf, who was of us all the most sensitively mindful in holding himself amenable to the Heraclean example in personal purity, scarcely ventured to cross the threshold, for among the hand-maidens Luocuratia had taken her place, but with her thoughtful face tinged with blushes shadowed from the dawning realm of unrevealed emotions. Her side-glances, timidly regardful in wondering perplexity, surveyed the object of her newborn attraction in thoughtful search for the evidences of reciprocal impression. But the educated society sophistications of M. Hollydorf's instinctive self clouded the frankness of his expression with the turmoil of impulsive excitement, that rendered him unintelligible and diffident in bearing. This sensitive shield of instinct baffled her longing search for the current impressions of assimilation, imparting to her hands a trembling uncertainty, plainly indicating that her devotions were not fully enlisted for the ritualistic perfection required for the shapely modeling of the oracular cakes intrusted to her leavening touch of purity. To our less enamored vision her touch seemed to impart chaste consecration, for not the slightest stain or discoloration from edible crudity, in preparation for the elaboration of fire, was retained by her hands, so that in contrast we were again inclined to revolt from ourselves. But with all our opposition, intrusive memory forced upon us, with prompted aid, the contrast of swinish priests administering their wafers of dough desecrated by their filthy hands for the unthinking drove specialities of the common herd. In verification of the common impression, the padre whispered to Mr. Welson, "I wish to goodness I dared receive one of those crumpets from her hands, for upon my soul I believe it would shrive me for a taste of purity?"

M. Hollydorf overhearing the padre's supplication cast upon him a grateful look of appreciation. Admonished by our feelings of grossness, we with reverence retreated beyond the charmed circle, but lingered within view screened by a hedge of rose and honeysuckle, through which our eyes paid worshipful devotion to the digital service of the kitchen nymphs. Without the aid of mystic conjurations, the scene seemed invested with a refinement of purity that exceeded the compass of instinct, raising our capacities for the realization of beauty, with a halcyon blending, for the perception of an enduring affection. Spell-bound within our enclosure, delightfully absorbed with our thoughtful contemplations, and nectarious impressions, varied with occasional voiceful melodies, concerted in time to the movements of busy hands and feet, we were startled from our reveries, and retranslated back to the grossness of appetite, by the exclamation, " Oh, for a Tobias sausage, well underlayed and flanked with gamey kraut, and a mug of foamy lager, for I am as hungry as a bear."

The body of Dr. Baāhar appeared in the rear of this hungry ejaculation, enveloped in flowers and cuttings bestowed by the teachers from the garden growths cultivated by the pupils. In a moment the carols of the kitchen celestas ceased, and side-long glances were directed to the hedge to detect the intruder whose guttural accents betrayed the profanity of his petition. The effect produced by this interruption may be truthfully likened to the hush imposed upon the twilight warblings of the water-thrush, swayed in tuneful measure upon the spray by the evening zephyr, and the rippling accompaniment of a flowing stream, when its evening carols are suddenly checked and silenced, for the night, by the croaking heralds of darkness from the sedgy confines of a neighboring bog. Even the padre, whose stomach had many a time and oft remonstrated with indigestive harshness against the introduction of crab salad, — sour-kraut's English and American cousin, — egg-noggs, brandy smashes, and like poetical compounds for its disposal, stood aghast at this profanation of the divinities' edible incantations. Finding himself unexpectedly subjected to an array of admonitory glances, his eyes sought through the openings in the hedge the cause of his cool reception, and with its revelation became aware of his invocation's apostate grossness. As he stood peering through the leafy screen, forgetful of his flowery decorations, he looked like a satyr wood-god of ancient devisement, in orchidian envelope, regaling his sight with a surreptitious view of the grove nymphs while adorning their persons for the festal mysteries.

Correliana, who came with the teachers to escort us to the refectory colonnade, with the desire of the scholars that we would test the relish of their food preparations, aided in disrobing the doctor of his flowery dress; this accomplished we joined the parents and children who were waiting to receive us in the vestibule. The tables were covered with cloths of tinted white interwoven from the fibres of the plantain and tree silk-floss, which produced a novel effect. This cloth was styled Tapalmtræ, a web of lighter texture being used for raiment. When seated, the Dosch addressing us from the platform of the tympano-microscope, which had been transferred from the prætor's table for the day, asked us to bestow our critical attention upon the cloth, to detect its conservative peculiarities for cleanly protection and rejection of corrupt attaint. The brightness, purity, and softness of the fabric, had not only attracted our attention as consonant with the characteristics of Correliana, on the occasion of our first interview, although reduced for the supply of others' necessities, to the limits of modesty, in extremity, but had with the scientific zest of curiosity been the subject of speculative investigation after our arrival in Heraclea. But since our introduction to the Manatitlans, it had only attracted our attention, feeling well assured that all in accruance for mutual benefit would in season be made known. "Its apparent peculiarities, in their partial perfection, we have been enabled to bestow upon the Heracleans," explained the Dosch, "for their advantage during the trials of the siege. Although, from the lack of material, and means of elaboration, imperfect in comparison with our attainments in its illimitable adaptation for the fulfillment of all material requirements for protection, it has subserved with them for the supply of a protective agent to their textile fabrics, conservative in transmissible durability and sanitary purity. Its special adaptative qualities are the extremes of mobility and immobility, and imponderability in degree sufficient for relieving the impressions of weight. These, together with a non-adhesive surface, with a capability for rendering it elastic and non-elastic to either extreme, and indestructible from exposure to the elements, have served as invaluable aids for comfort and their preservation. As an effective aid for increasing the durability of textile fabrics, you can judge when I state that the garments and cloths are heir-looms of centuries' transmission as well with the Heracleans as with our race; an electrical current keeping them repulsively free from impurity, they are to all intents new to each succeeding generation."

Padre. " Why, what a boon the art will prove to the world? especially to the poor, who will esteem you their benefactors forevermore."

Dosch. " It has, with many other attainments, been achieved by goodness for the perfection of purity; and as the miseries of your race are self-inflicted from the stupidity of over-indulgence, its bestowal upon them, in their present state, would prove an encouragement to evil, rather than for its abatement. From this consideration we do not intend to hold ourselves culpable by offering it as a premium for the cultivation of selfish greed and luxurious indulgence. The scientific improvements of your progressive race in the adaptation of vegetable, animal, and metallic productions for the development of their tiger instincts, is quite sufficient for the exemplification of their delusive aspirations, without prostituting the labors of affection for the encouragement of envious hatred."

Padre. "But do you arrogate to yourselves greater goodness in your decrees than God, who bestows sun and rain on the good and evil alike?"

Dosch. "Your distinction of Creative indications in the bestowal of gifts, is, in delusive appeal an assumption characteristic with sectarianism. It should be evident to perception, that Creative benefactions extend to the whole creation, to the reptile, and monkey, as well as to the higher grades of mankind. But the endowment of humanity with powers of discernment to distinguish between good and evil, is an indication of intention that directly implies the privilege of choice for securing the results of happiness or misery. In other words, if man prostitutes his privilege, and makes a brute of himself, he must expect the living void of bestiality, and incapacity for present happiness, with its affectionate premonitions of immortality."

When seated, the prætor, while acknowledging the superiority of knives and forks, drew from his hand its transparent glove, offering it as an apology for the use of their fingers in eating, by showing that it was repellant to adhesive matter. Although instructed in the use of chop-sticks, and knives and forks, they were not yet proficient in their use, and would prefer the use of their fingers with their silicoth gloves if the habit would not offend? This accorded, a maiden was self-assigned to each guest who adjusted Mappas (napkins) to their necks. Luocuratia, radiant with blushes and smiles, assumed the charge of M. Hollydorf, assisted by an Indian maiden of singular beauty. Correliana observing the curious interest excited by her presence and others of her race, introduced her by the name of Toitla, as one of their foster sisters of the Betongo tribe, taken when infants and adopted for a hostage education; their parents visiting them whenever an opportunity offered without attracting the notice of their savage allies, a swinging bridge having been constructed for the northern basin of the falls to facilitate their entrance and exit unobserved. "To their gratitude," she exclaimed with tearful eyes, "we are indebted for the food that preserved us from starvation, when the malignant river savages sowed caterpillars and other noxious grubs upon the wind, from the brink of the precipice, which destroyed our means of subsistence."

After the first course of maize and banana-bread, — styled by the padre crumpets, while under the moulding pressure of Luocuratia's fair hands, — the elder maidens seated themselves beside their parents, the little ones taking their places, their busy eyes watchful for an opportunity to render aid in supplying the wants of their parents and guests. So well versed were they in the language of eyes, tongues were rarely used. Our most skillful performer with the knife and fork caused them to stand on tiptoe with wonder, in view of their rapid alternations in the transfer of food to his mouth, although himself unmindful of special notice. Whether the pantomimic expressions evoked from their symmetrical hands, arms, and questioning eyes, were elicited from the quantity or facile speed in the disposal of food, we could not judge. At the close of the refection, the prætor remarked, that the impression of their debt of gratitude was accumulating so fast from an increase in happiness, they felt sensitively the poverty of their resources for making suitable returns. "But if you will only wait with confidence, our dispositions will find some method of recompense that will prove more acceptable than metallic gold?"

Mr. Welson assured him if true happiness could be considered a meed for equivalent reciprocation, the Heracleans had conferred far more by their example than they had received.

Dosch. "Then you must fain remain content with

each other, and bestow your mutual aid upon the less favored for the recognition of your source of happiness. As the day is drawing to a close, perhaps Dr. Baāhar will favor us, and the other children, with his impressions and ideas derived from his associations of the day ? "

The doctor, without apology, responded as follows: "During the day I have been so enchanted with the harmonizing voices of the parents and children, free from chiding, whining importunities and reproachful bitterness, common to our schools, both male and female, that I was often prompted to speak to you of the effect that has ever been accorded to harmony in musical concord, from the remotest antiquity; but checked myself from reverting to classical fables in view of the brighter reality of your example, which has impressed me with the reflection of a future, made glorious with the realization of your true affection, as the only abiding source of happiness. We feel ourselves novices in appreciation and capacity for reciprocation, as well as in the power of self-command, but will treasure your loving example for a clearer perception of our faults of omission and commission. Notwithstanding our gratitude has but recently emerged from its cocoon of selfishness, we feel that its rays are brighter, warmer, and more kindly in their influence and extension, and truly hope that we shall be able to reflect your example for the lasting good of the well disposed. If the possibility or probability of reducing a woman's tongue, young or old, of any race, to the limits of useful, witty, or consoling speech, dictated from thoughtful impressions for kindly reciprocation, had been advocated in my presence by the members of the royal scientific societies of London, Paris, or Berlin, I should have given less heed to their arguments in support of feasibility, than to the babblings of a brook. Or if in prophecy, the scenes of to-day had been foretold as a probable event likely to occur by any transition, I should have attributed its source to the fantastical chimeras of a fool. Moreover, if in thought suggestion the Manatitlan auramentors had substituted the idea that I could improve upon ancestral precedents, I should have thought myself, when free from their influence, subject to the freaks of insanity. Albeit not much given to respect in following advice, or imitating parental example in my youth, still both law and gospel forbade one to think himself wiser in his generation than his antecedents; from this prevailing authority we expected that our men would wield their swords, and the women their tongues, in opposition to their own promulgated ways and means of salvation, to the end. From the light of this morning's example I can realize, in view of the past, that inconsistency is the soul of instinctive selfishness, as well as the 'substance' of law and gospel, upon which we found our vaunted civilization. In addition, your system of education founded upon the practical adaptation of study to the requirements of life, makes me feel that I have used my brain as a store-house for the vile and useless lumber of past ages, which had better have been buried in the instinctive grave of oblivion. In fact, I have hibernated in common with the class styled learned men, in company with the corrupt bodies of a dead ancestry; and while subject to the winter gloom of instinct, have existed in ritualistic dependence upon the fancied nutriment derived from sucking my mental paws, while in truth exhausting my resources of vitality, and hopes of immortality. But whatever there is in me left of rational appreciation, capable of being cultivated in diversion from the baneful influence of the past, shall be devoted to the welfare of future generations, for the abatement of selfish greed which seeks to accumulate in excess of self-requirements to the detriment of others."

At the close of the doctor's declaration of faith, the padre quietly remarked to Mr. Welson, that he fully believed in the Manatitlans and their power of thought substitution. Then, after even-song, Correliana led in a hymn commemorative of Heraclean deliverance, of which the following is an imperfect rendition: —

> " Father Supreme, our guide and stay,
> When sore opprest for others' wrongs,
> In pity, Thou didst ope a way
> To save ; to Thee the praise belongs.
>
> " Guide those, to whom we owe the aid,
> Under Thy sole direction sent,
> That our paths of peace may be made
> Through them the sign of great event.
>
> " That instead of war brings goodwill,
> Preferring kindred love to self,
> That others' joy may prove their skill
> In place of hoarding useless pelf.
>
> "Nor deem it ill, that they can learn
> From Manatitlans peaceful sway,
> Love's power to bring like return,
> And bear from hate the palm away."

After exacting a promise that we would accompany their parents on their next monthly visit, we were permitted to depart, and, as the temple gates closed, held in review, with thoughtful silence, the scenes of the day, feeling within us that they were the index of future happiness for our race. Our thoughtful revery was broken by Lindenhoff, the corps' genealogical curator of sound, who expostulated: "It is strange that the Heracleans still continue to drone the old pæan cadences practiced by the Greeks four thousand years ago, after hearing the Manatitlan operatists; for they are really a wonderful people, and superior musicians, notwithstanding their lack of power for the expression of the deeper emotions of rage, love, and revenge, which are in reality the vitality and soul of our great master's compositions. They show but little versatility in fugue movement, which expresses the gliding power of musical intelligence; this certainly discovers a material lack of appreciation, however accomplished they may be in other respects. In fact, the Manatitlans would be esteemed as superior vocalists the world over, if they could register a little more volume to their voices. I would much rather undergo one of Mr. Welson's practical jokes than listen again to the droning of the Heracleans, for their execution was perfectly shocking, and they have far less capacity in the lower scale than the bumble-bee."

The music taster's criticism provoked a hearty laugh, but the padre, with warmth, exclaimed: "Upon my soul, for the life of me, I can't see any cause for fault finding with sound, when the words harmonized so well with one's feelings of grateful sympathy. A good heartfelt invocation from such voices, which were as beautiful as their faces, should not be questioned by our coarse natures! Why, man alive, if I had had the voice of a nightingale, it would have choked with kindly emotions from the harmony of their affectionate solicitations in our behalf! Faugh, man, your opera tral-la-la yells are as empty as the screechings of cockatoos and the croakings of frogs in comparison! The chord of sympathy they touched is beyond the reach of your Norma quirketizations."

All joined in hearty commendation of the padre's strictures on the hypercritical curator, Mr. Welson reminding him that the Maniculan choristers would have failed to impress his sensitive ears with their excellence without the magnifying aid of the tympanum. "In full chorus, to the unassisted ear, their music would have sounded monotonous, hardly reaching in volume the lisping chirrupings of an infantile cricket, heard from its home in a distant cranny. As with your registrations of impressions derived from its voice, the Heracleans would find Manatitlan instruction wanting in volume for successful imitation. But," he added, as Correliana overtook them, "here is the offending composer; we will now hear what she has to say in extenuation for neglect of opportunity for improvement in the cultivation of fugue flights above the reach of harmony."

Correliana, observing the quizzical expression of mirth that accompanied this appeal, inquired the cause. In answer, Mr. Welson rehearsed the criticism of the curator, to which she blushingly replied: "You will, I hope, consider in our behalf, when I acknowledge the justness of your criticism, that before your arrival we were constantly harrassed with troubles which required the active employment of our people's thoughts in the devisement of expedients for preservation. These kept us occupied with the full enlistment of our sympathies, so that we could only exercise our musical inclinations in the transmitted current of our original songs of thanksgiving. But in our greatest distress we longed for a harmonized extension of capacity, that you have supplied with adjuvantic aids, from which, in time, we hope that we may be able to render you satisfaction, with the evidences of industrious application."

The curator of sound was too much abashed for an apologetic reply; and the Dosch requested Mr. Welson to say, that for their evening's entertainment he would relate the circumstances that placed the "dulcetina" in the hands of Captain Greenwood for disposal.

CHAPTER XXVIII.

AFTER the evening song of salutation, on the day following the members of the corps' visit to the school of the "ninyetas" they accepted the prætor's invitation to join with his family in listening to the recital of the Dosch, which we transcribe.

During the ravages of the "coast" and yellow fever in Rio Janeiro in the year 18—, it made sad havoc among our provincial offshoots of Brazilian parentage, owing to a lack of means for provisionary precautions, so that I felt it a special duty and privilege devolving upon me to give my personal supervision for its arrest. The joint efforts of our Manatitlan corps of censors and nurses soon succeeded in rescuing our adherents from the deadly influence of the pestilence, affording us leisure to render succoring advice to the good of the Giga race. Among the foreigners, one had attracted my particular attention from the fact that he studiously avoided companionship with others, beyond the enforced necessities required for business relations. This, together with other singularities pertaining to his deportment, attracted a desire for an auramental investigation of the cause of his non-alliance with the herd. My first discovery after entering into auricular communication with his thoughts was, that his preference for communion with himself arose from a natural repugnance for association with men in form, whose instincts were degraded below the bestial capacity of the lower orders of animality. This, I soon learned, had its origin from the sympathetic impression of the animus of goodness revealed in desire. While studying his characteristics, as a key for after-thought substitution, I found that the intrusion of an indelicate impression from his own instinctive propensities, or in word reflection from others, gave him acute pain. Or when from natural promptings, induced from a genial disposition, he had been influenced to listen to or relate a humorous story, strongly tinctured with the passionate rulings of instinctive induction, for days afterwards he would subject himself to remorseful reproof. These sensitive traits, indicating a desire for the attainment of instinctive purity, although rare in the associations of Giga men, are by no means singular or unrealistic with the conceptions of the thoughtful. But a lack of discrimination in society association, subject to the arbitrary rule of money, blunts the perceptions of intelligent refinement, under the impress of the selfish policy it imposes for the successful enlistment of patronage. Vulgarity impairs the powers of inclination for refined perception, in like manner and degree with the action of foul odors upon the sense of smell, which renders it obtuse for the delicate appreciation of a well selected bouquet.

With this reflective introduction of our auramentee, we will ask you to picture him in meditative mood leaning against a huge pile of coffee-filled bags, waiting in the shadow they cast upon the wharf to witness the variegated effects of light imparted from the rays of the declining sun upon the beautifully environed waters of the harbor bay of Rio de Janeiro. The surface of the water, with its deeper blendings of green and blue, were tinted with the yellow light, while the rippled wavelets, gently moved by the waft of the evening breeze, sparkled in bright effulgence as their crests toppled and broke in foamy succession.

As the sierra peaks of the des Orgoaes began to

cast their long shadows over the distant foliaged and villa-fringed bay of Jurbajuba, he was attracted from his reveried meditations by the distant strains of music, in harmonious accord with his mood. The instrumental combination in trio was so blended in harmony that he failed to recognize their individual characteristics, until a near approach enabled him to distinguish the movements of the performers. While yet distant his attention was impressed with the beseeching undertone of melancholy that pervaded the apparently improvised variations of familiar melodies, as if in wailing supplication for sympathy. As the boat approached the wharf, within its shadow, the awning was retroverted to admit of the upright position of a harp, supported by a woman yet young, but the resemblance of her features to a boy and girl, sitting upon either side of the stern thwart, proclaimed the relationship of mother. The children were yet within their first decade of years, but had advanced to the stage that rules with its impressions the after course of Giga life, in act, for good or evil. Their instrumental prelude had attracted all within hearing to the wharf, for the unusual tones of sad sweetness proved alike irresistible to the troglodyte negro and more insensate sea-monster of brutality, the slave-ship's captain. The eyes of the mother, whose face was overshadowed by the broad brim of a Tuscan hat, moved with a quick glance from face to face of the gathered assemblage upon the wharf, while she directed the concerted movement of her children's musical appeal, from violin and dulcetina, by touching in timed lead the strings of the harp. When all accessible to her sight had been passed in review, her eyes became suffused with the sad mists of disappointment, which were imparted to her children's, upraised with hope. Drawing her veil to screen her emotions, she commenced a plaintive refrain, her fingers imparting to the strings of the harp an anguished tone of petition, so evident in its pleadings, that the uncouth negroes reverentially removed the turbaned bandas from their heads in recognition of the woful strains, and for the moment were raised above the grovelings of their debased condition. After the third repetition, the instrumental air was changed into an accompaniment for their voices, which in song preferred the following petition in Italian and English : —

> "Father dear, art thou near?
> Then listen without fear;
> We came not to reprove,
> But erring steps to soothe.
>
> " Italy, dear land of our birth,
> Though exiled, the choicest of earth,
> Truly, thou wast cherished for love,
> With only one object above.
>
> " But alas, how frail was my stay!
> Beguiled by a wanton away,
> These pledges of love now remain,
> To haunt me with loss, and the stain.
>
> " To save, I have sought every trace,
> A pilgrim to this distant place,
> Hopeless, I have come in despair,
> And now forlorn, breathe the last prayer."

When the refrain had been repeated for the fourth time without response, or sign of recognition, the mother sank back on her seat; the harp following, with its weight would have forced her backward into the water, but for the timely arrest of the padrone. In a moment her neck was encircled with the arms of her children, who bestowed, unabashed by the curious presence of the assemblage, the spontaneous promptings of their affection, in solace for the encouragement of hope. Never, in the course of a life devoted to auramental association with the Giga race, had I ever witnessed an influence that so quickly dispersed varied evidences of brutality in human expression, as from these manifestations of suffering in alliance with innocence, affection, and beauty, hallowed in preluded expression of emotions by instrumental and vocal music. The repulsive sensuality, so brutally prominent in the slave captain's and their " owner's " visages, which exceeded in the loathsome vulgarity of selfishness the hyena's, gave place to the shadowy reflection of sympathetic pity, as if from the impression of a reality retrieved from the dim memories of childhood. In default of tears, to the moisture of which their eyes had long been dead, they relieved their pockets of the last representative coins of sympathy, for bestowal "in charity" upon these wandering minstrels, who had recalled a flitting reminiscence of a mother's memory, which once entitled them to an alliance with affectionate humanity. In contrast, the black faces of the negroes glistened with moisture from eyes still open to the founts of primitive sympathy; those acting as boatmen collecting the coins with scrupulous honesty, deposited them in the sachels of the children.

The mother, aroused with the continued sound of falling money, for, as with the exampled impulse of panic fear in battle, and the gambler's reckless course in the downward path of fate, charity becomes heedless of self under the associate impression of congregated bestowal, made an effort to free her eyes from tears, that she might give expression to her thankfulness and stay the uncalled-for gifts of money. Then making known her desire to land, the padrone directed the boat to the stairway of the pier, the eyes of the children the while being engaged in a wandering search among the spectators, with a woful expression of loving desire. Ascending the stairway from the water, the motley crowd opened a free passage; the foreigners following the example of the negroes, removed their hats in token of respect. My auramentee had been greatly moved from the first sound of the instrumental prelude, but the appealing sadness of their voiceful invocation enlisted his sympathetic excitability beyond control. Unable, with his utmost exertions, to approach within speaking distance, he followed in the wake of the procession until he saw the padrone and boat's crew deposit the harp and baggage of the mother and children, at the street door of a house occupied by an *attaché* of the English consulate, in a court opening upon the Rua da Dereita. As their entertainer proved to be an acquaintance of the auramentee, he returned to his hotel well satisfied with the assurance of their congenial safety, which had fulfilled his kind intentions. On the second day after their arrival he obtained an introduction, and with an unobtrusive offer of service gained their confidence. When but partially recovered from the anxiety and fatigue of the voyage, they commenced their street perambulations as musicians, with a pecuniary success more than equal, to the exalted expectations of favorite opera singers, which to the credit of the Rioans was bestowed from the enlistment of true sympathy in their behalf, rather than in acknowledgment for their musical talent. The family of the emperor became interested from the universal expression of sympathy bestowed in recognition of their sufferings; although the cause was unknown, they extended to them their protection. Failing in their endeavors to dissuade them from the exposure of street concertizing, by the offer of a less laborious and more pleasing method of rendering their talent provident, they were content to aid them with their special protection and patronage. A week later, in a private interview, she gave them such reasons for the course she had chosen, that they used their power to facilitate the attainment of her object.

On the nineteenth day succeeding that of her landing, my auramentee was detained until a late hour in the evening at his place of business, and was hastening to pay a short visit to his *protégés*, when he was intercepted by a messenger from a friend who had been suddenly prostrated with an attack of the coast fever, who urged him to make haste as the symptoms threatened a fatal issue. We found the doctor in attendance on our arrival, who accepted a thought suggestion, and on the supposition that it was his own, adopted the recommendation, which served to relieve his patient from the fatal tendency, thereby relieving my auramentee from his apprehensions, in time to fulfill his first intentions. This fever scourge of Brazil differs from the yellow type of northern latitudes; as in Rio, during the first stages of accession, it is exceedingly erratic; suddenly appearing in one department to rage with deadly vigor for a few days, and then in apparent transfer, subsiding, to reënact in a remote district its fatal ravages. At a later period of its sway, when the partially exhausted venom has become more generally dispersed, it flits hither and thither with demon activity, fastening upon its prey without premonitory symptoms, perceptible to curative observation, devoted to empirical treatment, although distinctly visible in inceptive cause to our censors. Even with coincident cause and effect clearly exposed for detection in current transfer, the Giga physicians utterly ignore ante-investigation, for prevising the means of prevention. This observance of limits, overleaping adjoining, to locate itself in remote districts, gives plain indication of local infecting agency, and we discover that the fermentable cause was overlooked, and allowed to exhaust itself in putrefactive dissemination. With this hint, in recurring attestation of the fatuous fatalism that will ever attend the curative devisements of humanity, while they neglect the means of prevention, we will resume our demonstration in narrative vindication of the axiom, that remedy is inherent with the cause.

But a few minutes had passed, after the auramentee had reached his hotel, before he was summoned to the house of his Italian *protégés*. On our arrival we found the mother in the height of the febrile stage of the plague's accession, but calm and resigned in thought, although impressed with a premonition of the disease's fatality, which with our knowledge we felt that it was impossible to avert, still we suggested remedies for transient relief. With the morning's dawn, after soothing the anxious fears of her children, she expressed to them her desire to converse with the auramentee alone. Notwithstanding the unusual nature of the request, it was cheerfully complied with. She then related to him the cause of her husband's estrangement and desertion, affirming that her sole object in following him was for his rescue from self-inflicted wretchedness, as she had brought with her a feeling of fatality, that warned her that her own and children's days were numbered. This feeling had been confirmed in her mind by the strange sympathy which had been shown in her behalf, as the source of her sorrows was only known to an appreciative few. We used all our powers of persuasion to induce a more hopeful mood, by endeavoring to convince her that she was yielding to superstitious feelings unworthy of the courage which had sustained her through the trials of desertion, and her long search which had been continued in a manner humiliating to the affectionate pride of a mother in behalf of her children, exhorting her to bear up bravely until she had achieved the object of her mission. With a wailing sigh, quickly suppressed, she averted her face, while with choked utterance, scarcely raised above a whisper, she despairingly murmured, "I have seen him."

Surprise, mingled with an oppressive sorrow, held us speechless; for words of sympathy, however pure in expression, would have added to the pangs of her agonized affection, which seemed already struggling for liberation from the body, held back by her children's love; but divided, and bereaved of the sentient unity of her affection, grief overshadowed and dimmed her assurance of a happy immortality. A silence of many minutes followed, unbroken save by convulsed sobs, which she vainly tried to suppress; at last a flood from the fount of tears enabled her to regain self-command, but only to be borne back for the realization of deeper woe. Her children, with anxious solicitation for the revival of fond memories, had caught the reflection of their mother's lullaby, with which she had soothed them in dawning infancy, when with undimmed eyes she had breathed her affection in song. Then no cloud had arisen to darken with its gloom the joys of her wedded life. The daughter had been encouraged, with guided hands, to touch the strings of the harp during the period of toddling babyhood, when from feeble, faltering incertitude an answering response came to the mother's leading song. Soon her tiny fingers, instructed by a retentive memory, enabled her to render with remarkable accuracy the most difficult compositions within the compass of her reach. The sadly harmonious memorial that had opened with renewed anguish the fount of the mother's tears, was the sleep requiem early impressed on the daughter's dawning memory. Commencing with an imitative prelude, suggestive of childhood's hesitating touch, accompanied with her brother's violin, the various canzanatas were modulated with the far-off lisp of invocation, as if from dawning perception, intuitively increasing in volume until it reached the flowing harmony of present maturity. From the joyous expression of childhood's buoyancy, the strain suddenly changed into the sad wailing of uncertainty, improvised with mournful variations descriptive of their wanderings and disappointments. Again, in renewal, as if led by some inspiration beyond their control, they reached their present source of sorrow. The burden of the plaintive strains was frequently interrupted with sobbing outbursts, rendering their touch tremulous and uncertain, the efforts made for suppression being easily detected by hesitations, which they endeavored to cover with bolder movements. Recovering, as if with the sudden impression of hopeful assurance, there came a stream of melody of inconceivable purity, as from an echo of futurity bearing in waft joyful gladness. This change caused the mother to whisper, with tears fresh flowing over a sadly joyous expression, " I would have so, it is our requiem."

With the lullaby, that was improvised in quick succession, the mother again clasped her hands convulsively, while the spasmodic workings of her compressed lips and trembling eyelids bespoke the inward struggle made to suppress the gathering strength of her emotions. But with the rehearsal of the melodious symphonies of the halcyon days of united love, grief found vent in an abundant flow of tears, which called forth from the auramentee stifled throbs of masculine sympathy. But while the melodies were growing more earnest in the sad sweetness of their expression, the strain suddenly ceased with the startled cry of, " Father ! "

The mother sprang from the bed, but with tottering dizziness fell back, still retaining her consciousness with a placid expression, which despite the ashy paleness of the face bespoke the full consummation of earthly hopes. The children gently opening the door led in the wretched father, upon whose features were imprinted with haggard remorse the interwoven lines of despair. Blind with the searing touch of hopeless shame, he passed the auramentee unnoticed ; then pressed down by the remorseful revival of first affection, he knelt at the bedside and was enfolded in his wife's arms. Not a syllable had been spoken save the word, father, and the auramentee feeling that his

17

voice and presence would prove alike embarrassing, quietly withdrew.

Five hours later, while crossing the palace plaza on his return from a walk on the Botofogian beach, we met the husband hastening back to his house of refuge and partner in disgrace. Although evidently bracing himself for the utmost exertion of his powers of resistance and speed, in opposition to the foe whose seal was legibly visible in the ashy paleness of his face, the wavering uncertainty of his steps betokened speedy prostration. The natives, accustomed to the symptoms, detected the cause of his swaying progress, and held their course as far to the windward as possible, following his movements with eyes subject to the instinctive fascination that a person under the doom of deadly infection attracts. Becoming fully impressed with his condition from increasing weakness, and the fixed stare of the passers-by, who avoided him, his steps faltered and a momentary shadow of dismay caused a wavering of his eyes and lips; but in quick revulsion he again braced himself, with a determination that bespoke the energetic self-possession of the Englishman in extremity. Leaning against the palace wall, which he had reached, he hastily buttoned his coat to his throat, then drawing in his breath resolutely, he again started forward with a defiant stride. But he was in the deadly grasp of a foe who toyed with his mortal powers as relentlessly in sacrificial oblation, as he had with the ties of affection. This fact his tottering steps soon betrayed, for in despite of his desperate struggles he sank back in a half kneeling and leaning position against a pediment of stone, in transition for the tower of a neighboring church, while its priest passed by on the other side hastily crossing himself and muffling his face with aversion.

The imploring language of his eyes, which he cast around with beseeching entreaty for help, moved even the stolid pity of the natives to unwonted activity, causing them to start in search of the brotherhood in charge of the department. But the auramentee, forgetful of the unfortunate's great wrong, gave him supporting assistance, while urging with his voice the necessity for the utmost exertion of self-determination. Pointing to the Hotel des Estrangeiros he made an effort to take from his pocket a card partially in view. Understanding his wish, the auramentee took it, adding to its anticipated intention in his own handwriting, "sick with the fever," dispatched it by a kindly hand to its destination. Scarcely five minutes elapsed before a female form darted from the portal and directed her steps in wild dismay to the stricken one's side, and kneeling claimed the support of his head, while with a kiss she supplicated, " Oh, Edward, what can I do ? " A faint smile lighted his face at this appeal, as he whispered the ever abiding talismanic word, " home," so dear to the honest attachments of instinct, however much misused in collateral signification. The auramentee then entreated him to muster all the energy possible in aid of their support. Raising him with great difficulty to an upright position upon his feet, all his efforts to walk proved abortive, but a kind-hearted Frenchman who was passing, volunteered his aid to bear the doomed bodily to his hotel and bed. By profession a nurse, the Frenchman undertook the patient's charge, after he was placed in bed, but gave no hopes of his recovery ; on the contrary, with the coolness of a physician, urged him to use quick dispatch if he wished to dispose of anything by his will for the living advantage of others, as it was impossible for him to live longer than two or three hours.

A smile, with the answering words, " It is well," aroused the anguished despair of the being, who still ministered with all absorbing thought her tender care and caresses, bringing forth the expostulation,

"Oh, Edward, Edward, if you go, you must not leave me! for wherever you go, I must go with you!"

The dying man raised his eyes to hers with a look of unutterable fondness, then, with mustered energy, whispered: "Julia, it is hard to part from you, after so much suffering. But living, it would prove to us both a continued scene of remorseful misery, without the possibility of atonement, while dying, I have gloomy forebodings that there will be for us no future. Yet, whatever may come after death, it is better that I should die as the cause, than live as the renewed source of misery to others."

Such a look of despairing desolation as she cast upon her expiring lover I had never before seen depicted upon the face of Giga woman. Her beauty, surpassing, in fair unblemishing complexion those of kindred type, was refined by the hopeless anguish of its expression, which in its passionless void betokened, as with him, a reviving hope struggling for the bodily retrievement of an assured immortality. At the expiration of an hour, her arm that supported his head grew lax and nerveless; but his efforts to raise himself recalled her thoughts for his assistance. Perceiving that it was his desire to be left alone with her, while he yet retained his consciousness, the auramentee was prompted to inform him of his kindly attentions bestowed upon his wife and children, as it offered the opportunity of affording mutual consolation, by conveying to his wife and children some affectionate token or message. The announcement revived his energies, imparting to his "allovee" a kindred impression of desire. Beckoning the auramentee nearer, in earnest, whispered accents, he implored him to plead with his wife, Julia's forgiveness, as the "sin" of desertion was wholly his own. "Say to her," he continued, "that it was my own unencouraged infatuation; against which she, loving, did all that she could to resist my entreaties, striving earnestly in the toils to escape from me and 'love's' allurements. She is not wanton, but pure and devoted as a wife can be, although misguided. It is my own 'heart' that is divided, even in death, which makes me feel doubly thankful for its nearer rescue."

Charged with this message we left them, Julia courting the virulence of the malady with an assiduous intention that plainly declared her determination to share in death his grave, in opposition to his own and the Frenchman's vehement protestations. We reached the bedside of his wife in time to receive her last recognition, who answered with a smile and pressure of the hand her husband's last petition, and while passing away invoked, with the reviving spark of conscious vitality, the auramentee's guardian protection of her children, should they survive, as she was aware that they had been seized with the fever in the presence of their father, who had bestowed upon them his care with the intention of returning. After bestowing upon the children his affectionate care in the fulfillment of his accepted charge, he hastened as speedily as possible to the bedside of the doomed husband, and found the dying lovers supported in each other's arms. For Julia, in the short period of our absence, had excited the latent seeds of the infection, and was already nearing the confines of her desire. The husband, although speechless, still retained his consciousness, with the power of making known, with grateful expression, the consolation imparted from our tidings. Julia, in anticipation of death, placed her attendant in charge of the auramentee, desiring him to send her back to Italy, as she had followed her own misguided steps from affection. The auramentee promised the faithful discharge of all their wishes in the event of his own preservation. Then with a sorrowful farewell, in freedom from the bitterness of our first impressions, we left them with a sure remedy at hand for the cure of their self-inflicted unhappiness. Returning to the children, we bestowed upon them our personal care and affection until death relieved us of our charge; but the scenes that preceded their final departure from life are too harrowing for recital. Let it suffice, that on the morrow when the western hills cast their shadows over the city, under the upward halo of the setting sun, the father, mother, and children, with their cousin Julia, whose beauty was the sad cause of her own and their misery, were borne together, in their bodies' materiality, for burial far beyond the city's limits. The place of interment had been granted as a special mark of interest by the emperor, whose family were deeply affected by the tragic end of their *protégés*. The harp, violin, and dulcetina were retained by Captain Greenwood, the auramentee, as mementoes of the sad scenes described, and are held in "devout" estimation as pledges of affectionate remembrance.

Annette, the companion of Julia, while assisting in packing the instruments for shipment to Montevideo, displayed versatile accomplishments as a musician that astonished Captain Greenwood, and while playing some airs found noted in the satchels of the children, she was frequently moved to tears, and in explanation of the cause, it transpired in revelation that she was the daughter of Signor Pozzuoli, the inventor of the dulcetina, and early teacher of the children, a majority of the preserved musical annotations being of her own composition. On the day previous to the one appointed for the sailing of the steamer for Montevideo the captain proposed to introduce Annette to the consignee of a ship about to sail for Leghorn. She then declared her desire to accompany him to Montevideo, as she felt a disinclination to return to Italy, urging that her musical ability would prove amply sufficient for her support, if he would assume the character of guardian for her countenance and protection. From the mutual interest engendered from the scenes through which they had passed, the captain encouraged her decision, gladly assuming the charge of protector. In closing, the Dosch said, I have related the history of the dulcetina, with desire of enforcing the absolute necessity of the Manatitlan system of education, if the Giga race really wish to bequeath happiness from unity in the marriage alliance, as a memorial source of example to succeeding generations. As scenes of the kind are constantly increasing in an engendered series from degenerate inoculation, with thoughtful consideration its practicability must be apparent to the meanest capacity. The relation will also impress upon you the characteristic value of your late companion, when relieved from the influence of habit, as well as the discernment of Correliana, which penetrated beneath the crisp asperities of his outer husk. In the exceptions we are about to advise, you will recognize the prudence of our judgment. The "brides" will surely afford an invincible security from their incorruptible purity and goodness, which, with kindred beauty in personal endowment, would insure constancy in defiance of all the temptations that could be proffered by the most lauded belles of civilized society, even if the ages of their intended husbands were less by two thirds. The countenance of Correliana, during the recital of the Dosch, was a mirror of reflection for the grateful expression of her thoughts.

CHAPTER XXIX.

ON the third morning after our visit to the school of the ninyetas, the prætor and the tribune teachers with Correliana and her mother called at the quarters of the corps, to escort M. Hollydorf to the prætorial colonnades, as the husband elect of Luocuratia, for the fulfillment of his probationary term. After receiving the congratulations of adoption from the Heracleans, all joined in the matin song of thanksgiving in the lower fora. While the prætor and his wife were absent, aiding Luocuratia in her valedictory salutations, M. Hollydorf was entertained by Correliana and the Doschessa. In order that he might perfectly understand the premeditated process of transfer, and security achieved, Correliana stated : " The Dosch had auramentally learned your determination to make Heraclea your home three months ago, and suggested the apt adoption of your peculiarities to her disposition ; but until convinced of your constancy to our customs he advised the course we have pursued. The result of your trial has proved of happier import than we anticipated, as well as of Luocuratia's ready infilmentary adaptation for the unity of impression ; but now you can rest assured that her thoughts have already become interwoven in desire with your own, so that your example will be held paramount to ours. After the bewildering maze your presence caused was dispelled, her thoughts were directed for the shadowy investment of your image with her own as a prelude for more perfect realization, with a success which imparted a trust free from doubt or fear in question of its fulfillment; in this mood I left her, promising to visit her in the evening. In keeping with my appointment, I found her awaiting my coming in the garden, in full confidence that, with my aid, whatever there might be of mystery to her veiled comprehension would be cleared for her perception's perfect understanding. With an endearing caress, fluttering with the timidity of a newborn joy, her eyes drooping in tremulous expectation, were filled to fullness with happy anticipations, as she leaned her head upon my shoulder, invoking with attentive silence my aid for the full interpretation of her waking vision. That she might taste the cup of my own realized joy, without tantalizing prelude, I rehearsed your confided doubts and fears as the counterpart of her own, the while encircling her waist with my arm, in support of her head's nestling repose, that the body's medium of a sister's affection might more fully open to her the gates of revelation; then to the trama of her love I interwove, through the shuttled impulses of her ear, the vibrating threads of your affection, until they became involved with the stamina of your stronger nature; then the rustling sigh of relief bespoke the double investment complete in the unity of confiding reciprocation. This accomplished, inasmuch as the agency of my influence could represent the responsive source of sexual alliance, for the embodiment of affection, she became so deeply absorbed with sweet meditative reflection that she was unconscious of my departure. This ingraft of affection, in surety so propitious, should engender solicitude, on your part, in behalf of your race; for enjoyment ever lacks full maturity, when we feel that there are others with the prestige of purity and goodness, who are denied our privilege from the want of kindly direction."

To which supplication, M. Hollydorf replied: —

"Truly thankful for your pleading consideration, however little my faults merit your lenity, I must ask your continued forbearance; as you can scarcely imagine from the purity of your associations, the depth of insincerity that must ever oppress and haunt me with the bitterness of reflection for my unworthiness, in accepting the boon of an alliance that so far exceeds my present capacity for just appreciation. But if the neglected germ of good intention, brambled by evil example, can be redeemed to offer an equivalent worthy of your acceptance, it shall be my constant study to withhold your memory from the past, which is beyond the reach of extenuation, by the integrity of an exampled affection."

Correliana. "That you may feel to the full extent the confidence bestowed with Luocuratia, my father has left his written salutation for presentation, which with your permission, I will read, that it may convey to you the living warmth of a personal address.

"'To M. HOLLYDORF, *Director of the Heraclean Deliverers:*—

"*Carrissimus*, acting upon the information received with advice from the Dosch and his advisers, and your own confidence imparted to my step-daughter, Correliana Adinope, affording verification of our own observations, that there exists a unity of attachment between you and her twin sister Luocuratia, we offer you with unspeakable gladness our joyful congratulations, with the sum of our united affection. In bestowing our fullest sanction, we are truly happy in being able to contribute, from our Heraclean resources, the means of perfecting our ties of grateful reciprocation, and rejoice that we have achieved the privilege of calling you by the endearing name of son, as you were in anticipation wedded to our affection. In accepting our daughter for the cultivation and solace of mutual affection, you will have our assurance of her enduring devotion, which no mischance can abate; for with her the animus of goodness exceeds in thoughtful intention the power of expression. In her affection you will find an allied support all-sufficient for happy sustenance, and in its overplus an index of the homage outflowing in reciprocation from every Heraclean. In conformity with the happy experience of our ancestors, we herewith, in addition to our verbal invitation, proffer the formality of script, with the desire that you will become a member of our household, in domiciliation, during the three months allotted for probationary exemplification of congeniality in habits necessary for unity in affectionate reciprocation. Luocuratia will return to gladden our colonnades at the approach of noontide, then with your presence our joy will overflow in thanksgiving to the Source of direction, that devised the achievement of our deliverance through your instrumentality. ADESTUS."

Correliana, while reading her father's script welcome, watched with keen interest its effect, and was recompensed to an extent exceeding her expectations, by the warmth made manifest from the grateful emotions of the respondent. Fully satisfied with his relief from the sensitive reproaches of his disposition, for thoughtful diversion from his waiting suspense, she appealed to the Doschessa for encouragement in behalf of her meditated mission for the colonistic establishment of schools for the educational intuition of self-legislation, among the civilized Giga races. " Have you," in your auramental experience, which enables you to reach and advise the thoughts of Giga women, found them all so abjectly subservient to the trammels of society and its fashionable tyrannies?"

Doschessa. " The exceptions are simply modified, all worship at the same shrine of thoughtless fatuity, with a heedless tendency for the utter extinction of purity and goodness."

Correliana. " Alieu, woe is with our hopes, if to such a depth of desecration the animus of purity has befallen with our sex ! Surely, she must still retain the germ of her affectionate inheritance ; and with the lead of our example it must revive with the nourishing warmth of a mother's love."

Doschessa. " In truth, we have never failed to discover in women, when free from the actual vices of corruption, the latent spark of goodness that with exampled cultivation may be revived for truthful reciprocation. Dependent as we have been upon auramentation for the invocation of purity in thought, the impressions are as transient as the conjurations of a dream, which give place to the more tangible waking visions of sense. If the current of their superficial conversation could be stayed, for the silent inception of thought, your mission would be rendered easy. Perhaps the irresistible impression of your own and companion's beauty, will surprise from envious covetousness sufficient thought for the detection of an inceptive source, with the desire for its privileged bequeathment to their children ? For often in selfish lamentation we have heard Giga women supplicate in prayer for the abatement of their own scandalous dispositions. One of their formulistic invocations to the ' throne of grace,' offered as an oblation for the " contrite heart," I will repeat : —

" ' Purge from me hypocrisy, ere I from life depart,
And all deceptions, that belong to the lying art.
Then purify love, from thoughts of material sense,
And make me feel that goodness responds to future tense.'

This accusative conviction is by no means rare, and the purity of your personal appearance, in consonance with exampled goodness, might attract thoughtful consideration from its contrast to the degrading attrition of selfishness subject to the material influence of gold. Your example might lead them in train to adopt our dress, which is light and ' elegant,' subserving all the requirements of bodily freedom, and purity in protection; if so fortunate you would remove the embargo of oppression from their bodies, and the curse of talkative frivolity from their tongues."

Here the Doschessa was interrupted by the sound of light footsteps, quickly followed by the voices and presence of Cleorita and Oviata, who with an escort of Kyronese maidens came bounding into the triclinium with the joyful announcement, "He's coming!" Pale and breathless, Correliana, without waiting for farther words of explanation, sped forth, her feet with dainty touch kissing the earth with gladness, passing with the swiftness of an arrow in its flight those already hastening down the avenue of the latifundium, apprised of the near approach of Captain Greenwood. First, she passed Mr. Welson and Dr. Baāhar, then with graceful ease Mr. Dow, whose lank form and longer strides had distanced his associates; even the mayorong, inured to an active mountain life, and long journeys on foot, now fledged with the grateful remembrance of his people's preservation, was left behind. Then as Captain Greenwood, urging his mule to its utmost speed, caught her view as he entered the cinctus gate, the earth seemed to respond with elasticity to the touch of her feet, and before he could dismount from his quadrupedal conveyance, he was clasped in the frank embrace of her arms, and had received her kiss of welcome, while her face, eloquent with smiles and tears of joy, became radiant with beauty in contentful expression. Her hushed silence, from the fullness of happy enjoyment, was aroused from selfish-indulgence, by the salutation of her father to his already adopted step-son. Still in half embrace, as if loth to relinquish the body temple animated with the shrine of her devotion, she was not forgetful of the affectionate relation she held foreign to self. Her parents and sister, who had followed with equal steps, but had held anxious desire aloof, until the fullness of her first emotions of gladness had subsided, were first made known to each other with affectionate designations, newly fledged from the English idiom. In turn the captain received embraces of welcome from each, which unloosed his tongue from its accustomed reticent caution, Correliana's still encircling arm causing the grateful current of speech to flow in accord with her own emotions.

The family scene closed, M. Hollydorf, in freedom from rival jealousy, gave his cordial salutations of welcome, which were followed by the other members of the corps. Then the Heracleans and Kyronese claimed the privilege of expressing their affectionate gratulations. The mayorong, distrustful of his power for expressing the reverential emotions of his gratitude, for the deliverance of his people from their extreme peril, although second to the prætor's family, in greeting with his presence the captain's entrance, had allowed all the precedence. Approaching the captain, when beckoned by the prætor, hereditary impulse inclined him to prostration, but the humiliating act was arrested by an energetic embrace which relieved him from his embarrassment. While the prætor was gratefully presenting the Betongese for the captain's kindred recognition, attention was attracted by the musical call of children from his incoming train. Mr. Welson, recognizing the voices that were making the name, Don Guillermo's, melodious, on approaching, in quick transition, found his neck enwreathed with the arms of his little favorites, Lavoca and Lovieta, whose eager curiosity, after bestowing their kisses of welcome, inquiringly asked, in whispered accents, who the angel was that embraced the captain, and the other, and others ? Supporting with his arms, their bodies pendent from his neck, Mr. Welson carried the children to Correliana and Luocuratia, "and the others," to receive from their lips the much coveted welcome; which was given with such loving zest, that sweet surprise made dewy their jetty eyes, while their cheeks glowed through the olive tint, embrowned from exposure, with an exquisite blending that enhanced their rare infantile beauty. As all the Heracleans and Kyronese matrons and maidens claimed the privilege of surprised affection, in bestowing the salutation of welcome, it was long before they were restored again to the full possession of Don Guillermo, and then were so mazed with delightful impressions, and wondering gladness, that they were unable to give heedful answers to his inquiries. At length, after an apologetic round of besitos beneath his grizzled moustache, which caused him, laughingly, to utter the interjectional expostulation "bas-tan-te," they in rambling relation commenced the rehearsal of events which led to their transfer from parental care and their natural home.

Observing his inquiring gaze directed to a young woman, whose eyes were occupied with curious admiration in following the changing variations in the scene enactment from the loving outflow of affection evoked by the captain's advent, Lovieta and Lavoca exclaimed with united voices, "Oh, that's Annette, our governante, we love her very much."

Then with childlike simplicity, peculiar to Spanish infantas, they informed him that she was a nice beautiful teacher of music, and everything else, and Captain Greenwood's sister, but not in the regular way, although they were very fond of each other. "Father loves them very much, and when the captain told him he was coming to live here for good, she said, that with his permission she would go wherever he went, and make his home hers. This made him very glad. Then father seemed to be sad with thinking, and then he loved us so much while shedding tears, that when we could speak, we asked what made him so triste? He said, he wanted them to take us to Heraclea that we might be educated so that we would be always good, and could be present with him and mamma although absent in body, which would keep them from feeling sad and lonely. But we could see that mamma and he were very, very triste. This made us sorry. So he talked to us of all you had written of the happiness of the people here, because they were truly good and pure in their love toward each other, without selfish concealments; then we were glad and wanted to be with you. The mammatits you sent, who have been with us all the way, told us all about the school, and how loving the children were toward each other, which made them always very beautiful; but you, nor they, did n't tell us that there were angels here. Then they said that there were no dolls here, for the larger nines helped to care for the ninaquillas, — how very, very beautiful they are, — do they never grow old and ugly here, so that they have to paint their faces, and scold like grandma, because people don't like them? Then, as we were a going to say, mother don't feel exactly safe with herself, and becomes fearful, when grandma talks to her of her soul's perdition, which we can't well understand, only that it's padre Molinero's doings. So she wanted us to come here before we were too old to enter the school, that she and papa might visit us and learn how to be truly affectionate without talking too much; but we don't see, now, how that can be; ay, ay, pobre mamma and papa! But they said it was for our happiness that they wished to send us, which we can now see. How beautiful, nice, and clean everything looks! Is it always so? And papa said, that without us his home would be desolate. Then mamma looked at him wild like, but so pitiable, and choked so, then looked so sorrowful, that we hugged and kissed her and whispered we would n't go away and leave her; then she said, 'Pedro?' soft like, and papa took her in his arms, and we all cried together; but so happy, it did n't seem like crying, and could n't speak for ever so long a time, but then we felt so content, when we thought so lovingly, and said nothing. Then when mamma could speak, she said whisperingly and softly, so that we could scarcely hear her for our crying, 'Pedro, it is better they should go, and I will try to make you feel that your house and home are not desolate.' Then he kissed her and we all cried again for joy. But grandma made us feel so unpleasantly by saying that it was as good as throwing our souls away to send us where there were no priests and churches, we did n't feel any sorrow when we bid her good-bye. So we have come away from bad example to get souls that will make us live as though we had no bodies, for we were very much afraid of death. The mammatits said the Heracleans lived to make each other happy, so that each one was loved by all the rest, and in caring for one they cared for all the rest, so that there could be no grief and repentance for wrong-doing, for all were good, and cared more for others than themselves; and we can now see that you are all gladness, and were sometimes so triste when you were with us. We love father a little more than we do you, because you see we have always known him, and we have n't known you so long. But the gente pequenézas said you were so much changed in disposition we should hardly know you; and to be sure, now that we see you when you smile, we feel so glad; and sometimes, when you lived with us, your smile made us feel sad, as though we 'd rather you would not."

Lavoca. " Yes Lovieta, just think, did n't his smile at home remind you of the rose blossoms that look out from the old grated window of the claustro San Jaun; which seemed for all the world as though they wished to come out into the sunlight, but could n't, because they would n't confess it was sinful? But they say you never speak cross words here, for in loving to do so much for others, without money, you have nothing to scold for. You see we never told mamma all what the Manatitlans could do to make us good, as they said the priests would persuade her mother that we had dealings with the devil. Señor Arbitrator, the one that used to talk with father most, told him all about the schools, and how you live here. How queer it is that you can hear them talk when you listen to what they say in your ear, and can scarcely see them on your hand, let it be ever so clean, for they are very particular. Then their voices are so very small and chirruppy. But father says, that they are louder in proportion to size than the cicada's. Mother was very loving and cheerful after they came, but very much frightened; then we knew the gentle pequenézas had talked to her. So you see that we are here with Annette Pozzuoli, who has come to teach the Heracleans music, which you said in your letter they are fond of, and we have heard the mammatits chant their morning and evening songs of praise. Are you really and truly glad to see us, now that you are so good! Oh dear, what queer dresses, now that we see them! But how nice, sweet and clean they look! How very, very, beautiful! Do you think that they will love us if we are truly good?"

Mr. Welson, with the opportunity assured them that he was truly glad to see them, as he was certain that they would be loved and happy, and he was sure that mamma and papa would soon follow the lead of their affection, and in Heraclea forget that they had ever been unhappy.

The prætor and family had listened unobserved to the prattling relation of the diminutive maidens, and at its close bestowed upon them the much coveted caresses, then placing them in charge of Cleorita and Oviata, who could converse with them in their own language, they were subjected to the rights of the Kyronese bath, which excited their wincing but mirthful admiration; and their comfortable contentment was well assured when they found themselves invested in Heraclean raiment, which impressed them with the feeling of purified adoption.

After they had been placed in charge of the Kyronese maidens, the assemblage moved up the avenue toward the city. Mr. Welson, who had devoted himself to Annette, after his introduction by the children, was pleased to learn in more direct language the events which had transpired in Don Pedro's family, from the period of her return from a visit in the country, a few days subsequent to his departure on board of the *Tortuga*. From the description she gave of the children's thoughtful endurance and self-dependence during the river voyage, and journey from Amelcoy, it was evident that they had already entered upon their novitiate under Manatitlan direction; for they expressed a decided determination to take care of themselves for the relief of others, and exhibited so many traits of prudential foresight that they were a help rather than a burden. At the commencement of the voyage, she said, that Captain Greenwood and herself had felt great solicitude for the children, and was half inclined to look upon their exile as an inexcusable act of indifference on the part of Don Pedro. "But on the third day, when the poignancy of their grief had become consolable, they immediately evinced a desire to relieve us from anxiety. With permission and encouragement they took charge of their own clothes and personal purity, submitting themselves to our inspection for approval and direction; and have improved so much in foresight that we have found it hard to excel them in neatness. As you have seen them this morning, they have appeared throughout the journey, causing by their example a constant desire for cleanly renovation on the part of the muleteers, who were ashamed to appear in negligent garbs, of doubtful purity, to subject themselves to the reproving contrast. If you remarked their appearance, you must have observed that they are wonderfully clean and tidy in their department, and have been unusually attentive in rendering assistance to all, so that the trip has been one of unalloyed pleasure, from the exampled influence of the children. When we started from Amelcoy the captain took Lovieta before him on the saddle-bow, and I took charge of Lavoca; but on the second day they insisted that they could ride unsupported on the led mules, and their prudence had so completely inspired our confidence that they were allowed to make the trial, with such success that the mules on the fourth day exhibited such a manifest preference, that jointly with the children they declared their independence from the arrieros, and have since been recognized by mules and muleteers as especial favorites."

CHAPTER XXX.

THE process of ablution having been completed, before the sun reached its meridian, not only the new arrivals, with resident intention, but the members of the corps, appeared in the lower fora dressed in Manatitlan costume, which had been prepared for the occasion by the Heracleans in commemoration of full adoption. The effect produced by the change can be comprehensively expressed in the whispered announcement of Lovieta and Lavoca as they regarded with admiring eyes the improvement made in the *personel* of Don Guillermo when raised in his arms for affectionate congratulation. " Oh! Don Guillermo, you look, and we feel so nice and light, we could almost fly back to mamma and papa to make them glad with happiness." Then pointing to a group of Heraclean matrons they asked, " Do they ever fly? "

Don Guillermo. " Oh yes, in thought to make others happy, they are always in flight, and it is that which makes you feel so light and joyous."

Lovieta. " But shall we always feel so good, and grow to be like them?"

Don G. " Yes, we are certain that you will, because you are disposed to be glad for the happiness of others, and measure your desires with the wish that you may be useful in contributing to the welfare of others for the return of their affection."

Lavoca (thoughtfully). " But will it last, Querido Don Guillermo? At home we were sometimes so glad, and then [sadly] so very, very miserable."

Don G. " But you see before you those whose examples never change, but to grow brighter and happier. When you grow more thoughtfully considerate, you will feel that what you lack in attainment, others older and more experienced will impart from their affection. Then in grateful transfer you can assist those more inexperienced than yourself. Of one thing you can be certain, there will be no Padre Molineros here to mar your happiness with bead-prayers, exactions, and penances."

Lovieta. " Of course, we can't now understand all that you wish to have us know, but we shall try hard to learn, with thought, how to make others happy, that their love may teach us more than we know ourselves, so that you can see when you come to visit us that we have neither been idle or naughty. But now we can't make much of ourselves, we do so many things without thinking, and then are sorry after it can't be helped. But look, what are they bringing that table with the queer thing on the end of it, out here for?"

Don G. " The instrument that excites your curiosity, enables us to see and converse with the Manatitlans; but farther than that I am as much in the dark as you are. We shall soon see, however."

After a short pause occupied in arranging the tympano-microscope, the Dosch from the auricular platform said, " According to our custom, practiced from time immemorial, the sanction of parents in confirmation of the marriage unity of their children, has been deemed and proved sufficient for the affectionate realization of unity in fact. But as your race of enlightened progressives have substituted shadow for substance, as an act of conformity, in the lack of anointed priestcraft I have volunteered to act as an officiating sponsor. If the prætor sanctions my assumption, he can, with his wife, first bestow their daughters according to our custom, and then I will duplicate the gift with formulistic rites, so that there can be no question with regard to the 'orthodoxy' of the union."

To the glad surprise of M. Hollydorf, the parents bestowed Luocuratia to his keeping, and Correliana to Captain Greenwood's, in the same breath. Then, when the sun had entered upon its meridian radius of ascension, free from shadow, in the still hour of noontide when all nature was hushed for repose, without inductive explanation, after the prætor had placed his children in position, Manito and his choristers, in full chorus chanted the nuptial ceremony with impressive effect, the Heracleans joining in response.

> " Here beneath the vertical sun,
> Without shadow, you are plighted,
> And with us now, in love are one,
> And forever, 'soul' united.
>
> " With Creative sanction, this, we ever pray,
> May prove your present joy, and immortal stay.
> Hail, glorious noon! these, our notes of love prolong,
> And echo back with joy, this our nuptial song."

Although arranged by the Dosch and prætor as a surprise, known to all except the espoused, they quickly, with blissful perception discovered the intention, and joined with thrillful zest Manito's choristers, who made the tympanum reverberate with the following hopeful prophecy and refrain: —

> "'Old Lang Syne,' in bloody record rules the past;
> In the future, love and peace are now forecast.
> Blessings have source, and flow from power Supreme,
> Goodwill to all, now sounds the glorious theme.
> Through the smallest of the human race,
> Was delegated this act of 'grace.'"

At the close of the marriage ceremony, the Dosch with his family quickly regained the fossæ of Mr. Welson's ears, the sinuosities of which were made to resound with the prophetic responses, causing the eyes of the owner thereof to turn with an instinctive strabism toward the organ subject to the impression of Manatitlan vocalism. While his eyes were in their retroverted position they attracted the attention of Lovieta and Lavoca, who had recalled their wondering gaze for curious inquiry, causing them to exclaim in a voice : " O Don Guillermo, what makes you squint so?" Then, without waiting for an answer, they continued, glancing at the reflected figures of the Manatitlans in the microscopic field, "How much larger your little folks are than ours, and how beautifully they sing."

But Mr. Welson's attention was too strongly diverted to give more than an abstracted answer to his pets. At the close of the prophetic jubilee, the Dosch answered, from the interpretation of Mr. Welson's thoughts: " We do certainly feel, from the docility shown by the leading members of the corps, as if the wedge of rational thought had opened a passage through the cycled round of folly for the ejection of the many self-inflicted causes of misery, which have made life with your race a penal infliction. The educated substitution of thought for impulsive impression from the senses, which at present holds ruling sway, would interpose a shield to prevent the emblematic union of head and tail, for the vortex extinction of material civilization, and degradation for the reënactment of savage barbarism through a long series of dark ages. There is certainly a happy forecast inaugurated by this union, which reminds us of our provincial success in raising human Animalculans to become in reality Animalcumans; a distinction which our neophytes are emulous of having conferred from self-approving merit. If it was not for the selfish fighting disposition of Christian nations and sects, which inclines them to instinctive patriotism and holy wars, advantage might be taken of their superstitious reverence for things ancient, to attract them hitherward as pilgrims, for their own behoof."

Here the ears of the Dosch caught the subdued tones of familiar voices, which from their peculiar method of construing terms he quickly recognized, causing him to expostulate in this wise: " I hear the voices of your old sailor companions, Jack and Bill; is it fitting that their honest sincerity should alone be welcomed by their Kyronese admirers? If their quaint, unsophisticated bluntness has been able to recognize the truthful simplicity of Heraclean example, it will prove a greater acquisition for the encouragement of hope than your own; as they represent a class embalmed with the stupidity of erratic animal indulgence, which subjects them to the vilest servitude ever imposed by the arbitrary few upon the unthinking masses of humanity. Indeed, with the exception of the self-imposed penalties of your Giga women's vanity, they suffer more abjectly from the fiat of hereditary usage than the veriest slave that ever winced under the lash of the taskmaster."

Mr. Welson and Dow soon added a glow of grateful contentment to the weather-beaten faces of the two sailors, by extending to them a cordial welcome, which was increased to manifestations of "weakness," from the warmth of the Heracleans' affectionate reception. There was something so uniquely attractive in the instinctive attachment of these strange beings, who had wandered away from their element, that they had enlisted a strong interest long before the possible existence of a human Animalcuman had been conjectured; or an idea of the practicability of a preferred affection had been suggested, in exaltation above the instinctive type rendered "famous" by the fabled "friendship" of the legendary Damon and Pythias. Mr. Welson had tested the fealty of their attachment in a variety of ways on board of the *Tortuga*, with a constant result in confirmation of its disinterested integrity.

Finding themselves the centre of attraction, which

was seasoned with manifold tokens of affectionate sincerity, they were fain to have recourse to their "pocket-swabs, to clear the leakage from the run on scuppers of their eyes," with an occasional sounding of their nostril pumps, to divert the emotional overflow from its natural course. Correliana, in a transparent glow of radiant joy, for the relief of their "filling condition, that water-logged" the speech of the "honest tars," asked Jack in good English, if he did not prefer his present success in "drawing fire," to the method he adopted with the Indians? The question produced a sudden revulsion, which Bill seconded with a nudge and a whisper, that could have been heard in a gale: "Say, Jack, her leddyship had you there with the pint of a marlin."

Taken aback, Jack gave a short, subdued hitch to his waistband and mouth, and then replied, with the latter reefed into a smile: "You see, m'rm, your leddyship, w're now on a peaceful tack, for w've come to sign the articles and enter our name as landsmen, if he cares to ship us at a venture, and take us in tow for the v'yage of life, and mayhap for t' other." Then hesitating, to gather courage, which was gained by an extra hitch of his waistband, he resumed: "You see, m'rm, if so be your father would 'low us to splice, we'd like to port our helm in Heraclea for life."

Understanding the tenor of the sailor's petition, by the fond glances exchanged by himself and mate, with the two Kyronese maidens, who had attended upon them while acting as guards or gate-keepers, she addressed the Dosch in Latin, asking if it was agreeable to his judgment to have their request complied with; pleading her own assurance of their constancy from the disinterested affection they had shown toward each other. The Dosch not only expressed his full approbation, but desire that they should be immediately united. This decision receiving the prætor's sanction, and willing approval of the maiden's parents, and, above all, the blushing self-bestowal of the affiants, whose inclinations were consulted by Cleorita and Oviata, it was resolved to consummate the union at once, if the sun had not declined in its arc sufficient for the casting of a shadow. Placing them in position as quickly as possible, it was found that, with haste, the emblematic ceremony could be accomplished in a union without shadow. But when the sailors were asked for their family names, it was found that from long habit in using only their "Christian" names in addressing each other, they were at a loss in deciding to which of the surnames, Smith or Jones, they were personally entitled from parental endowment, although aware that one or the other had been inherited as a nominal adjunct to Jack and Bill. As they could not recollect to which they were separately entitled, the Dosch, from the urgency of the emergency, was fain to accept the only alternative of the dilemma, and in the formulistic style, peculiar to Giga understanding, consummated the ties by propounding: "You, Bill Smith or Jones, in mutual troth do plight your vows in constancy for life to Anonymosimia Doycymba, and you, Jack Jones or Smith, yours to Meerisia Abdosia?"

The hearty "aye, aye, sir" of the male respondents, and the softer modulated, but firmly expressed "ai toi" of dames Jones or Smith, closed the involved nomenclature of the ceremony. The Dosch remarked, after regaining the ear of Mr. Welson, that although the courtship had been conducted in the entire absence of an understanding speech communication, their auramentors were fully assured that there existed a stronger instinctive attachment, in nearer approach to an independence from bodily influence, than is usually attained by civilized reciprocants from the advantage of a common language, inasmuch as it restrained the tongue from its "yarned" propensity for exaggeration, peculiar to its use with sailors, and the more decided truthful negation by the votaries of fashionable society. The assembly, at the conclusion of the second improvised marriage scene, joined in the recitative invocation and song of thanksgiving subtranscribed: —

> "In gladness we to our Creator raise
> This grateful song, in everlasting praise,
> That through Manatitla's atomic life
> He has ope'd a way to end human strife,
> That in ' wedlock,' domestic joy
> Shall brighter glow and never cloy."

At the close of the hymn of invocation the Dosch dictated the advisory sanction adopted by the Manatitlans, which we give as rendered by the prætor: "In the full belief of your loving sincerity, we joyfully confirm this union with that of our children, hopefully believing that your affection will increase in fervor until death relieves you of your bodies' encumbrance for the full consummation of a joyous immortality."

The buzz and genealogical curators of sound were highly delighted with the harmony of the musical composition, declaring that its peculiar adaptation attested to the affectionate talent of a master spirit. The former, in enthusiastic approval, offered his warmest commendations to Manito, the Maniculan prætor; at the same time congratulating the Manatitlans in having possession of a musician of such eminent ability. Great was his chagrin and surprise, when Manito not only disclaimed the authorship, but stated that the merit of the poetical composition and musical adaptation belonged solely to his pupil, Mistress Correliana, of whose advancement and talent he was justly proud. The perturbed expression of Pettynose bespoke the revived memory of his former criticism, causing the padre to chuckle audibly from the recollection of the dogmatic snap he had received, when his suggested variation had abruptly closed the dulcetina improvisitation. The blushing attention of Correliana was too much absorbed with the admiring surprise of her husband to heed the professor's confusion. In explanation to him, she whispered that Manito had taught her how to use the dulcetina with the aid of the tympano-microscope, unbeknown to the members of the corps, and had also instructed her in the art of composition. Captain Greenwood had a strong passion for music, without vocal capacity for its expression. To compensate for his own deficiency, it had been his abiding desire to possess a wife with the talent he lacked, that she might impart its sympathetic solace for the relief of anxious care. This desire Correliana had intuitively discovered, which added a strong incentive for application, with the purpose of imparting her improvement to her people.

After the marriage confirmation by the prætor, Manito, through Mr. Welson, proposed to adjourn from the fora to the auriculum. On the way the Dosch passed to the ear of Captain Greenwood; his salutation caused a sudden start, with the motion of raising his hand, which Correliana detained; awáre of the cause from the divergence of his eyes, she asked: "Do you recognize the voice of an old familiar?"

Her husband's puzzled expression declared the nature of the communication, aside from his voiced expression as if in repetition, "Annette, harp, violin, dulcetina!" Correliana added to the sum of his perplexity by asking if the young woman having Don Pedro's children in charge was the sole survivor of the unfortunates who received his assisting sympathy while in the extremity of their distress in Rio? Startled by a question that implied her knowledge of a secret which he supposed was only known to Annette and himself, he answered, inquiringly, "Yes?" Receiving from his wife a fond kiss of benediction, she asked: "Do you wonder that I discovered the source that gained you Manatitlan approval, with an affection so fearless in its sympathy while imparting its succoring rays of goodness? You wonder at my ready acquisition of your language and the source of my information? Do you suppose, with an innate perception of the unselfish sympathy which prompted you to solace the sufferings of those forlorn beings, who had afforded me protection at the cost of their kinsfolk's lives, that I could remain content without perfecting myself for the full enjoyment of a languaged communion with your thoughts? The voice that startled your memory, was the prompting familiar's, who attended you in Rio through the sad scenes, that in termination bequeathed the harp, violin, and dulcetina, as mementos to stimulate your unselfish affection for the devisement of means for the future relief of your race from the cause of such calamitous hereditaments."

Tears glistened in the eyes of her husband as her loving sympathy brought back with graphic effect the scenes indelibly impressed upon his memory. Recovering from his emotions, he beckoned to Annette, who, attended by Mr. Welson, had held herself aloof from the newly wedded; quickly answering to his signal, she was introduced to his wife, who bestowed upon her a warm embrace as a prelude to more affectionate communion. Then in answer to his desire to listen in judgment of her proficiency, Cleorita and Oviata volunteered their service with Kyronese aids to bring the dulcetina, harp, and violin from the hospidoræ. When the harp was attuned to the dulcetina, Annette, with ready ear and touch improvised an accompaniment to the simple air of an anthem of Correliana's composition, at the same time watching her supposed self-taught success in the management of her father's instrumental conception. From the first Annette's pleasure became manifest, for Correliana retained the tenor of her composition independent of the harp. At the close of the instrumental duo, Annette highly commended her proficiency, giving her the desired assurance of capability for the attainment of unusual skill, both in vocal and instrumental music. Annette's skillful instrumentation and melodious vocalization caused Dr. Baāhar to observe, naturalistically, that there was a sensible diminution in the length of the buzz and genealogical curators of sound's musical horns. The prætor Manito was in ecstasies with the successful rendition of his pupil, and declared to his wife his intention of extending to Annette a kiss of welcome, out of his abounding love for her musical talent. Disappearing from our Giga eyes on the instant, but not from those of his wife, whose face mantled with a blush when she saw him in the very act of imprinting a kiss upon her lips, the recipient, with a vague impression, raised her hand and brushed it away. The Dosch remarking the effect produced upon Manito's wife, said, that it was an apt illustration of jealousy, for it never considered the relative disproportion of the exciting objects. Manito's next appearance upon the stage was under the lead of his wife's thumb and forefinger, attached to his ear, while with assumed tartness, in strong Giga accent, she upbraided him for the impudent infidelity of the act. But her curiosity getting the better of her assumed indifference, she tauntingly exclaimed, "I suppose you found the unreciprocated stolen kiss from a single Giga lip more than equal in sweetness to two of ours prompted by conscious affection?" Slipping from her finger's hold, he gave the flying answer: "Yes, truly, I found her lip as full in volume as the tones of her voice!"

Lovieta and Lavoca, who had witnessed this playful episode, whisperingly asked Mr. Welson: "Are these large ones related to the ear Manatitlans, and will they grow larger when they grow older?" But as his answer failed to satisfy their comprehension, they asked him, if they looked as pretty as the Mantitlans now that they were dressed like them? A caressing hand upon their heads proving a satisfactory reply, they declared it a *moda linda, facil, y agradable a la vista*, and in the Heraclean school, so nicely clothed, they felt sure that they should become good, graciosa, and would try to be as affectionate as the brides! In benediction, for the success of their good intentions, Mr. Welson bestowed commendatory kisses, and again placed them in charge of Cleorita and Oviata.

Anxious to read his letters, he, with the Dosch, retired to the quarters of the corps. After glancing at his formulistic letters of "friendship" and business, — which were closely interwoven, — under the supervision of the Dosch, who kept up a running commentary, in which he pointed out the prospective selfishness of each correspondent, in a manner so legible, their insincerity became so disgusting to the receiver that he laid them aside, wondering how he had allowed himself to be beguiled for a lifetime with such shadowy pretexts. The letter of Don Pedro Garcia, which he had reserved for the last, revived his hopeful trust in the latent goodness of humanity. We offer its chapter transcript for the benefit of the reader.

CHAPTER XXXI.

BUENOS AYRES, *November 5th*, 187–.

DEAR DON GUILLERMO: You will be very much surprised, notwithstanding the forewarning of intention, to find yourself unexpectedly greeted in Heraclea by your little favorites Lovieta and Lavoca. It has cost us a painful struggle to part with them; but we should have been unmindful of our privilege and duty as conservators for their future welfare, in joyous transmission, if we had consulted the disposition of our selfish feelings. Even if our household was of the most suitable description known to civilization for rearing children, I should not have hesitated a moment in imploring your influence for their admission into the Heraclean school. Consolata, with unprejudiced consideration, fully appreciates the advantages of the protective course, for conferring present and future happiness.

Realmente, the evidence appeals so directly to the affectionate understanding, the niñas fully comprehend the advantages that it will afford, which in thoughtful mood they expressed, by asking us if we did not think the Manatitlans and Heracleans from having such nice parents for so long a time were better than good? I am free to confess that I felt a wince at the touch palpable the question conveyed; tears glistened in the eyes of Consolata, as she enfolded the unconscious challengers in her arms, with a languaged embrace that impressed them with affirmative conviction, while it imparted a desire for their forgiveness, which trebly admonished me of my own unworthiness. It is strange how little use we now have for voiced words to give expression to our thoughts. This silent source of happy intercourse bespeaks with increasing flow our current perception of a joyous unity in affection, including in its circle of communion M. Baudois and our neighborly confluents. In our morning and evening walks for the exampled demonstration of happy regeneration, that in practice its source may be made known to others, reflections from objects are so similar in impression that a glance is sufficient for the conveyance of coincident thought comparisons. Last evening, while on our way to visit the "industrial" establishments of our foreign "citizens" (by invitation), who are constantly begging for governmental concessions for the encouragement of their enterprising "undertakings," we passed many of our black priestly scentipedes, and it was with conscious emotions of joy that I felt the confiding pressure of Consolata's arm, as she averted her face to avoid the necessity of hypocritical salutation. As my governmental position had evoked from Mr. Hogg an invitation to visit his distillery, I proposed to M. Baudois the advisability of taking the family, that we might observe the effect produced by the improvements, upon my wife and children. With his approval they were taken; and while we were cautiously picking our path along the causeway crossing of the slough that separates the brewery premises of Von Guzzledorf from those of the distillery, Mr. Hogg, in his obsequious desire to honor our visit with propitious attention, hastened to meet us with his ponderous body and jowled cheeks and neck in jellied tremor from the emotions of a waddling gait, in gloat of the expectant relish of selfish gratification. After a wheezing prelude, with a cough that in fitful gusts conveyed the foul odors of gamey engorgement, he lamented his asthmatic affliction, that refused to yield to remedies, although he had spared no expense to avail himself of the best talent that could be secured for "love" or money. But recently, he said, he had obtained great relief from the prescriptious of Dr. Bull, a member of the royal college of surgeons, who had recently come out from London, at his instigation; a man of eminent ability who had gained a great reputation for his prophylactic pills, a box of which every family should keep in their house, to be taken regular every morning. While engaged in this preliminary detail of health and its providential means of preservation, Lovieta and Lavoca with ill suppressed disdainful sniffs of comparison, allowed their eyes to alternate between his person and a family of his namesake's, who, with a strong personal resemblance and odor, were indulging in the luxurious contributions from the neighboring brewery and kindred establishments bordering upon the slough.

Our visits to the distillery and brewery were certainly not propitious for securing the desired grant of land, for the enlargement of their industrial facilities, notwithstanding the sacrificial offering of a box of Wolf's Scheidam Schnapps and a dozen of old London Dock Sherry, of 1824. When leaving, Mr. Hogg begged me for the welfare of myself and family, not to forget that Dr. Bull was for the present stopping at the Hotel del Mundo. On our return home, a bath was held in requisition for purification from the attaint of the visit. Under these contrasted impressions, our thoughts have withheld us from instinctive gratification by the force of repulsive comparison. Of course, our singularity has evoked the despiteful recognition of the evil disposed, but we feel an enduring recompense in the loving reciprocations of our household; for we no longer suffer from those transitions which of old subjected us in a day to mutations, that realized within the zones of affection, tropical heats, with terror motor accompaniments of tornado, thunder gusts, and intermediate alternations to the frosty stillness of the frigid.

With realized happiness, we are encouraged with the approving commendations of our Manatitlan sojourners to believe that a perfect eradication of the parental past will be accomplished in the memories of our children, under Heraclean tuition, for in the change they already recognize the happy cause. When a sufficient time has elapsed for the fulfillment of an end so desirable, we shall, with the prætor's permission, make Heraclea our permanent home. Then, with the privilege of monthly visits to our daughters, we shall enter upon an era of happiness that we feel assured will outlast the records of time; for we have already tested the efficacy of purity and goodness, in sufficient degree for an initial impression of immortality. In companionship with you and the members of the corps, under the direction of the Dosch, and the example of the Heracleans, I shall endeavor to amend my own crotchety humors, which, if we except the superstitious devotion of instinct to rites and ceremonies, and herding predilections, are in no great degree above those I affect to despise, in the matter of stability. But I am fully impressed with the belief, that Creative wisdom devised purity and goodness, as the simple creed of self legislation and approving test of immortality.

I will now offer for your remorseful consideration, a counterblast. Have you reflected upon the responsibility you have incurred in leading us astray from the glorious system of rewards and punishments offered for the encouragement and correction of indulgences, by the "good" old mother church? Why did you not leave us in ignorance of the audacious Manatitlan system of education, that dares usurp with self legislation the infallible promulgations of law and gospel for the correction of excessive indulgences. The substitution of direct responsibility for salvation by saving grace, deprives instinctive humanity of its barter privilege of using the gold of its god as a compromising compensation for indulgence, and the consolations of confession and absolution, causing us, especially, to hold them in extreme contempt as subterfuges begot from the vileness of hereditary cause and effect. The instinctive reverence we once paid to priestly mediums of heavenly assurance against the devices of hell, you have turned into the bitterness of shame, from the reflection of past humiliation imposed by their impudent assumptions of delegated divine authority.

Our present feeling of responsibility for perfection in purity and goodness, revolts at the thought that our perceptions were ever so weak as to believe there ever existed in the pampered bodies of priests, other than a swinish divinity endowed with an instinctive audacity that prefers deception to honest labor. We now avoid church and street mummeries of host processions and masses, as a profanation of good instinctive common sense, as our presence would implicate us in the blasphemous degradation. To our chance acquaintance with you, fostered with affectionate sympathy, I must charge this defection, also the retrospective pangs of shame engendered from sensitive regrets for my stupidity, which suffers from revival with every rattling peal of church bells. Not content with casting these heretical shadows across our path, you have devised to rob us of our niñas, our abiding source of love. Bethink you of the accountability you have incurred, with your chances of self-forgiveness.

With a falcon script in acknowledgment we shall feel assured that you will bestow upon our niñétas a father's care and love. . . . . P. G.

Mr. Welson joyfully exclaimed, as he closed the letter, "It is really reviving to hope, to hear a provincial, of Spanish descent, express himself with such perceptive clearness, in freedom from the thraldom of sensual embargo, which has rendered our race the phantasmal victims of stomach indulgence, from time out of date!"

The Dosch, in confirmation of Mr. Welson's train of thought, urged, " It has required our utmost efforts to hold our feelings of contempt subdued that they might not altogether usurp the rule of discreet generosity, in view of the obstinate stupidity of civilized Giga races; who, with star gazing, and moon conjectures, forget their preservative accountability to Creative indications designed for the attainment of a living realization of immortality. How it has been possible for the races of civilized humanity to exist in individual communion with self, and the advantage it affords for deductive comparison, aside from the mutations of attrition, through so many cycles of re-degradation, without discovering the active cause and remedial source, puzzles our comprehension. The ridiculous abstractions of your philosophers, which in thoughtless aberration leads them to lose the substance while pursuing the shadow, was aptly illustrated by some of our frolicsome dames, who decoyed Dr. Baāhar, in successful test of his lack of discernment, to follow the shadow, more vehemently, from its supposed evasions, than he had previously bestowed upon the gossamer substance of a butterfly. While amusing themselves with a morning volant airing upon a beautiful azure-tinted specimen of the L. Matutinal in its wafting sips from the flowers of the latifundium, it caught the covetous gaze of the doctor, who was on his way with his net for a hunt beyond the walls. After pursuing its doubling variations in flight, which from graceful composure seemed void of evasive intention, to the verge of vexatious anxiety, while his back was to the sun, it suddenly soared, substituting its shadow, which the dames made still more attractive by prismatic colors rayed from their silicoth mantles. His efforts grew more frantic from the apparently miraculous escape of the coveted prize through the meshes of the net; which caused the infatuated pursuer to suppose that its illusive power indicated a new species. To our surprise the illusive chase continued with increased avidity as the prismatic colors were varied, the shadow being kept just in advance of the net's swoop, until after an hour's pursuit, the excited naturalist fell exhausted upon his knees in the formulistic attitude of prayer. In this position, with upraised eyes and hands, he watched imploringly the lessening shadow of his morning's devotions. Although surrounded with evidences, which should have led to an impulsive detection of the cause of the tantalizing movements, previous to substitution, there never was for a moment the least hesitation that indicated suspicion or doubt of the limn's substance reality.

In character, nothing has surprised us more than the insensibility of Gigas, who claim the disciplined aid of collegiate education, to the effects evolved from their own experience. Yet, notwithstanding the contempt you personally feel for the selfish enactments of your past life; your relapse, in partial degree, would not appear strange if subjected to reversed example. But with the Heracleans, from hereditary usage the impression has become immutable. The contrast will lead you to realize the great difficulties that will attend the inceptive stages required for reversing the progressive position of your people; as they have been accustomed to advance with their backs to the future from time immemorial, the change will prove embarrassing for many generations. These contrasted facts, which expose in extremity the habits of usage, with the inveteracy incurred for good or evil, has caused us to view the act of self-denial on the part of the parents of your pets, as a trustful deviation that exceeds any in our former experience, and it will certainly insure a harmonious transmission. If the mother's affection had been void of delusive infatuation, permitting her to act in accordance with the promptings of its natural expression, the devisement of the children to the Heraclean school would not have appeared strange, as unprejudiced by delusive agency and its evil tendencies the love of maternity extends to the future. In exemplification of the evil tendencies of your delusive legendary book of creeds, that has defied the efforts of ninety-five thousand human commentators to render it comprehensible, with accumulating legions of preaching expounders, I will relate an event that transpired during the inceptive period of my auramental labors.

The dramatis personæ in the triangular scriptural duel I am about to relate, for exemplifying the utter perversion of intelligent affection, wrought by this precedental tramway to discord, were in scenic representation, a grandmother of Scotch extraction, derived from the amiable clan McGregor, her daughter, and granddaughter, who in passionate exacerbation occupied counter positions in domestic antagonism. The *locale* of the scenic enactment, for your more perfect understanding of its mouthpat *entente cordiale*, we will award to Londonderry, Ireland. Having, by a strange fatality peculiar to auramentation in youth, been led by diverse circumstances to become a member of the household, I was forced to witness the repulsive drama in all the progressive stages to culmination. When in the morning avocations the three "fell out" to disagree, from a flux of hereditary passion, they would exhaust their store of word provocations, and then have recourse to prayer, after the mid-mother had read a formal challenge from the creed "omnium gatherum." These formulistic rites concluded, and the male members dispersed to their employments foreign to the house, the three antagonists would enter the " sitting-room " to decide their quarrel, each taking a corner of the room with their book of missals in hand. With the room appropriately darkened, they would each, with hasty avidity, hurriedly search for some virulent passage of " scripture " of innuendic import, and when found it would be hurled in venomous recital against her adversaries. Often the voices of the three would be intermingled in the combined discharge of offensive similitudes, each vying in the melée encounter for ascendancy in loudness of report, and precision in the diabolic aim of their denunciations. Slight wounds, of Jezabelic imputation, were borne heroically, but the grandam of eighty when severely hurt by thrusts into old wounds rankling with the personal reflection of her accountability for whatever was amiss from lack of amiability in the tempers of daughter and granddaughter, would seize her plaid, and, in defiance of weather, would seek an asylum in the house of her son, distant two miles from the scene of action. The flight of the grandam would add new vigor to the vituperative discharge of invective quotation between mother and daughter, which generally resulted in a drawn battle, leaving the envenomed cause to smoulder in belligerent tendency, until their magazines were replenished with holy war munitions, sufficient for the adventure of another trial of anathematizing strength. When one of these encounters had proved fatal to the grandam, from a severe cold caught from wading through the snaw broo in her retreat from the battle-field, her grandsons invoked from day the fall of darkness, to stay the portended renewal of hostilities, in opposition to the commands of Joshua, but coincident with the prayer of the modern battle hero of Waterloo, for Blucher or night. These belligerent domestic inconsistencies incited by creed incongruities, and indigestible food, and lack of bodily exercise, are by no means rare among the civilized peoples of the old and new world. As with the Scotch, singed sheep's head, haggis, and whiskey, were the inciting cause of religious intolerance and border warfare, it will be found that like causes rule as a source of provocation for distempered aberrations of every kind.

"We will now seek your infantile protégés, and see how it fares with the newly united."

"First resolve me of my doubts with regard to your consistency," urged Mr. Welson. "How do you reconcile the hasty unions you have sanctioned, with your 'invariable custom,' that requires three months probationary test of compatibility before the full consummation of unity."

"You should be aware of a distinction that it would be impossible for us to reconcile in your marriage adoption," replied the Dosch. "You are strangers to our system of education, so that we are obliged to accept an alternative for your tests; as you must realize that the quarantine of a lifetime would not render you compatible according to our acceptation. But nominally all have complied with the probationary requirements; even M. Hollydorf, as Correliana proxied her twin sister. Still our chief dependence is in the incomparable beauty and goodness of the bride, which will render disagreement impossible. But I perceive that you have still another indigestible example that you would have reconciled. We have claimed that our 'system' of education renders the unity of affection between the marriage affiants indivisible; yet in seeming contradiction, your thoughts refer to the second marriage of the prætor with Correliana's mother. We could have explained to you this apparent discrepancy, but for our wish that you might discover from the promptings of your own perception the admissibility of an association designed for mutual solace and companionship in bodily representation. The prætor Adinope when premonized of death's approach, preferred the tribune Adestus, who had lost his wife, to the prætorship, as the chosen companion of his wife, under the temporal privilege of correlative correspondence in the body. With the dying prætor's sanction, Adestus assumed the charge of the household and prætorial advisorship while the husband yet lived. You have in thought questioned this as an impeachment of the unity we profess in the assimilative fulfillment of our first affection. But I can assure you there is neither divorce or abatement in the troth unity of the first allegiance. In fact, they become more perfectly wedded in thought with those who have preceded them to the current realms of immortality; and in vicarious communion, commend without stint or prevarication the ever present manifestations they enjoy with their beatific spouses, and longings for the speedy consummation of a disembodied reunion. If Heraclea could furnish wives for Giga representatives, as well endowed with reason and as free from prejudicial taint as you are, the labor of educational induction in its inceptive stage would soon be accomplished. For in wifely Heraclean example, the joyous brightness would be reflected with such purity that it would irresistibly attract assimilative reciprocation, in thought, from all within reach of its influence, causing vanity, with its promptings for adornment, to become an exile beyond the reach of material redemption. With the current of your women's affections once emancipated from the shallows of personal ornamentation, the clear depth of the stream would purify itself from the undertow of man's grosser instincts, casting the refuse of precedental habits and customs back upon themselves, and the cycle shores of the past, with their memorial odors of instinctive corruption."

Mr. Welson. "As you have answered these problems, which opposed themselves to our understandings, as stumbling blocks preventing our full appreciation of your wisdom's infallibility, in a manner so practically agreeable, will you apprise me of the method you propose for reducing the appetites and passions of Giga humanity to an initial accord with the Heraclean standard? This request I proffer under the privilege conferred by your maxim, 'that we should never cavil or criticise without being practically able to amend.'"

Dosch. "Although our maxim in application to your race lacks, or has hitherto lacked, the secondary power of example required for practical efficiency, we will answer your inquiry by holding your example as our prospective means of introduction. As an initiatory step for reciprocal purification, in prefatory advisement for the introduction of our protective system of education, and its inauguration of self-control and legislation, we shall auramentally propose an international dietary congress, for the studied adaptation of food in quality and quantity, for the healthy requirements of the body. With this as a basis for thought direction, we shall propose a method for the inductive substitution of a common language, free from sectional prejudices, by the introduction of international schools, kindred to our own, which in hostage reciprocation will eradicate the seeds of instinctive jealousy. Husband and wife are to be held as a unity for representative expression in congress; with the special proviso that the voice of the man shall alone bespeak the unity of intention in public assembly. Under the ruling of the dietary congress, stomach codes could be established for the mouth rejection of all indigestible and incompatible compounds, in solid or liquid form. This would emancipate the stomach from the arbitrary tyranny of individual hodge-podgery, relieving the body and brain from the incubus imposed by the unreason of ages. The prestige of a single generation's restful rendering of these intuitive examples of reason would result in the utter abolishment of such songs as the 'Watch on the Rhine,' so characteristic in guttural expression of the indigestible philosophy of a German diet, and the more musical, battle-inspiring Marsellaise, instinct with the lighter French national régime, suited to the Zouave accompaniment of 'Leap on, leap on!' while in substitution there would arise blendings of song, and salutations replete with joyful gladness in the new-fledged accents of affectionate reciprocations. This innovation would effectually liberate the German language from the bondage of nose and stomach, and the French from the frothy sibillations of vanity, causing them to harmonize in peaceful goodwill, with contributions from every tongue, until special idiomatism would become involved in the sympathy of universal accord."

The discourse of the Dosch was here interrupted by the voices of Lovieta and Lavoca, calling for Don Guillermo, who gave an answering invitation for them to come in and see where he lived. This brought them to his knees in full chorus for the rehearsal of the marvelous impressions they had received. But in the rapid scan their eyes gave to the alcoved apartment they caught a view of the Dosch and Doschessa, with other Manatitlans reflected in the field of the table tympano-microscope, which hushed their voiced exuberance into regardful silence. The Dosch, after watching for a few moments their curious awe, reminded Mr. Welson that his wife was specially anxious for a personal introduction to his children in trust. This given, the Doschessa soon won their confidence, and imparted to the eager germ of Giga curiosity some of the winning traits of affectionate reciprocation encouraged in the Manatitlan schools for the enlightenment of thought perception. Her success was soon evident from the gathering mists that sparkled in rayed mementos of affection from their eyelashes to be resolved into tears, as an accompaniment to the plaintive vocals, "mamma, papa." As the dew of inborn memories yielded to soothing direction, natural affection expanded, until it included, with the "extreme unction" of goodness, the infantile query of possibility for the redemption of Padre Molinero from self. In questioning expression, from the impressions of memory, they asked with a toddling perception of cause and effect, if it would not change him if he ate and drank less, so that his mouth wonld not make a noise so porcuno? Then, as if in thought consultation tracing the effect of renovation, they asked if a priest could be made as loving and respectable as Mr. Welson and Captain Greenwood by removing his hat, gown, and fat?

The Dosch laughingly replied, that if he and his kind would adopt the first restrictions mentioned, it would certainly indicate a desire to become respectable in self-estimation, and show a disposition to merit the confidence of others.

When well ingratiated in their affection, the Doschessa asked by what token they wished to be made sensible of her watchful care? This seemed to puzzle their ingenuity for the devisement of a tangible method of communication. But Lavoca, after demure consideration, said, that she thought it would be easy to kiss and embrace, if she could manage to continue as large as she then appeared.

The Dosch then explained to their ready comprehension, that their reflection in the field of the microscope was like the vanity of personal adornment shadowed in a mirror, which when removed left nothing but its vague impression for the delusive gratification of self. But the Doschessa said, if they wished to retain a lasting impression of her as she then appeared, they must keep themselves free from passion by bestowing their thoughts upon others;

then she would be ever present with them to be kissed and embraced in thought, which was a reality that with goodness would last forever. They promised that they would always try to be good, but hoped if they sometimes forgot, she and the teachers would forgive, and let them try again. "Because," Lavoca urged, "our people have not been good like yours, and we have n't learned how to be always the same." She assured them, that with all their disadvantages, if they tried to make their associates and teachers happy, they would forget their own selfishness, and feel that the merited affection of others would always make them joyous with gladness. Perceiving that they were still anxious in thought for an intuitive token of her affection conveyed in the language of a kiss, she proposed to comply with their wish, but cautioned them to be gentle in their reciprocation when they felt her pressure upon their lips. First to Lovieta, and then with an ear premonition to the more impetuous Lavoca, she imparted the loving thrill that ever attends the reciprocal blending of instinctive sense with the animus of goodness. Both were exultant in declaring that her kiss, although exceedingly small and tiny in its touch, was larger in making them feel more happy everywhere than any they had ever felt of their own kind, and were certain they should know whenever they were kissed by a Manatitlan. After this happy introduction of the novecetas to the Manatitlans, the Dosch and Doschessa accepted Mr. Welson's invitation, and occupied their accustomed seats on the tragus of his ear; and then with the escort of Cleorita, Oviata, Lovieta, and Lavoca, started in search of the newly unionized, who were found enjoying the cool shade of the tamarisks on the terraced descent from the summit to the basin of the falls.

The Dosch, while the presence of Mr. Welson's party was yet undiscovered, called his attention to the unity of expression exhibited by their faces as they gazed in thoughtfully silent meditation upon the fantastic sprays of falling water, whose misty vapor, bearing perennial freshness in dispersion to air and vegetation, represented in similitude their own thoughtful desires for the extension of their glad happiness to others. Mr. Welson's face became subject to regretful shadows, as he passed in review the instinctive follies recalled to his memory in contrast by the constantly recurring variations in manifestation of the happy influence transmitted from hereditary self legislation. In thought he expressed thankful praise that his life had been spared to witness scenes which in truthful representation realized more of bliss than had ever entered into his most sanguine conceptions. In thoughtful admiration of the unionized beatitude expressed in the silent flow of current reciprocations, stimulated by the stentorian promptings of the Dosch in the lulls of the wind waft, he resolved to avail himself, without delay, of an example so pregnant with current joy. With lingering desire he motioned away his escort, then withdrawing himself without disturbing the mystic harmony of the wafting ingraft of affection, he sought within himself for an assurance of hopes that had surprised him while visiting the school for nynetas.

At the descending junction of the avenue with those of the basin and incrematium he met the prætor and his wife, who were accustomed on the occasion of a marriage to visit its sweet scented groves for communion with their current selves in purification from the body's probation. They were quick to detect in the subdued but hopefully eloquent expression of Mr. Welson's face, an undefined longing, and were not surprised when he unburthened to them his desire for their censorial consultation, and judgment after an explanatory intercession with the object of his premised thoughtful affection. With warm commendations in support of the wisdom of his choice, from her special adaptability, they immediately entered upon the eliminary negotiations required for a verdict of relief; the result of which will be detailed in a subsequent chapter.

CHAPTER XXXII.

CLEORITA and Oviata, with intuitive susceptibility, detected the new phase of attraction to which Mr. Welson had become subject, and assumed the entertainment of his adopted children until evening song, although they were inclined to hold padre carita in attendance on the plea that they should not see him from to-morrow for a whole month. With the canopied shadows cast by the sun's decline over the city, the eyes of Lovieta and Lavoca, under the reactive weight of unusual excitement, began to flicker, and they were dismissed to their beds while expressing in drowsy accents the happiness they anticipated from the warmth of their adoption by the niñétas de escuela. The twilight chirps of the suffragian court sparrows from trellis-vine perches, in preparation for their song of praise, had scarce betokened the break of day, before Cleorita and Oviata aroused with fondlings their nycephas from their deep sleep into the dreamful slumbers that precede with herald prelude the instinctive mood of ruling impression. While yet hovering with suspended wings in the balance of sleeping and waking impressions, conscious affection murmured in pleading accents, "Mamma, papa, poco mas sućño?" But with the renewal of the rosy salutations, the waking twilight of perception began to dawn, and then, with the first sunlight rays of memory, their eyes quickly opened to receive the fond greetings of the Kyronese maids with tokens of kindred affection.

After a responsive mediation with arms and lips, in grateful acknowledgment for the service rendered by their loving monitors, who were self-delegated aids for dress adjustment, in preparation for joining in the morning song of praise, they made quick dispatch, and, unassisted by the proffered hands of the maids, accomplished their first self-introduction into Heraclean costume.

At the conclusion of the matin salutation of praise, offered from all the thresholds of the inhabited homes of the city and latifundium, in which they enthusiastically joined with infantile zest, they exclaimed, "How beautiful and neighborly kind!" Then joyfully asked, "Why, padre carita, how did you learn to sing? Will papa learn when he comes? It will be so nice for him and mamma to join together in the cantata viva of morning and evening, when all is right, without a wrong, to make them unhappy."

After the morning meal, Lovieta and Lavoca, with Annette in special charge, under the care of Correliana and Luocuratia, were escorted by the Heracleans, Kyronese, and members of the corps to the gate of the nymphatasium, and were there fondly received in the matriculating arms of the teachers and censors, and their future associates, with such tokens of affectionate sympathy that home longings lost their poignancy, causing their eyes to overflow with genial joy, watering the smiles evoked from greetings of self-forgetfulness, with the balm of affection. Annette's reception was one of equal warmth, but timed for the Giga reserve of dignity conferred with the distinctions of age. Before her entrance, she had gratefully accepted the proffered adoption of childless Heraclean parents, so that she could look forward to the forthcoming monthly day of visitation, with the certainty of receiving their unselfish endearments.

In the farewell greetings were mingled the novecetas' assurances of their determination to merit warm approval for their improvement at the next monthly visitation; of this all felt assured; for with other adaptive regulations the hours of sleep were so well timed for the requirements of healthy recuperation, that even the dream shadows of morning liberation were curtailed of their precedental phantasmagoric impressions of instinctive hallucination.

CHAPTER XXXIII.

For some weeks subsequent to the arrival of Captain Greenwood, the daily avocations of the corps were assimilated with those of the Heracleans, if we except the erratic disposition of Dr. Baāhar, which seemed to have become more enamored with entomological pursuits. In apology, he said, that the great beauty and ephemeral existence of the butterfly declared its special intention for the accomplishment of a transient purpose; and as angels' terrestrial visits were few and far between, he had come to the theoretical conclusion that they were intended as relief vehicles for their conveyance during their earthly visits. For the verification of this theory he had increased his vigilance, with the hopes of catching an angel napping, which would recompense his trials from the jeers of an unbelieving world.

After the morning salutations, four hours were passed in the cultivation of the garden allotments in the latifundium, by all except the padre, curators, and artist, the former assisting the Kyronese in renovation in his vocation of carpenter; the latter named preferring pastoral occupations as more consonant with their instinctive affinities. From nine to eleven the time was occupied in the auriculum in conversational consultation for the exposition of Manatitlan usages, applicable for initiatory adoption by the Giga races. Thenceforward until the noon-day hours of meridian heat, devoted to repose in the shady colonnades, each individual employed his or her time in rendering neighborly aid or solace. When the shimmering heat shadows were reflected in gleams from the falling water indicating the sun's decline, a slight refection was served.

From thence until evening song the time was occupied in associate consultations contributing to amusement and projective goodwill, embracing in scope devices for penetrating the armadillo shell of civilized vanity and selfishness. The ever changing novelty of thoughtful inventions suggested by these associations, were in moments of reflection a fruitful source of wonder to the members of the corps, from the constant increase of real enjoyment afforded, in contrast with the vague pursuits of instinctive pleasure followed with the routine regularity of the kitten's pastime, by the civilized races. In the cultivation of associate worth they derived such abiding satisfaction from the increasing reach of happy perception, they were at times inclined to doubt their real identity as personal actors in the delusive scenes reflected from memory.

The self-imposed absurdities reflected from the accumulative worriments of business pursuits, and sensual gratifications were truthfully illustrated by Jack and Bill, in the quaint relation of their experience; who declared that their bodies had been launched and shipped with just sconce enough to eat, drink, scrub, chew, splice, smoke, and reef, under the old gaff, without the flutter of a sky-sail's worth of thought more than what they were bid to do. "But thanks to Captain Greenwood, we've been saved from a dive into Davy Jones' locker, where we once expected to be keelhauled in brimstone scaldings by Old Nick, without ever being able to take a squint beyond. Homsoever, now with the Dosch for a skipper, we've taken soundings, and know our bearings, so with a clear look ahead we can see a smooth surface in the channel without a ripple, or a scud aloft to take us aback from our portage."

Notwithstanding the constancy of the sailor's ruder perceptions, the thoughts of the padre and Dr. Baāhar were often auramentally caught revelling in past visions of instinctive indulgence, so that it became necessary for the auramentors to remind one of his medical society, which held its stated meetings for the correction of ethical correspondence between its members in a beer cellar; and the other of a condition, in which he argued with his wife the propriety of retiring for the night with his boots on. Mr. Dow would, in like manner, be occasionally surprised in a mood of covetous calculation in anticipation of conferred honors and titles likely to be bestowed by potentates and societies in reward for his persevering merit, which had led to the discovery of the Kyronese, Heraclean, and Manatitlan races. But the slightest lisp of his first honors obtained for the discovery of a new species of crab, which was christened "Cancer Doweri," restored him to a conscious appreciation of Heraclean example. M. Hollydorf and Captain Greenwood were proof to the lure of selfish thought.

The visit to the nymphatasium had been eventful, under the direction of the Dosch and Doschessa, in attracting an assimilative sympathy between Mr. Welson and a maiden teacher, Cæluiformia by name, the daughter of the pastor Corycebæus; this, through the intercession, or mediation, of the prætor and wife, had been matured for a surprise. The pastor had set his house in order for the return of his daughter, and the probationary reception of Mr. Welson. When the arrangements were perfected, the unwitting brideswick was greeted at the portals of his thalmia when emerging for matin salutation, by the prætor, tribune censors, Kyronese, and Betongese, who escorted him after the morning song of praise, accompanied by the entire population to the pastoriza. At the portal the happy mentor received the embrace of welcome from the pastor, and one of equal zest in the expression of sincerity from the prospective "mother-in-law," who introduced the blushing Cæluiformia, radiant with affectionate anticipations, to the arms of her betrothed.

This consummation was the signal for the waiting choir of Manito, who made the tympanum resound with an anthem prepared for the characteristic expression of the Scotch instinctive type. Correliana, initiated into the proemic espousal dedication, directed the measure from Manatitlan lead. We give a rendering of the words in translation below: —

> "From Scotia's lock'd inlet shores,
> Rough highland crags, and sombre glens,
> Where heather glints o'er boggy fens,
> And shivering, sighs lonely plaint;
> In misty tears the lowland saint,
> Of bracken braes, that rise from moors.
>
> "With love, we hail the herald sage,
> Who dares disdain the bogle chain,
> Of myth-bound sects and all their train,
> Whose fenny thoughts in muirk arise,
> To obscure love's creative skies,
> With miasmatic hate and rage.
>
> "All hail to his love's perpetual vows,
> That Cæluiformia's now espouse."

At the close of the salutatory greeting, the parents bestowed upon the current unity of affection, in espoused accession, their joyful benediction, introducing them with a glad welcome to the freedom of their household colonnades. After their installation the assemblage dispersed to their daily avocations.

With Mr. Welson's departure, the "quarters" of the corps seemed to have lost its active principle of vitality, and its members were to be seen in daily attendance at the house of Corycebæus, after the morning salutations. Indeed, the transfer was so complete that the tympano-miscroscope followed in train, from the proposed consent of all, the Dosch remarking, that in their course they followed the universal "law" of attraction, that recognized the lead of strength, for self-control, as the predominating source of power for the control of others. This axiom you will find amply verified in all the motor relations of animate and inanimate matter, as well as in all the votive enactments of life. The sun, as the supreme source of effulgence and heat, attracts the lesser luminaries within the pale of its orbit, and as the revivifying source of vitality, force, and motion, it receives from instinct worshipful reverence; while in mundane expression, its effects are instinctively preëminent in the attractive power of the preacher, lecturer, and democratic leader, for the control of the unthinking herd, as the oratorical expositors of sound. In your own relations you were controlled among your own people by precedental habits and customs, accepting them, without a questioning thought, as well approved by the ordeal of time. Away from your precedental theorisms, in enactment by the controlling majority, you were attracted by the influence of Correliana's happy example over the Kyronese, and for the first time, with the majority, your thoughts were directed to facts for deduction and analytical comparison, which with the leading influence of Heraclean example has happily called forth into active life your latent appreciation of goodness. Following in its lead, after liberation, it has harmonized and rendered subservient your instinctive tempers, so that with the ascendant portion precedental argument is unknown, and politic prudence controls the less appreciative minority, even when opposed by the aggravations of material rebuttal. In apt illustration of the power of self command achieved by the pastoral members of the corps, while engaged in Olympic sports with the herd under the lead of the pastor Corycebæus, Dr. Baāhar, the most pertinacious, politic, and irascible imitator of antiquarian revelations among you, having unwarily allowed his stronger passion for butterfly hunting to intrude upon the portion of the day set apart for the entertainment of the flocks in field gymnastics, was surprised while stooping to disengage a gaudy victim from the meshes of his net, by a disjunctive butt, in the rear, from the censorial horns and head of a precedental guannaco, which caused a cycle revolution of his body. Regaining his feet, he in wrath unthinkingly opposed himself to the sportive cause of his mishap, who was collecting his energies with blind zeal for the renewal of his " good old times salutation." But with quick perception the doctor subdued his reactive wrath, and while the sportive ram was poising his head to follow up the advantage he had gained in reversing precedental ideas of naturalistic progression, he wisely concluded that diplomatic discretion would, for the occasion, be the better part of valor ; acting upon the suggestion, with bipedal advantage, he dodged instead of opposing his body fatuistically with the adaged shield, " what has been, will be." Notwithstanding his " presence of mind," shown upon this occasion, he obstinately continued to pursue his predilection for fly catching, with increased zeal. Often in the midst of the most alluring conversation, devised for the reciprocation of instruction by Correliana, with a refrain of notes from woodland songsters to the musical tones of her voice, he would start wildly up, with his net raised " rampant " for the catch, with his eyes absorbed for the detection of the species and order of a butterfly attraction. When assured of rarity, he would rush forth with eyes and net upraised for the capture of the tempting lure. Gentle expedient, and every form of pleading inducement had been exhausted, that could be suggested for exampled persuasion, when an incident occurred which appeared in coincident similitude, like a conjunctive interposition, for the cure of his malady.

On a morning which had been freshened with

night showers, betokening the approach of the winter solstice, Corycebæus led forth his flocks, attended by all whose inclinations were not stayed with the occupations of gardening and household employments. Conspicuous above the happy throng, whose voices were melodious with song and mirthful repartee, made vivacious with bantering chase, was raised the pennon net of Dr. Baāhar. But for the contrasting halo of exuberant gladness, the bevied groups, as they passed beneath the cinctus portal, might have been taken for actors in some memorial scene enactment, expressive of festive gayety in historic commemoration of ancient ceremonial rites. Nathless, upon nearer inspection it would have been readily discovered that instinctive pleasure, from anticipated indulgence, bore no part in the joyous emotions that flowed in sportive current from affectionate association. Even the pennon net, borne aloft in naturalistic ardor by the enthusiastic fly hunter, had received its characteristic "fields" of red, scarlet, blue, and yellow, from a peaceful Kyronese dye pot, under the baptismal hands of the mirth loving sisters Cleorita and Oviata. After their arrival and dispersion among the hill glades, selected for the grazing of their flocks, Dr. Baāhar, apparently forgetful of the net staff, supported on his shoulder, was imparting to a bevy of matrons the secrets of vegetative propagation and fruition, when his words were suddenly arrested by the shadow of a butterfly of large dimensions cast by its interception of the sun's rays upon the flower of his speech demonstration. A glance upward, with an exclamation of enraptured covetousness, and all his impressions and energies were concentrated for the capture of the resplendent andean queen of butterflies. Rushing from among his pupils, heedless of apologies, instinctive gallantry, and masculine courtesies bestowed in deference to the weaker privileges of the sex, he started under queenly lead down the incline of the hillock, with eyes upturned, fixed upon the rainbow glints reflected from the swaying waft of the andean regina's wings, which were radiant with cerulean tints, as if in blending to proclaim her ethereal source. Like the ancient falconer, who with frantic gesticulation was accustomed to wave his luring staff to attract the attention of an eyas gaffling, who in freedom soared after striking his quarry, the doctor, with outstretched arms, pursued the tantalizing evolutions of his intended prize, which were sustained just beyond the reach of his net, — when, lo! while in full career, from an opposite direction, the king appeared, and a sudden concussion followed in quick succession, causing the doctor to drop his net staff, and in reciprocation enclose with his arms the object he had encountered, which, with the impulsive instinct of woman's self-possession in dangerous emergency, embraced with her arms his neck. With faces in near approximation, the objects of this strange conjunction in wondering surprise held emotional consultation; then, in freedom from the reflection of modest embarrassment, which would have caused sudden release, the right shoulder of the doctor became clothed in raven tresses, intermingling with his own flowing locks, his right arm having fallen instinctively to the waist for the support of the fair possessor's yielding form.

Forgetful of his net, and the vanished object of his first pursuit, he, in "good" Germanic Latin, free from the guttural ingesta inflection of saur-kraut, lager bier, sausage, and tobacco, offered apologetic consolation for the shock he had unwittingly occasioned, to which she replied in equally good English, "Pray, don't mention it;" while with lingering fondness, her sighs and steps were made eloquent in responsive continuation, as he led her back in half-reclining mood to her parents. The prætor, who had witnessed the scene with a peculiar smile of satisfaction, explained the predisposing cause of the encounter, — inasmuch as it was appreciable to ordinary observation, — that it might not be thought an act of premeditation on the part of the female respondent, or her relatives.

"Our Heraclean marriage alliance is so closely interwoven with instinctive impression, hallowed by the unity of an affection independent of the body, that the rupture by death of either of the coaptive sexual individualities, leaves a void, from the material deprivation of functional reciprocation, so desolate to the female in its impression of lonely isolation, that instinct conjures some gentle hallucination, to supply the broken threads of sympathy in the weftage of the severed ties. This illusive visionary substitution is held as a consecrated indication of continued affectionate unity, for the survivor's material direction in the body. Indeed, all our bereaved experience in some form this impression, in translation to some memorial object presented to view in the agony of instinctive disseveration. Isolita, the daughter of our cremator, who is now reclining in the support of Dr. Baāhar's arm, had her attention attracted, while in the anguish of separation, by a superb andean butterfly, which floated over the body of her expiring husband, and with his last sigh settled for a moment on her head with wafting wings, as if by invocation to inspire her hopes in bereavement of a material emblematic source of communication and direction; then, from the court colonnades, soared directly upward until lost to view in the blending tints of ethereal azure. This scene impressed us all with its omenic signification, so that we could scarcely wonder that Isolita in her great sorrow received it as a presage of vehicular translation, to be treasured as a token of animus visitations from her departed unity in the flesh. Without doubt, she will hold the conjunctive act you have witnessed this morning as an intimated sign of direction for the selection of a scocius, or companion, for the completion of her earthly term of sojourn. The confiding trust, evinced from her retained position, already betokens her belief in the consummated fulfillment of delegated substitution. In like verification, you will observe that the doctor has abandoned his net, and the winged vanity of his pursuit, for the realization of a more happy and abiding achievement."

In confirmation of the prætor's prognostication, but a few moments had been numbered with the past ere a procession had formed headed by the cremator's family, in hopeful conformity with the ceremonious rites they were disposed to accord in recognition of the instinctive liberalisms of sense which had been fostered by the doctor's precedental education. Being obliged to pass the scene of encounter in their passage up the dale, the prætor's face grew anxious as they approached the discarded net, but assumed an expression of gladness when the doctor passed it within the measure of a footfall, and without wincing saw it trodden under foot by the mother of his prospective affiancee. Relieved of his fears by the disdainful look cast upon it by the captured fly hunter, the family group of the prætor moved downward to meet the symbolical procession, and greet the advancing victor of self. While bestowing their congratulations, the fanatical fatuity, inherent with the expression of the the doctor's face, became broken and dissipated, as with mist clouds under the genial rays of the morning sun. In answer to the doctor's application for the required sanction of his betrothal with Isolita, the prætor expressed his warm approval, with the hope that he would soon be able to derive his happiness from the prospective good his example would confer upon future generations.

"Still," he continued, "without a clear knowledge of Manatitlan coöperation, in directing the wisdom of the 'choice,' I might have questioned the prudent propriety of the betrothal, from your pertinacious adherence to precedental habits, in defiance of the constant increase of self-inflicted misery. Especially, as I have learned from auramental source, that it has been the custom of the Germans, practiced from time immemorial, to render their wives servitas of convenience, rather than for the fulfillment of Creative intention, designed for the perfection of unity. From this isolating peculiarity of self-indulgent German instinct, it might be well for me to question, even now, whether in thought you treasure selfish desire that would detract by indulgence from the socius companionship of bereaved affection. Although naturally endowed with a strong instinctive predisposition, Isolita is in no way derelict in her full appreciation of an affection, matured in purity, independent of the body's functions. Bethink you, in answering, of your deposed net?"

In reply, the doctor said, "My net has subserved its purpose, in fulfilling its destiny of prestige; for, as you well know, I have expressed my full belief in the especial design of the butterfly's vocation, from the unrivaled beauty of its embellishments, which indicate the celestial transport, in previsemental aid of angelic visits. This morning I have received satisfactory evidence of the fact, and for the future have no farther object for its use; or, as we might say in quotation, 'sufficient for the day is the evil thereof.' If you can assure me of Heraclean reciprocation in the bestowal of the angelic capture I have made this morning, I will endeavor to discard, with my net, precedental pursuits."

The ingenuity of the insect savan's reply bespeaking the sanity of his self-possession, the prætor repeated to him the peculiarities of Isolita's widowed hallucination.

Still himself, the doctor replied, " I feel confirmed

in my impressions of her angelic nature, from your acknowledgment of the fact, that as a woman she harbors but one hallucination, and that I have been preferred as an equal for association with her, a privilege which has yetafore been awarded, in civilized society, solely, to her sex's insatiate unabridged vanity by the cajoleries of man."

With this additional evidence of the doctor's consciously sane appreciation of the happy conjunction his morning's encounter " foreboded," the espousal received general approbation. The prætor suggesting the efficient aid Isolita would be able to confer in systematizing his botanical labors, from the thorough knowledge of her acquirements necessary for fruitful vegetation, they departed upon their first united essay in botanical research, and were not seen again until the herdsmen sounded their calls for their return to the city; they then appeared crowned with floral decorations in overture anticipation of united reciprocations.

Of all the returning train the padre's face alone remained subject to the fitful indications of thoughtful sadness. The conjurations of the day had separated him from his last mythological hold upon instinct, raising a happy barrier between him and the familiar confriality of genial gossipings in the language of talk. Returning to the desolate quarters of the corps, after indulging freely in chiremoya and milk, he became subject to the indigestive broodings of instinct, barren of thoughtful resources for occupation. In this condition, disconsolate, he paced the deserted colonnades long after Mr. Dow and the curators of sound had retired to rest. But the Kyronese, with sympathetic consideration for his lonely plight, busied themselves in the court and cochina, ostensibly in preparation for the duties of the morrow, until, with the impression that he would prefer solitude for the melancholy nursing of his ruminations, they yielded to the drowsy influence evoked by the approaching midnight hour; and unaccustomed to the vigil unrest of anxiety, begot by the dismal forebodings of dread from a belief in mythological rewards and punishments, their eyes were sealed with such sudden surprise that little choice was permitted for the selection of easy positions for repose. Of late, mindful of others' comfort, he saw these sympathetic vigalantes overcome with sleep unheeded. Even Corycæus intermitted his thought auramentations with the solace of an occasional nap, and with the padre still waking, and walking in a mood of increasing nervous excitement, he at length sank into a dreamless sleep.

The darkness which gathers its deeper pall of blackness, in reversion to the brightness in vivid glow of the dying spark, had merged from the palpable coldness of its impression into the murky gray of the shadowy dawn, when there came a change so sudden and peculiar in the outward sway of the hammacas of the auramentor's family, — suspended from the vibrillæ of the tragus across the fossæ to the antetragus,—that the forward lurch awoke the occupants. Curious to know the cause of a motion so unusual, Corycæus hastened, with the recollection of the padre's condition, to take an observation, in which his wife joined with sympathetic alacrity. They found the padre kneeling and bowing before a rough-hewn statue of an ancient Heraclean mother, with a child, which she supported in her arms, the while counting with a " vociferous " whisper the beads of the rosary presented to him by Fraile Gallagato, alternating his devotional manipulations by cross " cuts " on his forehead and breast with his index finger. The scene was so ludicrously absurd, in evidence of the superstitious revival of his religious instincts, that the auramentors passed to a neighboring branch to watch his motions and hear his prayers engendered from selfish fears, wrought by indigestion and sleepless innervation, aided by the changes of the night. The image had been closely veiled with a vine embossure of iriditrope, which had been noted for its close resemblance to the sculptured statues of the immaculate virgin, without being aware of the model beneath. By some coincident freak, combined with fear, mist, and muirk, confounding with the incertitude of vision-fancied resemblance, he had discovered the statue beneath, which tended to raise his phantasmic emotions to a pitch of fanatical devotion. Impressed with the belief that it was a special revelation, designed as a reproof for his "backsliding" departure from grace, and neglect of his opportunities for the conversion of the Heracleans, he ventured to unveil the miraculous discovery, before seeking inspiration through the celestial gates of bead prayer. Notwithstanding the impression made upon the family of Corycæus by the ridiculous farce, there was a weird instinctive effect that reminded them sadly of the benighted condition of his race, who still made themselves blindly miserable with selfish labor, to the utter perversion of affectionate ease imparted from the current equality of self-legislation to the Heracleans.

After an hour's devotional exercise with hands, and mumbling prayer dronings and enumerations, wearied nature closed the scene with sleep, and he sank forward with his body and face prone upon the virgin bed of vine, in dreamless oblivion. In this condition he was found, as the ruddy beams of day began to dispel the lingering misty light of dawn, by the mayorong, who in sad fright made the courts and colonnades resound to his calls for assistance. Fearing that the vital spark had forever fled from the prostrate form of the kind-hearted padre, who, in despite of his incertitude, begot from his thoughtless reliance upon instinctive impressions, was alike the cherished favorite of the Heracleans, Kyronese, and

Betongese, the mayorong made no effort for his resuscitation. The shrill, wailing cry, reverberating in anguished appeal, reached not only his own people who were preparing for morning salutation, but the Heracleans, who hurried in the greatest consternation to the quarters of the corps to learn the cause of the fearful outcry. Proof to the mayorong's mournful cry, hastening footsteps, and exclamations of the excited throng, the padre continued unconscious, the gathering assemblage regarding his prostrate body with blanched faces and horror-struck gaze. When at length their surprised emotions had subsided into thoughtful sadness, " presence of mind " revived under the impression of regretful sympathy, which caused Cleorita and Oviata to kneel and raise the padre's head, and with the assistance of their grandfather to turn him upon his back. As gentle hands withdrew the dank hair that enshrouded his eyes, the fall of tears upon his face brought forth a deep sigh, as if conscious of the source from whence they came ; this, with a muttered ave, was followed by a quivering stretch for relief from the stiffness of his limbs, significant alike of retained vitality and reviving consciousness. Then, as if under the herald impulse of a dream of dread, he, with a spasmodic start, suddenly raised his head from the pillowed lap of Cleorita, bringing his nose in abrupt contact with the toe of the figure that projected over the pediment of the statue. This brought forth, with tears, his accustomed ejaculation, " My goodness gracious! " while he administered to it extreme unction with the soothing touch of his hand. The grimaced accompaniment, in revulsion, brought forth, in contrast from the depth of sadness, an irresistible outburst of laughter, from the late mourners whose eyes were yet moist with the tears of sorrow.

Starting up, amazed at his own unaccountable position, and the assemblage of faces that bestowed upon him their gaze, with mingled expressions of grief and mirth, the padre's fingers sought his hair for the disentanglement of his bewildered impressions. Failing in his attempt to recall the causeful events, his looks appealed to Cleorita and Oviata, whose eyes were glistening with gladness through their misty veil of tears, like the sun's rays through the celium of rosebuds sparkling with dew drops; but the anxious inquiries of new arrivals diverted explanation from them. Evil tidings are ever quick in spreading when borne by the scandalous impulse of gossiping tongues intent upon marvelous impression, but with sad sympathy the alarm had spread from portal to portal, with tongueless celerity, heralding the source of affectionate bereavement.

Among the nearest, and earliest to be apprised of the padre's supposed demise, was a young widow named Madonnasta, who resided with her parents without the oppidum gate, in the racept of the latifundium. Her husband had been taken by the savage besiegers when returning from a forage sortie; their hatred against the whites had been embittered by cruelties practiced for intimidation while the Jesuits were endeavoring to found missions among them for subjective utilization and the ruling advance of instinctive religion and civilization. In woful ignorance, they accredited their civilized foes with a united faith in a common form of worship, designed for the immolation of all unbelievers. Prompted by revengeful defiance, the unfortunate captive had been stretched and bound to a cross, the sacrificial emblem of Christian faith; and in that condition had been suspended over the brink of the precipice, in view of the besieged, who were forced to witness his agonized struggles, under the scorching heat of the sun, increased by the absorption and reflection of the basaltic rock, aggravated by the pain of his bonds, and the gnawings of hunger and thirst; but not without witnessing the desperate sallies made by sympathy for his rescue, in which with a wife's devotion Madonnasta had engaged. When at length death relieved his mortal torments, and the vultures, with time and the elements, had severed the cords that bound the bleached skeleton to the crucial framework, and it to the precipice, it fell to a resting place beneath; then a successful sortie was made for their recovery, and they were cremated with the wood of crucifixion; but a portion was retained by the widowed wife, who with great care and ingenuity formed it into an emblematic cross, corresponding in memorial form with the one upon which her husband had suffered; this she suspended from her neck, as an instinctive memento of the sad scene of her mortal bereavement. Her devotion to the relic soon imparted to sorrowing emotions the hallucid impression that the crossed pieces of wood were enacting the part of a spiritualized medium of communication with the animus of her departed husband, and were consulted at certain hours of the day for direction. As the hallowed memories clustered around the waking hours of the morning, when from repose the grateful impressions of thanksgiving had been revived for affectionate reciprocation, she was ever the first in readiness for the orison hour of morning greeting. In these moods, the fervor evinced by her reverential endearments plainly indicated the instinctive lapse of her faith into the implied belief of material transubstantiation eucharistic for imparting the hallucination of actual presence for the renewal of connubial felicity. These impressions, which from their sincerity involved consolation, in no wise impaired the sanity of her thoughts and acts in matters pertaining to the rational employments of her bodily existence in purity of intention. On the contrary, it strengthened the outflowing tide of her affection, so that its tangibility was imparted with a perceptible thrill from touch, voice, and presence, to all within the sway of purity and goodness.

It was the good fortune of the padre, on the morning succeeding to that of his first Heraclean advent, while yet subject to the relict baneful effects of whiskey, tobacco, and their habitual hereditaments of impurity, to be attracted by the beneficent fervor of the widowed Madonnasta's pitying glances of sympathy, while passing the portals of her father's house. The effect of these interviewing glances became immediately reformatory, for he sought a retired spot, where he "devoted" himself for an hour to the rapid chewing of his remaining tobacco, — supplied from the the limited store of his friend the doctor, for a stipulated butterfly consideration, — the while ruminating upon the incomparable charms of his inconuistic discovery. After fully expressing its narcotic power to the offaled dregs, he, in the vernacular phrase of instinct, incontinently swore off, while from a fountain in the crematium he thoroughly abluted his mouth; then returning past the house of Madonnasta, he paid her his reverence free from the actual impression of defilement. Afterwards, whenever he contemplated a visit to the predisposed object of his adoration, he subjected his mouth to a thorough purification with the chloride of lime, recommended by his "friend" as an excellent deodorizer for the correction of effluviums. This politic course partook of in advisorial advocacy, and exampled acceptance, the ostrich's fatuity, who in closing or concealing the eyes to self-reflection "supposes" its material body is rendered invisible to others. With the passage of time and his reproof pilgrimage to Amelcoy, he gradually became impressed with the mishaps attendant upon self-indulgence, and under the direction of goodwill he had obtained with her greetings manifestations of affectionate approval, which inclined her to study his language with rapid achievement and understanding success. These interallusions will afford the reader an understanding impression of antecedent and subsequent passages, elicited from the eventful singularities of the morning's transpositions.

When the padre's forlorn or dead estate was announced by the mournful cadences of the mayorong's call, Madonnasta was among the first of her people who had flown upon the wings of sympathy, to realize with her own eyes the truth of the startling rumor that knelled the second bereavement of her hopes.

The padre, at the moment of her approach, was endeavoring, with his right hand in his hair, to establish an equilibrium for the use of reflective thought, while his eyes wandered from face to face in search for the cause of their congregated anxiety, manifested in his behalf. Observing the roseate flush of gladness that quickly succeeded the pallor of dread anticipation in Madonnasta's face, when she found that he still lived, the padre essayed to address her in his own language, but upon the instant of his first articulation she caught sight of the cross dangling from his neck suspended by its chain of beads. Suddenly raising her hands in the clasped attitude of thankful surprise, she uttered the exclamation, "Al han espousita directicio!" (It is by thy fond direction!) and springing forward fondly clasped his neck in a joyfully conscious swoon. This episode proved fortunate, else she would have discovered his great trepidation and lack of glad reciprocation, which would have sadly chilled the realization of her transubstantial vision of predilected reunion dedicated for enactment through the padre's substituted mortality. With his usual tardy perception, dulled from the renewal of superstitious impression, he gave only mechanical support to the form of Madonnasta, resplendent with the charms premised from prospective reduplication in the body.

Cleorita observing his perplexity and evident abashment, pointed to the cross of Madonnasta, and his own, then with eucharistical fervor he bestowed upon her lips a baptismal kiss, while with a blush of shame he concealed the pendent emblem, suspended from his own neck, beneath his vest. This devotional exercise and symbol, recalled to his memory the events of the night, with a circumstantial impression that Madonnasta had by miraculous interposition been converted to the Christian faith, which led him to exclaim with enthusiastic ardor, " Upon my conscience' sake, it's a miracle, how she has kept the faith among pagans ! With pity and admiration, he again administered the baptismal rite of instinctive communion, which served to revive the lapsed faculties of his incumbent burthen. As his somewhat tardy tenderness revived the waiting perceptions of his angelic godsend herald, sighs, like the rustling flutter of leaves stirred in the stillness of noon-day by the advance of a shower, betokened the restoration of vital energy, with the genial accompaniment of joyful tears. When at length the rosy eyelids of Madonnasta began with trembling vibrations to unfold, the padre's features in waiting expectation flickered with the *ignis-fatuus* expression of catholic zeal, in the full belief of miraculous intervention for the preservation of the ordinances of revealed religion under the fructifying influence of saving grace. As with a convulsive shudder the full orbs of Madonnasta's wondering eyes were unveiled, and made glorious with the expression of delegated affection, the padre's face became illuminated with the propagandic light of zealotry, causing him to seize and bestow upon her cross an emblematic kiss of reverence. This act fully revived the pervading strength of Madonnasta's hallucination, causing her, with a look of fond recollection, peculiar to widowed grief, to embrace his neck, while with her lips she realized to his fanatical zeal the confirmation of faith.

The wonderingly amused spectators of this pantomimically enacted scene of mutual hallucination, with this act of consummation opened a passage for Mr. Welson and the prætor, who had been attracted from the house of Corycebæus by the hurrying excitement within the city portal. A glance sufficed for the assurance of a provisional "wedding" crisis, and the prætor was about to add his sanction, but the moment the padre observed his intention, he started back objuringly in the greatest alarm, muttering an interposing exorcism, at the same time exposing his own and Madonnasta's crosses as shields of protection. His impetuous array startled the prætor with the fear that the padre was in reality instinctively mad. But M. Hollydorf explained to his adopted father that the padre's disarray of thought had undoubtedly been occasioned by an unusual conjunction of circumstances, recommending an adjournment to the ordinarium of the corps for an investigation of facts, and a mutual understanding, under the sanction of advisement. When convened Corycæus related all that had transpired within the scope of his waking knowledge, which extended through the devotional vigil of the padre. It was easy to trace from subsequent events the source of Madonnasta's and the padre's coincident delusions. She had recognized in him the transubstantiated form preferred as a substitute by her crucified husband, from the cross attachment to his rosary; and he, from the bias of an instinctive Christian education, had supposed from coincident impressions that she was a miraculous convert appealing to him for a husband's protection. The Dosch advised that the padre should be made acquainted with the circumstances attending the death of Madonnasta's husband, and her consequent monumental delusion from the derangement of her instinctive perceptions occasioned by affectionate solicitude. Then, if he chose, in prospect of their incurability, to solace her with his companionship during their allotted terms of mortal sojourn, their union should receive Heraclean approval, upon the plea of like illusive adaptation.

The padre was greatly abashed when the facts in

plain demonstration were confirmed by Correliana and her mother. Mr. Welson then urged him to accept the coincidence as an omen of happy premonition. He then gratefully received the rites of sanctioned betrothal without demurring, after Madonnasta had been offered a like privilege of revocal by a statement of his mythological delusions derived from the ghostly precepts of a Christian education. Both with firm adhesion retained the bias of first impressions, and were escorted with joyful mirth to the house of Madonnasta's parents, where, with parental welcome, the padre assured the assemblage that he felt himself proof against lonely relapse into the mythological haunts of instinct. After the padre's betrothal and domiciliation, Mr. Welson with the Dosch returned to the house of Corycebæus.

While the prominent eccentricities of the last two conjunctions were the subjects of mirthful explication, Mr. Welson abruptly addressed the Dosch and Doschessa with an inquiry, which, like the shadow of a cloud passing over a landscape made humorous by man's instinctive invention, caused gleams of joyous transition from the reflection of absurdity, which, with authority, we are permitted to report.

Mr. Welson. "You have found it necessary in coupling the doctor and padre with yoke mates, to adjust their ruling infatuations with like characteristic hallucinations of will-o'-wisp affinity! will you now expound to me my own, that led to the direction of my choice? For I will frankly acknowledge, that with studied aid afforded by Cæluiformia's reflection, I have only been able to discover a lack of equality from my own instinctive imperfections."

Dosch (with a joyous accompaniment of laughter). "You have an old ritualistic proverb of more than ordinary worth, recorded among your mythological oddities and traditional 'saws,' for the expression of Giga infatuation, which we will reverse for your especial benefit in aid of perception for the explication of the enigma you ask us to solve. The reading your experience confirms, in quotation, should be rendered, Sufficient for the day is the good thereof! and with us, Sufficient for the day is the exampled proof thereof!"

Their mirthful inclinations were stayed by the entrance of the padre and Madonnasta; the countenance of the former having reassumed the vacuity of expression peculiar to the fanatical rule of instinctive fear and prejudice, which in language we will allow him to express.

Padre (addressing the Manatitlans). " You must know that it is not my wish or intention to be ungrateful; but then one must have a care for the preservation of his soul; for what is the whole world to a man if he loses his own soul. I am certain you could not have failed to see by my actions, all along, that I had qualms of conscience that all was not right with me. Not that I would wish for a moment to question the motives of Heraclean example, or ever have, for I know that in purity it's above my reach. But works, you know, are as nothing in the balance with one's soul without faith, which works wonders. Neither can I blame you in any way, except that you reject the light, confident in your own good works; and I greatly fear that the sources of happiness you suppose to be real are the delusions of the devil, who goes about like a roaring lion seeking whom he may devour. For what says Father Jaen, ' Good works are of no effect without saving grace administered under the seal of confession!' Well, after the strange marriages of Captain Greenwood, M. Hollydorf, Jack and Bill, with the sun overhead, — which I suppose is a pagan fashion, — and begging your pardon for expressing the truth of my mind, were no marriages at all, being extraordinary, without the sanction of the holy rites of the church, anointed under priestly seal, in sign manual of registry in heaven, which prevents divorce. (Addressing Mr. Welson.) Then you were espoused in another strange way, which shows that there is no regular sanctified method, as there should be. But yesterday, when Dr. Baāhar and Mrs. Isolita came together in such an extraordinary way, my eyes were opened, and I could not sleep, so I prayed to the virgin and her child fervently, which led to the miraculous discovery of her image, and only begotten son, just as the light was dawning, and while praying to her I was suddenly overcome with a slumber so peacefully sweet and deep, that I awoke to find myself dead in the belief of you all, at least those that saw me in that condition. In my vision I saw angels in tears, who seemed to express sorrow for the death of my body without the salvation of my soul by confession. Now, perhaps I was dead, for I found Cleorita and Oviata and the mayorong and his people and the Heracleans weeping, and felt uncomfortable in my body, as though I had just risen. Hardly had I begun to think, and had just bethought myself how I was overtaken, when I heard caress me (carissima!) spoken in an imploring way, at the same time found that Madonnasta had fainted in my arms, in an embarrassing way, which again bewildered me, until Cleorita and Oviata pointed to our crosses; then a light burst upon me, for I saw that she was a Christian among pagans, miraculously interposed for my reproof and her salvation, before I had sinned away the day of grace! All that I have said is true, and much more, if I could recollect it, which you would have seen if you had had faith like a grain of mustard seed. At any rate, I feel that the immaculate virgin and her holy son are my guardian angels, and the Manatitlans acknowledge that they are human, and depend upon good works for happiness, which is against the fathers and Scriptures, and I cannot, upon second thought, bring my mind to submit to your rites of marriage, which I fear are but little better than concubinage, that would endanger my soul and that of Madonnasta. From this you must know how anxious I am to depart, that Madonnasta may receive the rites of baptism and consecration for adoption into the bosom of our holy mother church. Then, after our regular marriage, we may return to assist in your conversion, if I am found worthy of confirmation in holy orders."

It would be hard to express in language the mixed emotions of those of the assemblage who understood the padre's interpretation of his waking visions of the morning, bred in emergence from sleep to the impressions of reality. The face of Mr. Welson assumed an expression of humorous admiration, seemingly gratified with the padre's revived superstitious simplicity, which gave encouragement to his playful disposition for quizzing inquiry. The microscopic reflections in like manner appeared to enjoy, for the moment, the "tutored" dismay evinced by the rambling impressions of the padre's rehearsal, incoherent with the precedental intuition of faith expressed in his memory of words. Madonnasta's face, although apprised of the padre's mythological delusions, "underwent" the varied changes of curiosity, puzzled for the want of a clear interpretation of emotions so foreign to the affectionate current of sympathy. Mr. Welson and the Dosch were alike dreaded by the padre, when the tenets of his religion and its instinctive incongruities, supported by faith in impossibilities, were rendered farcical by the contradictory absurdities of his questioned exposition of the law, prophets, and revelation; for, with a few interrogations, they invariably made him feel the ridiculous mist of his self-involvement. His incoherency had been increased by the proboscidial waggish indications of Mr. Welson's nose, which he felt was searching for a tender point beneath the superficial flow of his religious faith; so he mustered all the dignified acerbity possible for repelling attacks made for the exposure of his gullibility.

Mr. Welson. "You have preferred your former desire to leave Heraclea, and propose to take Madonnasta with you for sacramental confirmation and marriage. After the explanation you have heard of the cause that led her to adopt the memorial emblem, would it not be well to question her farther, that you may learn whether her disposition inclines her to the course you propose?"

Padre. "You know very well, Mr. Welson, that I cannot speak the Latin language, neither can she understand the English sufficiently well for the full comprehension of my wish. But what is there under the sun more evident than the common language of the cross, commemorative of our Saviour's crucifixion? Why, my goodness gracious, man, can't you see that its use in her husband's death, was the inscrutable means used for her conversion? Then what led me to discover the virgin and her child, — which you had passed hundreds of times without noticing, — when I was in the greatest need for their intercession from the want of sleep? I know that you say it is the statue of an ancient mother of the household, reverenced for her " virtues," but this, as you well know, would not account for the effect produced on me, when my prayers were directly supplicating repose?"

Mr. Welson. "Our faces undoubtedly show what we cannot deny. But our smiles are not provoked by a scoffing disposition; on the contrary they are more inclined to sadness than derision, for it is hard for us to conceive the incomprehensible nature of an obstinacy so void in perceptive appreciation, although by the reflection it forces upon memory the perverse insensibility and difference of our past lives to the true source of happiness. That you, of us all the most highly endowed with the natural manifestations of goodness, should prove so dull as not to realize the source from which the happiness you really feel is derived, bespeaks an infatuation that exceeds the measure of our comprehension. But as you have determined to leave us, it is proper for you to understand the true interpretation of your betrothed's feelings in prospect of her removal from home."

Dosch. "Before she is questioned, I would have the padre fully comprehend the true nature of the alliance he would assume with Madonnasta, for she, in common with the Heracleans, has a realizing perception of the unity still existing between herself and former respondent in the flesh. Her true impressions, excited by the symbol in your possession, were that your sympathies flowed in unison with hers toward the severed reflection of her own identity, and that you were the preferred successor chosen for the representative solace of her sojourn in mortality. To disabuse her of this gentle hallucination, imparted from the severed ties of instinctive association, would, if possible, be cruel. Still if your prejudices are over strong against soothing her partial preference in the interchange of proxied solace, it would be better for you to depart alone. If, on the contrary, you can reciprocate in substitution her instinctive affection, you will find her a constant source of happiness, that will advance your perception to an earthly realization of the joys imparted from a foretaste of immortality, through the current reciprocation of goodness. Now that my wife has explained to her your multiplied delusions, founded upon the sounding words, faith and saving grace, with their attendant instinctive inducements for the patronage of gross indulgences, I will state to her the motives of your intention which prompt you to leave Heraclea, also your desire to have her body undergo the ritualistic manipulations of the priests of your sect, for the salvation of its instinctive soul, and recommend that you bestow upon her during the relation your regardful attention."

Madonnasta, during the recital, devoted her attention to the close study of the padre's personal peculiarities, which were described to her as a prevailing index of corrupting effect with the Giga civilized races. With the mention of tobacco and distilled liquors, that could not be disguised to his ear, his face assumed the scarlet hue of shame, while with downcast eyes and tremulous folding of hands, he pleaded in thought parental example and the encouragement afforded by priestly absolution. Quick to appreciate his regretful sufferings, she was attracted to his side, and with the soothing action of her hands imparted sympathy for his self-inflicted misery. Shamefaced from the constantly recurring examples of his heedless lack of purpose, he made no attempt to renew his promises of constancy, but remained silently submissive to the reprehensive admonition of the Dosch. "If," continued the Dosch, "you and your race would give heed to the warning impressions of your bodily functions when oppressed, your perceptions from memory would soon act as a guard against incompatibles and excess. Functional experience as an example for good and evil, in provisional guard for the welfare of digestion and healthy assimilation, is better by far than the theoretical tests of chemical analysis and the empirical counteractives of the doctor, which only serve to exhaust vitality, distempering in the process protective mental power designed for the corresponding elimination of instinctive purity and goodness. Our bodily perfection, which ignores in age the artificial aids of plaster and paint, for the concealment of living depreciation from unnatural causes, has been attained by the thoughtful provision of ancestral example, which constantly held in view, with themselves, their responsibility to future generations, with reactive profit to their own happy correspondence with material self. It should appear, from the example of your last experience, which has rendered you phantasmally mad, that judgment should be trained individually and collectively to recognize in representation the adaptability of food in quality and quantity for healthy support. The cherimoyer as a fruit, and milk as a vegetable production from animal elaboration, are each separately, with a recognition of time and quantity, well adapted for the nutriment of the human body. But you, as with your race, have paid the penalty of heedlessness, and in relative degrees can realize from self-experience the origin of war, gospel, law, and medicine, with their legions of phantasmal abettors, which renders life a waking nightmare of miserable variations in opposition to happy realization. In the calm quiet of Heraclean life, with associate correspondence in purity and goodness, your impressions and desires have been so occupied with happy realities, that even in reflection, from memory, the gala day celebrations that attracted your instinctive passions of sense with evanescent beguilement have proved an aversion to thought. Picture in impression your emotions, if in the distance you can imagine a scene so abhorrent to the realizations of affection, you saw a procession passing up the now peaceful avenue of the latifundium, heralded with deafening shouts, cannon, Chinese crackers, bombas, the clangor of cymbals, obstreperous shrillness of fifes, screechings, groanings, and dronings of bagpipes, the monotonous boom and clattering roll of drums, a procession with banners borne by soldiers in the popinjay "uniforms," glittering swords, bayonets, and like pharaphernalia of vanity and death! Or the horror that would suffocate your tender hopes inspired for the increasing purity and goodness of future generations, if the temple schools of germination should be usurped to give place to the stable ritualisms of priestly compostors! When, with the study of your personal requirements, you seek to make your habits inoffensive and agreeable to purity and goodness, you will be able to avoid the humiliating impressions evoked by your morning's exposure, which were solely attributable to a heedless lack of attention to your former experience and advice of Anticipator, who warned you of the effects you provoked. From the effect produced upon your involuntary powers from indigestion, you can judge of the living nightmare freaks of insanity which have been provoked from ages of conceptive indulgence to give birth to hallucinations of your present progressive civilizations. Once entered upon the realities of self-legislation, in its current form of affectionate solicitation for the welfare of others, the germ of goodness will expand for reciprocation, until in revivication it embraces not only the human race, but in instinctive effect and degree the lower orders of animality.

The padre feeling the justness of the direction, and kindly sympathy manifested by the Manatitlans and Heracleans, could not withhold his eyes from giving misty manifestation of emotional appreciation. This "weakness" caused Dr. Baáhar, who had with politic diplomacy conformed, in outward appearance, to Heraclean usage, to become cynically provoked, openly urging that his childish tears accounted for his mistaking the rough-hewn Heraclean statues for the Christian prototypes of his creed. Notwithstanding the padre's regretful humiliations, from a lack of thoughtful consideration, he could not withhold a retortful reminder from his old noli me tangere opponent, of his more flagrant assumption; after a moment's hesitation, he replied, "I claim but a limited knowledge in genealogical matters pertaining to mythology, but I think I was not more daft in my judgment when I mistook the statue in the misty morning light, for the virgin mother and child, than you was in judging the Heracleans politic worshipers of one of your old Sclavo-vendic deities, because you found a statue garlanded with vine-disguised Kyronese mouse-traps."

This ever ready repartee, and apt for the occasion, served to dispel the reproachful shadows, that in impression hung over the padre from his listless predisposition to lapse into his old fatuous rulings of instinct. The admonitions of the Dosch had also aroused in him a reproachful fear that his example would serve to impair the confidence of new arrivals in the effective permanency of Heraclean example; which awakened in him a determination in his own mind " to make his calling and election sure," by a thoughtful avoidance of precedental inclinations.

CHAPTER XXXIV.

DR. BAAHAR, acting in accordance with the suggestion of the Dosch, given a few months previous, had devoted his attention to the cultivation of fruit-bearing plants, shrubs, and trees, but his success from a lack of objective constancy and discriminative judgment, was inclined to be enigmatical in practical results. Instead of studying the practical adaptation of productive vegetation for the requirements of healthful subsistence, he was quite content with transplanting rare growths, obtained from the surrounding country in the latifundium, without anxious regard for the development of fruitful utility, often introducing those that it had required the labor of years to exterminate, when sowed upon the wind from the brink of the precipice by the Indian besiegers. Fortunately his democratic ideas, which reverenced the rights of naturalization in freedom from adaptability, and rapid succession in office, gave his citizen plants but little time to take root, except those of the most worthless description that live upon the blight of the fruitfully good. Yet with all his inadvertencies, accident occasionally favored a useful result, as many of the fugitive growths which had in seed-flight adapted themselves to congenial soil proclaimed their trans-atlantic origin and capability for life sustaining reproduction with provident forethought in cultivation. His botanical ambition found ample satisfaction in tracing their genealogical relationship without testing their fruitful capacity, except in chimerical conjecture, founded upon precedental arguments advanced by the most ancient writers.

Under the affectionate tuition of necessity, Isolita's instincts had been trained for the consistent conservation and advancement of vitality, and her knowledge, despite the disadvantages of siege, had extended with a wide reach beyond the cinctus walls. With cultivated attainments for the discernment of cause and effect, she had with the dependent emergency of her people upon a continued supply of vegetable products become a practical botanist, capable of tracing at sight the natural life-sustaining affinities of fruits and roots, although ignorant of their technical classification into generas, orders, and species. Visiting the embryotic garden of the doctor, shortly after their espousal, she was surprised to find the only thriving plant the noxious venoseminata, the evil genius of fruitful vegetation, which when once allowed to take root, in new soil, offered hydra resistance to the efforts bestowed for its eradication. With her quick perception she discovered the danger incurred from its heedless cultivation, not only to the plot of her adopted Socius, but to the neighboring plantations, which with full exampled growth would become subject to its contagious encroachments. Quick in preservative action she seized a dibble, and before the technical precedentalist could arrest her practical intention, the malignant parasite was uprooted, and hung dependent from the branch of a tree exposed to the full rays of the sun. Too late for expostulation, the theorist stood aghast at her audacity, but kept silence lest from her skillful use of the dibble she should trace the noxious thrift of the plant to his jesuitical cultivation, despite the warnings of his neighbors. Recovering, when he saw her raise plant after plant, consigning them to the same fate, and in process exposing others to remove from their roots the fatal tentacles, he remonstrated; but she still

continued her labor, the while congratulating him
when she discovered that none of the diffusia had
trespassed beyond his limits. At length convinced
that no stray fibre remained, she carefully gathered
every leaf, branch, and tendril to be united in the
fate of the parent stalk. Completing her search for
the garroting quirls of the venoseminata, that strangulated above the surface with an effect as deadly as the
wide spreading roots beneath the surface, she silently
replaced those promising fruition worthy of cultivation, then standing in a smilingly questioning attitude
of graceful solicitation, she waited to learn the measure of her Socius' approval. To which he answered
in words, with eyes fitfully glancing askance, half
with shame, inwrought with furtive displeasure, " To
be sure I understood the nature of the plant, but I
wished for the others to grow in company with it,
that they might improve upon its evil example, in
vindication of our theatrical enactments which portray
sensational evil, that they may show the shadow of a
surviving moral, for it is the duty of the good to
shame the evil; for what says one of your old Roman
poets?

" 'In evil company you should ever show,
That purity can protect itself, and ever grow.' "

Isolita. "But did you not see that it was destroying all within its reach?"

Socius. "But as in war, evil eventually exhausts
itself; and by furnishing more hardy growths I should
have overcome it in time."

Isolita. "But it would have soon extended itself
beyond your limits. Besides, of what avail the cultivation of your ground if your useful plants were condemned to be constantly devoured by this parasite
without reaching fruition. In permitting evil to grow
and expand under your hand for neighborly infliction,
when in the beginning you have the power of suppress-

ing it at ease, to perfect extinction, would make you miserably culpable as an abettor."

Socius. "But your Manatitlan advisers advocate the practical good of their school of hypocrisy, that their scholars may be fore-armed by being fore-warned."

Isolita. "Yes, but the professors are as harmless for evil and injury as those that I have hung in the sun to indure the scorching noonday heat, with the fruitful soil beyond their reach. Besides the human venoseminatas serve as a warning to their kind, and in their professorial speciality of ingratitude are detained from propagating their deadly example."

Socius. "But your Manatitlan advisers advocate the practical good that comes from exposing hypocrisy; and their arguments sustained by example, are equivalent to preaching, and our theatrical entertainments founded upon precedental enactments, which appears to be a distinction without a difference in reality."

Isolita. "As you are aware, the Manatitlan school of hypocrisy was an ulterior resort, forced upon them by the ritualistic duplicity of their Mouthpat neighbors; which, aside from the beneficial result derived from exposing the deceptive incongruities that entailed constantly increasing misery upon the races of mankind, afforded thoughtful stimulus to the graduating novices for suggesting the means of auramental direction, in their aural correspondence with the civilized Giga races."

Socius. "You are speaking as a Manatitlan, under direction. Is it well for you to submit to the prompting of third parties in your intercourse with me? I have been taught that the marriage alliance should be held sacred as a body corporate united in its parts for communion with self."

Isolita. "If we consult our mutual advantage, it is not from extending injury but help to others, and with the flow of recurrent reciprocation, we in turn are filled to overflowing with grateful emotions of joy. We certainly should not disdain good instruction from any source, which offers experienced advantage for aiding our endeavors in perfecting the attainment of a happy union. As with us of Heraclean lineage, you have acknowledged the near approach of the Manatitlans to happy perfection, and should unite with us in grateful expressions of joy that they are pleased to devote their experience for our advancement in happiness. Their exampled experience in goodly purity, revived in current reciprocation from disembodied affection, affords us a more perfect realization of Creative intention, through the indications of perceptive endowment. If we live dependent upon the vitality of others, without reciprocation, to the exclusion of confidence imparted from the joyous trust of unselfish goodness, we should in fact enact the part of the venoseminanta that destroys useful vegetation with the growth of its own grasping evil propensities, which yields to itself a destructive existence in compensation for the injury it inflicts upon the fruitful beneficence of its neighbors."

Socius. "Your language betrays the Manatitlan philosopher, rather than the wife; who according to our creed should obey her husband in all things. We have a proverb in Germany, that says 'Two literary philosophers can never agree in a common household;' and another that reads, 'It is better to have a wife submissively weak in intellect, than strong in mind.' So you will perceive that in sequence it logically follows that children born from united strength will become heterodox to ancestral faith, unless left early to the example and correction of a surviving parent."

Isolita. "With the indwelling sanction of purity and goodness, we accord to the Manatitlans a better interpretation of Creative indications from practical knowledge, and are grateful for their aid prompted by well tested experience devoted to an enduring perception of our immortal privilege of living in life away from the gross control of instinctive desires, which in confluence with united parental example lives ever with us proof to bereavement."

Socius. " I certainly wish to understand you, and better still, I would have you comprehend me without other aid than I am able to impart. For as I have been taught, it is esteemed absolutely necessary for a wife to reverence her husband as a director from acknowledged superiority, with a submissive affection contentfully obedient in affording a guarantee for the peaceful assurance of the household. Law and order, under the ruling control of the husband, are as essential for the preservation of domestic discipline as public.

Isolita (smiling sadly). " Can union abide with the superiority of one part above another, that with assumption dictates subserviency in the place of equality? To love, with us, is to be loved; and, as you have experienced, we have no jarring discords from selfish indulgence, for in recognition of the unprivileged specialities of brute instinct, in contradistinction we are enabled to consult the body's requirements for healthy support, in appropriate degree for the healthy manifestation of affectionate equality, in check of the cravings for excessive gratification that with material clog is the pampering source of all the woes of the Giga race."

Socius. " So, so, I see that a Heraclean wife includes the dictation of a Manatitlan bride."

Isolita. " It is not our wish to 'dictate or argue,' for we have been well informed of the dissentious meaning of the words in exampled use with the Gigas. But you must be well aware, that unless confidingly united in sympathy our union is void, and our example would impart evil rather than good. It is from the consciousness of grateful appreciation derived from imparted affection that we obtain our impressions of happiness, and in extension of immortality."

Socius (abashed). " You make me feel from your affectionate solicitation, in self reproof, for my repellance, like a father who has dictated to his children, by recommendation and example, politic hypocrisy, sword exercise, and dancing, as passports for the enjoyment of life, with the expectation of affectionate reciprocation ; but will honestly acknowledge that the prepossessions of my instinct oppose concessions ; still will try to make myself subservient to your affectionate direction in all things, for I am fully impressed with the fallacious follies incubated from my conceptions."

Isolita. " It is not my desire's wish to have you subservient in any respect, but affectionate in the reciprocation of purity and goodness, so that our companionship may never admit of selfish deviation, but experience in daily appraisement the novelty of new ardor in loving returns."

The voice of the padre, tremulous with mirthful enjoyment, interrupted the doctor's grateful reply to Isolita, by calling upon him to act as umpire in deciding a question that had arisen between him and Madonnasta with regard to the germination of the bean. In answer to his call the doctor and Isolita advanced to his plot, where they joined in the voiceful mirth of Madonnasta, who had surprised her espoused while engaged in reversing the supposed resurrection of some Indian beans he had planted, which in germination had been forced in division above the surface. The first salutation of Madonnasta, when she discovered his unnatural occupation, was a look of startled inquiry directed to his face, to detect whether his employment was prompted by humorous suggestion from the delusive effects of his Christian education, which inculcated a material resurrection of the interred body, or simplicity. His pantomimic expression of disgust, at the supposed trifling of vegetation in returning to him his labors in material bifurcation, put to flight her first impressions that he was facetiously endeavoring to show his willingness to uproot old prejudices founded upon precedental habits and customs. His ludicrously despiteful gesticulations, evoked from nature's supposed practical joke, quickly dispelled her shudder of dread by making apparent the real cause of the reversion. The simplicity of his ignorance caused a depreciating flow of mirth, as she stayed his hands from farther sacrilege. Readily understanding from her movements and facial indications, the cause of her remonstrance, he expostulated with her by signs and words to respect his judgment. But unable to stem the current of her mirthful resistance to his labors, he appealed to Dr. Baāhar for judgment upon the cause of the bean's bodily resurrection. Upon learning the cause of the doctor's summons, Isolita was unable to resist her inclination to join in the mirthful peals of Madonnasta. But her socius with dignified seriousness, natural to the scientific professor, proceeded to give an elaborate opinion from authenticated authority founded upon the technicalities of order, genera, and species, in conjunction with a supposititious analysis of the soil; which caused the padre to interrupt him with the exclamation, "My goodness gracious, doctor, I called for light, and you give me darkness! The fact is we might as well acknowledge our ignorance at once, and submit gracefully to our betrothed's Heraclean direction; for even your maiden aunt, upon whom you was mainly dependent for your theoretical ideas, illustrated the virtue of necessity in like emergency, by an example of celibacy, and the value of our opinions tend to the same end."

To which the doctor replied, "If you are to be

made the pons asinorum for the passage of Manatitlan wit, we had better adjourn to the auriculum where we can have a more direct rendering."

In response to an implied allusion to his brogue, the padre, in retort, urged that the doctor's name gave indication of an instinctive alliance equally remote with his own, although more simple and less prolonged in vocalization. These repartees, although civilized in evolution, were void in chivalrous results, as each party held themselves amenable to kindly goodwill in the revival of their ancient badinage in the presence of their betrothed.

CONCLUSION.

HAVING, with advisorial aid, completed the historical part of my delegated labor, designed for the initial elaboration of Manatitlan habits and customs in design for Giga adoption, I am directed to urge for any lack of perspicuity, in addition to my own defects, the limited variety of words and terms embodied in the languages and idioms of civilized races, for the expression of affectionate purity and goodness, with the impress of reality, independent of the selfish distinctions imposed by the arbitrary rule of meum and tuum. As a reflecting pharos for the Manatitlan rays of affection, I have endeavored to render from their dictation with truthful impartiality, rarely offering comments or suggestions of my own. Still, I am fully aware, that, as the medium, I shall subject myself, as a target, to the defilements of instinctive stigmas, which so abundantly replenish in Giga vocabularies the lack of words endowed with affectionate expression; but feel myself so well protected by the initiatory silicoth-garment of Manatitlan adoption, that the omniscentiferous capacity of humanity for the ejection of odors from mouth and pen, will prove as harmless in effect as if in aim directed to my dictators.

With the assurance of affectionate reciprocation from the good of septs and nationalities, I shall, with the grateful solace of their sympathy, rest content in freedom from annoyance, although assailed with odors, grunts, and growls vented from mouthpat instinct. If, peradventure, the future of Giga races may be withheld by the adoption of the Manatitlan system of education, in devisement for the attainment of legislative self-control, from thoughtless submission to the mouthed and written precedentalisms, which have served to render misery the chief object of life rather than happiness, it will prove an ample source of recompense for the untoward contributions bestowed in opposition to creative indications by progressive instinct.

Yet with all the hoped for joy anticipated from the grateful confluent reciprocations of goodness, we acknowledge a selfish grief foreboded in our exile from Heraclea into the civilized world of instinctive strife, although consoled with the auramental presence of our Manatitlan familiars.

R. ELTON SMILE, *Proscriptor*.

In testimony of joyful authenticity, we the undersigned members of the Teutonic corps of the R. H. B. Society subscribe our names in verification of the Historiographer's correct interpretation of the events transpiring under our observation.

GIGANTEO XL.,   ADESTUS,
Dosch of Manatitla.   *Prætor of Heraclea of the Falls.*

M. HOLLYDORF,
Director of the Corps.

LEPIDOPTERUS BAAHAR,
Entomologist.

OCTAVE PETTYNOSE,
Buzz Curator of Sound.

FALSETTO LINDENHOFF,
Stridential Curator and Recorder of Genealogical Sounds.

Honorary Addendas: GUILLERMO WELSON,
Mentor.

DIEGO DOW,
Naturalist.

TRULY RURAL GREENWOOD,
Expeditionary Aid.

PADRE SIMON,
(*Under protest*) *Mythological Curator of Souls.*

JACK and BILL SMITH or JONES,
Sons of Neptune, and volunteer Aids.

www.ingramcontent.com/pod-product-compliance
Lightning Source LLC
Chambersburg PA
CBHW051844300426
44117CB00006B/260